Birkhäuser

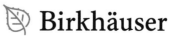

Applied and Numerical Harmonic Analysis

More information about this series at http://www.springer.com/series/4968

Michael Ruzhansky • Sergey Tikhonov
Editors

Methods of Fourier Analysis and Approximation Theory

Birkhäuser

Editors
Michael Ruzhansky
Department of Mathematics
Imperial College London
London, United Kingdom

Sergey Tikhonov
ICREA Research Professor
Centre de Recerca Matemàtica
Barcelona, Spain

ISSN 2296-5009 ISSN 2296-5017 (electronic)
Applied and Numerical Harmonic Analysis
ISBN 978-3-319-27465-2 ISBN 978-3-319-27466-9 (eBook)
DOI 10.1007/978-3-319-27466-9

Library of Congress Control Number: 2016932897

Mathematics Subject Classification (2010): 41-XX, 42-XX 32A-XX, 65C-XX, 49K-XX

Printed on acid-free paper

This book is published under the trade name Birkhäuser.
The registered company is Springer International Publishing AG Switzerland
(www.birkhauser-science.com)

Preface

This volume consists of a collection of papers originating from two events devoted to areas of harmonic analysis and their interplay with the approximation theory.

The first event took place during the 9th ISAAC Congress in Krakow, Poland, 5–9 August 2013, at the section "Approximation Theory and Fourier Analysis".

The second event was the conference on Fourier Analysis and Approximation Theory in the Centre de Recerca Matemàtica (CRM), Barcelona, during 4–8 November 2013, organized by the authors. It continued the successful tradition of workshop series: Osaka University 2008, Imperial College London 2008, Nagoya University 2009, University of Göttingen 2010, ICREA Conference 2011 in CRM, and Aalto University 2012.

The topics of the conference include: Fourier analysis, function spaces, pseudo-differential operators, microlocal and time-frequency analysis, partial differential equations, and their links to modern developments in the approximation theory.

London, UK
Barcelona, Spain
November 2015

Michael Ruzhansky
Sergey Tikhonov

Contents

Some Problems in Fourier Analysis and Approximation Theory

Michael Ruzhansky and Sergey Tikhonov

Abstract We give a short overview of some questions and methods of Fourier analysis, approximation theory, and optimization theory that constitute an area of current research.

Keywords Approximation theory • Fourier analysis • Harmonic analysis • Optimization theory

Mathematics Subject Classification 42-06, 42-02, 41-02, 41-06

1 Introduction

In this review we present a short survey of topics related to the two conferences (see the preface to the volume), with brief introduction to more extensive presentations contained in the papers of this collection.

2 Fourier Analysis

2.1 Parseval Frames

Chapter 1 by L. De Carli and Z. Hu is "Parseval Frames with $n + 1$ Vectors in $\mathbb{R}^n$". A frame in a finite-dimensional vector space is a set of vectors that

M. Ruzhansky (✉)
Department of Mathematics, Imperial College London, 180 Queen's Gate, London SW7 2AZ, UK
e-mail: m.ruzhansky@imperial.ac.uk

S. Tikhonov
ICREA, Centre de Recerca Matemàtica, and UAB, Campus de Bellaterra, Edifici C, 08193 Bellaterra (Barcelona), Spain
e-mail: stikhonov@crm.cat

M. Ruzhansky, S. Tikhonov (eds.), *Methods of Fourier Analysis and Approximation Theory*, Applied and Numerical Harmonic Analysis,
DOI 10.1007/978-3-319-27466-9_1

contains a basis. Frames (or redundant systems) have become increasingly popular in applied mathematics, computer science and engineering; representing signals in terms of a set of a redundant set of basic frequencies ensures resilience against noise, quantization errors and erasures in signal transmissions. Appropriate frame decomposition may reveal hidden signal characteristics, and have been employed as detection devices. The references are too many to cite, but see [9], the recent book [10] and the references cited there.

Parseval frames are nontrivial generalisations of orthonormal bases and have important applications. Vectors in a Parseval frame are not necessarily orthogonal or linearly independent, and do not necessarily have the same length, but the *Parseval identities*

$$v = \sum_{j=1}^{N} \langle v, v_j \rangle v_j$$

and

$$||v||^2 = \sum_{j=1}^{N} \langle v, v_j \rangle^2$$

still hold for every $v \in \mathbb{R}^n$.

In recent years, several inquiries about tight frames have been raised. In particular: how to characterise Parseval frames with N elements in $\mathbb{R}^n$ and whether it is possible to scale the vectors of a given frame so that the resulting frame is Parseval.

In the paper by De Carli and Hu, the authors give a method for constructing Parseval frames with $n + 1$ vectors in $\mathbb{R}^n$ that contain a given vector w with $||w|| < 1$. Precisely, they construct a Parseval *triangular frames* (i.e., a frame $\mathcal{B} = \{v_1, \ldots, v_n, w\} \subset \mathbb{R}^n$ such that the matrix with columns $v_1, \ldots, v_n$ is right triangular) and they show that $\mathcal{B}$ is essentially unique, in the sense that if $\mathcal{B}' = \{v_1', \ldots, v_n', w\}$ is another Parseval triangular frame that contains w, then $v_j' = \pm v_j$. Scalable frames with $n + 1$ vectors in $\mathbb{R}^n$ are also discussed. The method can be used to construct Parseval frames with $n + k$ vectors in $\mathbb{R}^n$ that contain given vectors $w_1, \ldots, w_k$ (this work is still in progress).

2.2 *Hyperbolic Hardy Classes and Logarithmic Bloch Spaces*

Chapter 2 is the paper "Hyperbolic Hardy Classes and Logarithmic Bloch Spaces" by E. Doubtsov.

Let $H(B_n)$ denote the space of holomorphic functions on the unit ball B_n of $\mathbb{C}^n$, $n \geq 1$. Given a holomorphic mapping $\varphi : B_n \to B_m$, the composition operator $C_\varphi : H(B_m) \to H(B_n)$ is defined as

$$(C_\varphi f)(z) = f(\varphi(z)), \quad z \in B_n.$$

As indicated in the monograph [34], the main problem in a study of composition operators is to relate the operator-theoretic properties of C_φ and the function-theoretic properties of the symbol φ. So, Ahern and Rudin [1] posed the problem of characterising those holomorphic mappings $\varphi : B_n \to B_m$, $m = 1$, for which the composition operator C_φ maps the Bloch space

$$\mathcal{B}(B_m) = \left\{ f \in H(B_m) : \sup_{z \in B_m} |\nabla f(w)|(1 - |w|^2) < \infty \right\}$$

into the classical Hardy space $H^p(B_n)$, $p > 0$, or into BMOA(B_n), the space of holomorphic functions having bounded mean oscillation.

First major results related to the Bloch-to-BMOA problem were obtained by Ramey and Ullrich [32]. In particular, quite surprisingly, every Lipschitz symbol φ has the required property for $m = 1$ and $n \geq 2$. Also, Ramey and Ullrich indicated that the Bloch-to-BMOA property is related to BMOA$_h(B_n, B_m)$, the hyperbolic BMOA class introduced much earlier by S. Yamashita for $n = m = 1$. In fact Yamashita used the following formal rule: replace a function $f \in H(B_1)$ and the Euclidean metric on $\mathbb{C}$ by a holomorphic mapping $\varphi : B_1 \to B_1$ and the hyperbolic (Bergman) metric on the disk B_1, respectively. He investigated hyperbolic analogs of various classical spaces, in particular, those of the Hardy spaces $H^p(B_1)$; see [42]. For $m = 1$, a solution of the Bloch-to-Hardy problem in terms of the hyperbolic Hardy classes was obtained by E.G. Kwon in a series of publications (see, e.g., [16]).

The paper by E. Doubtsov is motivated by the above problems for $m \geq 2$. Standard arguments are not applicable in this case, so the first technical idea is to use test functions with appropriate sharp lower estimates of their moduli instead of lower estimates of their derivatives. It turns out that such derivative-free approach gives a complete theoretical solution of the Bloch-to-BMOA problem for all $m, n \geq 1$; see [13]. Also, Kwon [16] argued that various hyperbolic classes introduced by Yamashita naturally arise in the descriptions of related bounded composition operators defined on the Bloch space $\mathcal{B}(B_1)$. So, the second idea is to consider logarithmic perturbations of the standard weight $1 - |w|^2$ and to show that the Bloch space $\mathcal{B}(B_m)$ is not an exception. Namely, given $\alpha < \frac{1}{2}$ and $0 < p \leq 1$, the hyperbolic Hardy class with parameter $(1 - 2\alpha)p$ is related to the bounded composition operators from the logarithmic Bloch space $L^\alpha\mathcal{B}(B_m)$ to the Hardy space $H^{2p}(B_n)$.

2.3 Logan's and Bohman's Extremal Problems

"Multidimensional Extremal Logan's and Bohman's Problems" by D. Gorbachev is Chap. 3. A.M. Odlyzko formulated the extremal problem related to zeros of Dedekind zeta functions. Let f be an arbitrary positive definite function such that $\operatorname{supp}\hat{f} \subset [-1, 1], \hat{f}(0) = 0, f(0) = 1$, where

$$\hat{f}(s) = \int_{\mathbb{R}} f(t)e^{-2\pi ist}\,dt$$

is the Fourier transform of f. The Odlyzko problem is to determine the quantity $T_* = \inf T$ provided that $f(t) \geq 0$ for $|t| \geq T$. This problem was solved by Logan in [22]. He proved that $f, t^2 f \in L^1(\mathbb{R})$, $T_* = 3/2$, and

$$f_*(t) = \frac{\cos^2(\pi t)}{(1-(2t)^2)(1-(2t/3)^2)}$$

is an extremiser.

The periodic case of Logan's problem was considered by Chernykh (1967, 1979). He proved sharp Jackson's inequality in the space $L^2(\mathbb{T})$ with the optimal argument. A generalisation of the Chernykh results for the multidimensional case was given by Yudin (1981). For an arbitrary convex 0-centrally symmetric body $C \subset \mathbb{R}^n$, Yudin constructed a positive definite function $F_C \in L^1(\mathbb{R}^n)$ such that $\operatorname{supp}\widehat{F}_C \subset C$, $\widehat{F}_C(0) = 0$, $F_C(0) = 1$, and $F_C(x) \leq 0$ for $\sigma_1(C)/\pi$, where $\sigma_1^2(C)$ is the first eigenvalue of the Laplacian Δ in C. Using properties of the function F_C, Yudin also obtained several results in analytic number theory and in algebraic lattice theory (1996). In particular, he got the asymptotic estimate

$$R(L)\lambda_1(L^*) \leq \frac{n}{2\pi}\left(1 + O(n^{-2/3})\right), \quad \text{as } n \to \infty,$$

where L is a lattice in $\mathbb{R}^n$, L^* is the dual lattice to L, $R(L)$ is the covering radius of L, $\lambda_1(L^*)$ is the first successive minimum of L^*. By a more complicated argument Banaszczyk (1993) also derived this result.

The multidimensional analogue of Logan's problem was stated by Berdysheva [6]. Suppose that $U, V \subset \mathbb{R}^n$ are convex bodies, and let $f \in L^1(\mathbb{R}^n)$ be an arbitrary positive definite function such that $\operatorname{supp}\hat{f} \subset V$, $\hat{f}(0) \geq 0$, and $f(0) = 1$. The problem is to determine the quantity

$$T_* = \inf T$$

provided that $f(x) \leq 0$ for $\|x\|_U \geq T$. Berdysheva solved multidimensional Logan's problem for the l_p^n-ball $U = B_p^n$ and $V = \tau B_\infty^n$, where $1 \leq p \leq 2$, $\tau > 0$, and B_∞^n is a cube. For $\tau = 1$, it was shown that an extremiser is Yudin's function $F_{B_\infty^n}$ and $T_* = \sqrt{n}/2$. To prove this fact, the multidimensional Poisson summation formula was used. Using one-dimensional Bessel quadrature formulas, Gorbachev (2000) solved the problem for the Euclidean balls $U = V = B_2^n$. It turns out that Yudin's function F_V is a unique extremiser and $T_* = q_{n/2-1}/\pi$, where $q_{n/2-1}$ is the smallest positive zero of the Bessel function $J_{n/2-1}$.

Logan's problem is closely related to extremal Delsarte's problem on finding of $\inf f(0)$ provided that $f \in L^1(\mathbb{R}^n)$, $\hat{f} \geq 0$, $\hat{f}(0) = 1$, and $f(x) \leq 0$ for $|x| \geq 2$. Using the solution of Delsarte's problem, one can derive the upper bound for the sphere packing density of $\mathbb{R}^n$, as shown by Gorbachev (2000) and by Elkies and Cohn (2001). Using an approximate solution of Delsarte's problem and analogue of first Logan's problem, Cohn and Kumar (2004) proved the optimality and the uniqueness of the Leech lattice in $\mathbb{R}^{24}$.

The main goal of the paper by Gorbachev is to generalise the methods used by Logan and Berdysheva. Theorem 2.1 in Chap. 3 asserts that if $\lambda_1(L^*)_V = 1$, then

$$R(L)_U \leq T_* \leq \pi^{-1}\sigma_1(V) \sup_{|x|=1} \|x\|_U.$$

The proof is based on the generalised Poisson summation formula and Yudin's function F_V. The above estimates are sharp in the case of $L = \mathbb{Z}^n$.

Logan's problem is closely related to Turan's, Bohman's and others extremal problems of harmonic analysis. Turan's problems were investigated, for example, by D. Gorbachev, C.L. Siegel, V.V. Arestov, E.E. Berdysheva, V.I. Ivanov, M.N. Kolountzakis, Sz.Gy. Révész.

In the second part of his paper, Gorbachev solves Bohman's extremal problem. The problem is to determine a minimum of the second moment

$$M_* = \inf \int_{\mathbb{R}^n} |x|^2 f(x)\, dx$$

for a nonnegative function $f \in L^1(\mathbb{R}^n)$ such that $\operatorname{supp} \hat{f} \subset V$ and $\hat{f}(0) = 1$. For $n = 1$, Bohman [7] showed that $M_* = 1/4$ and $f_*(t) = \frac{8\cos^2(\pi t)}{\pi^2(1-(2t)^2)^2}$. Bohman proved the optimality of some numerical integration methods of probability density functions. Later, Papoulis (1972) also proved this result by means of variational methods in connection with a determination of minimum-bias windows for high-resolution spectral estimates. Periodic Bohman's problem for applications to approximation theory was studied by Korovkin (1958), Yudin (1976), and Ivanov (1994). Yudin constructed the admissible function G_V which is a generalisation of one-dimensional Bohman's function. In the case of the Euclidean ball $V = B_2^n$, Bohman's problem was solved by Ehm et al. [14]: $M_* = (q_{n/2-1}/\pi)^2$ and $f_* = G_{B_2^n}$.

Theorem 2.2 in Chap. 3 asserts that if $\lambda_1(L^*)_V = 1$, then

$$R^2(L) \leq M_* \leq (\sigma_1(V)/\pi)^2.$$

In the case of $L = \mathbb{Z}^n$ and $V = B_\infty^n$, one has $M_* = n/4$ and $f_* = G_{B_\infty^n}$.

2.4 Weighted Estimates for the Hilbert Transform

In Chap. 4, "Weighted Estimates for the Discrete Hilbert Transform", E. Liflyand deals with the classical topic of boundedness properties of the Hilbert transform.

Being somewhat apart of the mainstream, the Paley-Wiener theorem asserts that for an odd and monotone decreasing on $\mathbb{R}_+$ function $g \in L^1$ its Hilbert transform is also integrable, i.e., g is in the (real) Hardy space $H^1(\mathbb{R})$. The oddness of g is essential, since by Kober's result, if $g \in H^1(\mathbb{R})$, then the cancelation property holds $\int_{\mathbb{R}} g(t)\, dt = 0$.

In [19], the monotonicity assumption has been relaxed in the Paley-Wiener theorem, and in [18] weighted versions have been obtained for both the sine and cosine Fourier transforms of functions more general than monotone ones. The periodic case has also been covered in [18] in the same manner. One can find historical background relevant to these problems in [18].

The goal of Liflyand's note is to prove the weighted analogues of the Paley-Wiener theorem for odd and even sequences, that is, for the discrete Hilbert transform. More precisely, the estimates are given in the weighted ℓ^1 spaces, the space of sequences $a = \{a_k\}_{k=0}^{\infty}$ endowed with the norm

$$\|a\|_{\ell(w)} = \sum_{k=0}^{\infty} |a_k| w_k < \infty,$$

where $w = \{w_k\}$, $k = 0, 1, 2, \dots$, is a non-negative sequence, or a weight.

The author uses the definition of general monotone sequences [38, 39]. A null sequence a, that is, vanishing at infinity, is said to be general monotone if it satisfies the conditions

$$\sum_{k=n}^{2n} |a_k - a_{k+1}| \le C \sum_{k=[n/c]}^{[cn]} \frac{|a_k|}{k}, \qquad n = 1, 2, \dots,$$

where $C > 1$ and $c > 1$ are independent of n. See also the survey [19].

In particular the author shows that for a weight $w_k = k^{\alpha}$, the discrete Hilbert transform is bounded in $\ell(w)$, when $-1 < \alpha < 1$ provided that a is an odd and general monotone on $\mathbb{Z}_+$ sequence, or, when $-2 < \alpha < 0$ provided that a is even and general monotone on $\mathbb{R}_+$.

Thus, assuming general monotonicity of a allows one to obtain analogues of the results of [18] to sequences; recall that the estimates in [18] were, in turn, generalisations of Flett's, Hardy-Littlewood's, and Andersen's results to the case $p = 1$. These results can also be considered as generalisations of the known results for sequences in [3].

2.5 Q-Measures and Uniqueness Sets for Haar Series

Chapter 5 by M. Plotnikov is "Q-Measures on the Dyadic Group and Uniqueness Sets for Haar Series". In 1870 G. Cantor proved the following result: *if a trigonometric series*

$$(TS) = \frac{a_0}{2} + \sum_{n=1}^{\infty} a_n \cos(nx) + b_n \sin(nx)$$

converges to zero everywhere on $[0, 2\pi)$ *except possibly on a finite set, then* (TS) *is the trivial series, i.e., all* $a_n = 0$, $b_n = 0$. Cantor's theorem starts the theory of uniqueness for orthogonal series, which nowadays is well-developed (G. Cantor, P. du Bois-Reymond, A. Lebesgue, Ch.J. Vallée Poussin, W.H. Young, F. Bernstein, D.E. Men'shov, A. Rajchman, N.K. Barí, J. Marcinkiewicz, R. Salem, A. Zygmund, I.I. Pyatetckii-Shapiro, etc.).

A set of uniqueness is one of the basic notions of the theory of uniqueness. Let $\{f_n\}$ be a system of functions on $[a, b]$. We recall that a set $A \subset [a, b]$ is called a *set of uniqueness*, or U*-set*, for series

$$\sum_n c_n f_n(x), \qquad c_n \in \mathbb{R} \text{ or } c_n \in \mathbb{C}, \tag{1}$$

if the only series (1) converging to zero everywhere on $[a, b] \setminus A$ is the trivial series.

The Cantor theorem means that every finite set $A \subset [0, 2\pi)$ is a U-set for trigonometric series. The next theorem is one of the most remarkable results in this direction: *let* F_ζ *be the Cantor symmetric set with a constant ratio* ζ*; then* F_ζ *is a* U*-set if and only if* ζ^{-1} *is a Pisot number* (N.K. Barí, R. Salem, I.I. Pyatetckii-Shapiro, R. Salem and A. Zygmund, 1937–1955). In particular, *the Cantor middle thirds set* $F_{1/3}$ *is a set of uniqueness*. So there is a fascinating connection between sets of uniqueness of trigonometric series and number theory.

In general, the problem of a characterisation of U-sets is open even for closed sets. Moreover, in 1980s A.S. Kechris shows that this problem can not be solved constructively. The theory of uniqueness of trigonometric series is treated for example in [15].

Nontrigonometric sets of uniqueness were systematically investigated since 1960s. The *Haar system* $\{H_n(x)\}_{n=0}^{\infty}$ is widely used in harmonic and functional analysis. Note that every series on the Haar system can be represented as some finitely additive set function (so called *quasi-measure*). Moreover, the set of all Haar series and the set of all quasi-measures are linear isomorphic. It was shown in for example [27, 36]) that convergence and behavior of the partial sums of any Haar series is associated with differentiability, continuity, and smoothness of an appropriate quasi–measure.

In contrast to trigonometric series the empty set is the only set of uniqueness for Haar series. Some authors (Skvortsov, Wade, Talalyan, Mushegyan, Gevorkyan, Tetunashvili, Plotnikov) have studied sets of uniqueness for various subclasses of Haar series (*conditional sets of uniqueness*). A relationship between such sets and Hausdorff measures was established in [26]: *A set* $A \subset [0, 1]$ *is a U-set for Haar series satisfying*

$$a_n = O\,(n^{p-1/2}), \quad 0 < p \leq 1, \tag{2}$$

if and only if A contains no perfect subsets of positive Hausdorff $(1-p)$*-measure. Let* F_ζ *be the Cantor symmetric set with a constant ratio* ζ*. Then* F_ζ *is a U-set for Haar series under condition* (2) *if and only if* $p < 1 + \frac{\log 2}{\log \zeta}$*. In particular, the Cantor middle third set* $F_{1/3}$ *is a U-set for Haar series under condition* (2) *if and only if* $p < 1 - \frac{\log 2}{\log 3}$. So such sets of uniqueness can be described only in terms of metric characteristics.

In Plotnikov's paper, he extends the family of subclasses of Haar series and uses so called Q-measures for description of appropriate sets of uniqueness.

2.6 Off-Diagonal Estimates for Calderón-Zygmund Operators

Villarroya's Chap. 6, "Off-Diagonal and Pointwise Estimates for Compact Calderón-Zygmund Operators", deals with operators T satisfying an off-diagonal estimate on $L^p(\mathbb{R}^n)$, i.e., when there exists a function $G : [0, \infty) \to [0, \infty)$ vanishing at infinity such that

$$\|T(f\chi_E)\chi_F\|_{L^p(\mathbb{R}^n)} \lesssim G(\mathrm{dist}(E, F))\|f\|_{L^p(\mathbb{R}^n)},$$

for all Borel sets $E, F \subset \mathbb{R}^n$ and all $f \in L^p(\mathbb{R}^n)$.

These inequalities play an important role in analysis because they enclose all the orthogonality properties of the operator required to prove its boundedness. Their interest in harmonic analysis renewed with the proof of the T(1) Theorem based on wavelet decompositions [11]. This approach involved estimates of the form

$$|\langle T(\psi_I), \psi_J\rangle| \lesssim \left(\frac{|J|}{|I|}\right)^{\frac{1}{2}+\frac{\delta}{n}}\left(1 + \frac{\mathrm{dist}(I, J)}{|I|^{\frac{1}{n}}}\right)^{-(n+\delta)} \|\psi_I\|_{L^2(\mathbb{R}^n)}\|\psi_J\|_{L^2(\mathbb{R}^n)},$$

for all cubes $I, J \subset \mathbb{R}^n$, with $|J| \leq |I|$, under the appropriate hypothesis on the operator T, the parameter $\delta > 0$ and the functions ψ_I, ψ_J. Similar bounds were also crucial in the solution of the famous Kato's conjecture [4] about boundedness of square root of elliptic operators and are nowadays extensively used in the study of boundedness of second order elliptic operators (see [5] for example).

In [41], Villarroya developed a $T(1)$ Theorem to characterize not only boundedness but also compactness of Calderón-Zygmund operators. Now, the paper *Off-diagonal and pointwise estimates for compact Calderón-Zygmund operators* provides off-diagonal bounds for singular integral operators that extend compactly on $L^p(\mathbb{R}^n)$. Under the right hypotheses, the estimates can be stated as follows:

$$|\langle T(\psi_I), \psi_J \rangle| \lesssim \left(\frac{|J|}{|I|}\right)^{\frac{1}{2}+\frac{\delta}{n}} \left(1 + \frac{\operatorname{dist}(I,J)}{|I|^{\frac{1}{n}}}\right)^{-(n+\delta)} F(I,J)\, \|\psi_I\|_{L^2(\mathbb{R}^n)} \|\psi_J\|_{L^2(\mathbb{R}^n)},$$

for appropriate bump functions ψ_I, ψ_J well localized on cubes $I, J \subset \mathbb{R}^n$. The last factor $F(I,J)$ is a new element, completely absent in the analog inequalities for bounded singular integrals, which encodes the extra rate of decay obtained as a consequence of the compactness properties of the operator. Therefore, the focus of the work is placed on obtaining a sharp and detailed description of this new extra factor in terms of the size and location of the cubes I and J.

3 Function Spaces of Radial Functions

3.1 Potential Spaces of Radial Functions

Chapter 7 by P. De Napoli and I. Drelichman is "Elementary Proofs of Embedding Theorems for Potential Spaces of Radial Functions". Embedding theorems play a central role in the theory of nonlinear partial differential equations, in particular when applying variational methods for proving existence of solutions. In many cases, the problem and the associated functional are symmetric under the action of the orthogonal group, hence one is led to the study of radial solutions.

Moreover, since the pioneering works of Ni [24] and Strauss [37], it is well known that better embedding theorems can be obtained when restricting to subspaces of radially symmetric functions. Indeed, Strauss' inequality implies that a radial function in the Sobolev space $H^1(\mathbb{R}^n)$ necessarily has a decay at infinity like a negative power of $|x|$, and Ni's inequality gives a bound for the behaviour near the origin of such a function. As a consequence, one can recover compactness of the embeddings, which is an essential feature for the success of variational methods and which does not hold for arbitrary functions in $\mathbb{R}^n$ due to the translation invariance of the Sobolev norm. These results were generalised by Lions [21] in several directions, including Sobolev spaces of fractional order.

The paper by De Napoli and Drelichman surveys some known results in the theory of radial Sobolev and Bessel potential spaces. The authors give two simple proofs of a version of Strauss' inequality for potential spaces, one for $p = 2$ (using the Fourier transform) and another one for general p (using a weighted convolution theorem for radial functions by the authors [12]). Another section of the paper is devoted to embedding theorems with power weights for radial functions, which include a generalisation of Ni's inequality in $\mathbb{R}^n$. Then they analyse the compactness of the embeddings, both in the unweighted case (giving an alternative proof of a theorem by Lions, avoiding the use of complex interpolation) and in the weighted case. Finally, they discuss a generalisation of Ni's inequality and embedding theorems for potential spaces in a ball.

The proofs presented are elementary in the sense that they avoid the use of interpolation theory and sophisticated tools such as atomic or wavelet decompositions.

3.2 On Leray's Formula

In Chap. 8, "On Leray's Formula", E. Liflyand and S. Samko suggest a new direction in a study of the multidimensional Fourier transform of a radial functions. It is well known that if f is an integrable radial function, that is, $f(x) = f_0(|x|)$, the n-dimensional Fourier integral $\hat{f}(x) = \int_{\mathbb{R}^n} f(u)\, e^{-ix\cdot u} du,\ u, x \in \mathbb{R}^n$, reduces to the one-dimensional integral:

$$\hat{f}(x) = \hat{f_0}(|x|) = (2\pi)^{\frac{n}{2}} \int_0^{\infty} f_0(t)(|x|t)^{-\frac{n}{2}-1} J_{\frac{n}{2}-1}(|x|t) t^{n-1}\, dt,$$

where J_μ is the Bessel function of first type and order μ. This makes sense also for radial L^p-functions with $1 < p < \frac{2n}{n+1}$: in that case $\hat{f}(x)$ is everywhere continuous away from the origin. However, working with Bessel functions is quite tedious business sometimes, and certain attempts are known to simplify this formula towards obtaining a "genuine" one-dimensional Fourier transform of some function related to f_0; let us mention [28] and [20].

Leray (see [17, 33]) proved that if

$$\int_0^{\infty} t^{n-1}(1+t)^{\frac{1-n}{2}} |f_0(t)|\, dt < \infty,$$

the following relation holds

$$\hat{f}(x) = 2\pi^{\frac{n-1}{2}} \int_0^{\infty} I_{\frac{n-1}{2}} f_0\,(t) \cos |x| t\, dt, \tag{3}$$

where the fractional integral I_α is defined by

$$I_\alpha f_0\,(t) = \frac{2}{\Gamma(\alpha)} \int_t^{\infty} s f_0(s)(s^2 - t^2)^{\alpha-1} ds.$$

This formula seems to be very attractive, however much is hidden in the fractional integral of f_0. Liflyand and Samko change (3) in such a way that it reduces to the one-dimensional Fourier transform of a function "closer" to f_0.

4 Approximation Theory

4.1 Approximation Order of Besov Classes

"Order of Approximation of Besov Classes in the Metric of Anisotropic Lorentz Spaces" by K. Bekmaganbetov is Chap. 9.

In this paper the author deals with the anisotropic Lorentz spaces $L_{\mathbf{q}\theta}(\mathbb{T}^{\mathbf{d}})$ endowed with a norm

$$\|f\|_{L_{\mathbf{pq}}(\mathbb{T}^{\mathbf{d}})} =$$

$$\left(\int_0^{(2\pi)^{d_n}} \cdots \left(\int_0^{(2\pi)^{d_1}} \left(t_1^{\frac{1}{p_1}} \dots t_n^{\frac{1}{p_n}} f^{*_1,\dots,*_n}(t_1,\dots,t_n)\right)^{q_1} \frac{dt_1}{t_1}\right)^{\frac{q_2}{q_1}} \cdots \frac{dt_n}{t_n}\right)^{\frac{1}{q_n}} < \infty,$$

where $f^*(\mathbf{t}) = f^{*_1,\dots,*_n}(t_1,\dots,t_n)$.

Using this definition, he defines the Besov space as follows:

$$\|f\|_{B^{\alpha\theta}_{\mathbf{pr}}(\mathbb{T}^{\mathbf{d}})} = \left\| \left\{ 2^{(\alpha,s)} \|\Delta_{\mathbf{s}}(f)\|_{L_{\mathbf{pr}}(\mathbb{T}^{\mathbf{d}})} \right\}_{\mathbf{s}\in\mathbb{Z}^n_+} \right\|_{l_\theta} < \infty,$$

where $\|\cdot\|_{l_\theta}$ is a norm of the discrete Lebesgue spaces with mixed metric l_θ and

$$\Delta_{\mathbf{s}}(f,\mathbf{x}) = \sum_{k\in\rho(s)} a_{\mathbf{k}}(f) e^{i(\mathbf{k},\mathbf{x})}.$$

Here $\{a_{\mathbf{k}}(f)\}_{\mathbf{k}\in\mathbb{Z}^{\mathbf{d}}}$ are the Fourier coefficients of the function f with respect to the multiple trigonometric system and

$$\rho(\mathbf{s}) = \{\mathbf{k} = (\mathbf{k}_1,\dots,\mathbf{k}_n) \in \mathbb{Z}^{\mathbf{d}} : \left[2^{s_i-1}\right] \le \max_{j=1,\dots,d_i} |k^i_j| < 2^{s_i}, i = 1,\dots,n\}.$$

Bekmaganbetov obtains sharp order of approximation of functions from the Besov classes $B^{\alpha\theta}_{\mathbf{pr}}(\mathbb{T}^{\mathbf{d}})$ in the Lorentz space metric, that is, he finds two-sided estimates for

$$\sup_{\|f\|_{B^{\alpha\tau}_{\mathbf{p}\theta}(\mathbb{T}^{\mathbf{d}})}\le 1} E_{\gamma'\mathbf{d},N}(f)_{L_{\mathbf{q}\theta}},$$

where $E_{\gamma\mathbf{d},N}(f)_{L_{\mathbf{pr}}}$ is the best approximation of $f \in L_{\mathbf{pr}}(\mathbb{T}^{\mathbf{d}})$ by polynomials with harmonics in hyperbolic crosses.

The obtained results are related to the previous investigations by K.I. Babenko, Ya.S. Bugrov, E.M. Galeev, Dinh Dung, V.N. Temlyakov, and the recent results by Akishev [2] and Nursultanov and Tleukhanova [25].

4.2 Ulyanov Inequalities for Moduli of Smoothness

Chapter 10 by M.K. Potapov and B.V. Simonov is "Analogues of Ulyanov Inequalities for Mixed Moduli of Smoothness". The authors study an important problem in approximation theory on sharp inequalities between moduli of smoothness in different metrics (L^p, L^q), $1 \le p < q \le \infty$.

In the one-dimensional case, if $f \in L_p(\mathbb{T})$, $1 \le p \le \infty$, the modulus of smoothness of fractional order is given by

$$\omega_\alpha(f,t)_p = \sup_{|h|\le t} \Big\| \sum_{\nu=0}^{\infty} (-1)^\nu \tbinom{\alpha}{\nu} f(x+(\alpha-\nu)h) \Big\|_p .$$

Sharp Ulyanov inequality [35] provides an optimal estimate for moduli of smoothness in (L^p, L^q), $1 < p < q < \infty$:

$$\omega_\alpha(f,\delta)_q \le C \Big(\int_0^\delta \big(t^{-(\frac{1}{p}-\frac{1}{q})} \omega_{\alpha+\frac{1}{p}-\frac{1}{q}}(f,t)_p \big)^q \frac{dt}{t} \Big)^{\frac{1}{q}} .$$

This inequality also holds if $p = 1$ and $q = \infty$ but not if $p = 1$, $q < \infty$ or $1 < p$, $q = \infty$. In these cases, one has [40]: for $p = 1$, $1 < q < \infty$,

$$\omega_\alpha(f,\delta)_q \le C \Big(\int_0^\delta \big(t^{-(\frac{1}{p}-\frac{1}{q})} (\ln 2/t)^{1/q} \omega_{\alpha+\frac{1}{p}-\frac{1}{q}}(f,t)_p \big)^q \frac{dt}{t} \Big)^{\frac{1}{q}}$$

and for $1 < p < \infty$, $q = \infty$,

$$\omega_\alpha(f,\delta)_q \le C \int_0^\delta t^{-(\frac{1}{p}-\frac{1}{q})} (\ln 2/t)^{1-\frac{1}{p}} \omega_{\alpha+\frac{1}{p}-\frac{1}{q}}(f,t)_p \frac{dt}{t} .$$

For periodic functions on $\mathbb{T}^2$, the mixed modulus of smoothness of a function $f \in L_p(\mathbb{T}^2)$ of orders $\alpha_1 > 0$ and $\alpha_2 > 0$ with respect to the variables x and y, respectively, is defined as follows:

$$\omega_{\alpha_1,\alpha_2}(f,\delta_1,\delta_2)_p = \sup_{|h_i|\le\delta_i, i=1,2} ||\Delta_{h_1}^{\alpha_1}(\Delta_{h_2}^{\alpha_2}(f))||_p ,$$

where $\Delta_{h_1}^{\alpha_1}(f)$ is the difference of order $\alpha_1 > 0$ with respect to the variable x and $\Delta_{h_2}^{\alpha_2}(f)$ is the difference of order $\alpha_2 > 0$ with respect to the variable y. For the main properties of the mixed moduli of smoothness see [30]. Sharp Ulyanov inequality for the mixed moduli was obtained in [31] for $1 < p < q < \infty$.

In their paper, Potapov and Simonov prove a sharp Ulyanov inequality for the mixed moduli in the case of $p = 1$ and $q = \infty$.

4.3 Approximation Order of Besov Classes

Tleukhanova's Chap. 11 is "Reconstruction Operator of Functions from the Sobolev Space".

In this work, she deals with the reconstruction problem of periodic functions from the spaces W_p^α with dominant mixed derivative by values of a function at a given number of nodes. Let $X = W_p^\alpha[0,1]^n$ and $Y = L_q[0,1]^n$. The author studies the decay order of the orthogonal diameter given by

$$d_M^{\perp}(X,Y) = \inf_{\{g_j\}_{j=1}^M} \sup_{\|f\|_X=1} \|f - \sum_{j=1}^{M} (f, g_j) g_j\|_Y,$$

where the infimum is taken over all orthogonal systems $\{g_j\}_{j=1}^M$ from $L_\infty[0,1]^n$.

It is shown that

$$\sup_{\|f\|_{W_p^\alpha}=1} \|f - F_m(f)\|_{L_q} \sim d_M^{\perp}(W_p^\alpha, L_q),$$

for $1 < p \le 2 \le q \le \infty$ and $\alpha > \frac{1}{p}$, where $F_m(f)$ is given by

$$F_m(f;x) = \sum_{\substack{|k|=m \\ k\in\mathbb{N}^n}} \frac{1}{2^m} \sum_{0\le r<2^k} f(\frac{r_1}{2^{k_1}}, \ldots, \frac{r_n}{2^{k_n}}) \phi_{k,r}(x_1 + \frac{r_1}{2^{k_1}}, \ldots, x_1 + \frac{r_n}{2^{k_n}}),$$

and

$$\phi_{k,r}(x) = \sum_{0\le\nu\le k} (-1)^{\sum_{j=1}^{n-1}(r_j+1)sgn(k_j-\nu_j)} \sum_{\mu\in\rho(\nu)} e^{2\pi i\mu x}.$$

5 Optimization Theory and Related Topics

5.1 The Laplace-Borel Transform

Chapter 12 is "Laplace-Borel Transformation of Functions Holomorphic in the Torus and Equivalent to Entire Functions" by L. Maergoiz.

The Laplace–Borel integral transform is an important concept in the theory of entire functions of one and several variables. It has numerous applications in functional and harmonic analysis.

In the beginning of the twentieth century, Borel [8] considered the interrelation between the series

$$g(z) = \sum_{k=0}^{\infty} a_k z^k , \tag{4}$$

which converges in a neighborhood of the origin and the following entire function of exponential type

$$f(z) = \sum_{k=0}^{\infty} \frac{a_k z^k}{k!}. \tag{5}$$

He proved the following theorem. *Let*

$$G(z) = \int_0^{\infty} f(tz)e^{-t}dt, \quad z \in B,$$

where B is the interior of the set of absolute and uniform convergence for the integral $G(z)$. Then $G(z)$ determines an analytic extension of the series g (see (4)) *to the domain B. Moreover, $B = \{z \in \mathbb{C} : h(z) < 1\}$, where $h(z) = h(te^{i\theta}) = th(\theta;f)$, and*

$$h(\theta;f) = \overline{\lim_{r\to\infty}} \, r^{-1} \ln |f(re^{i\theta})|, \quad \theta \in \mathbb{R}$$

is the indicator of the entire function $f(z)$ given by (5).

In 1929 Polya [29] introduced a notion of the *indicator diagram I_h* of the entire function f. The latter is the compact convex subset of $\mathbb{C}$ such that its support function is equal to the indicator $h(\theta;f)$ of f. G. Polya studied the Laurent series

$$G(z) = \sum_{k=0}^{\infty} \frac{a_k}{z^{k+1}} \tag{6}$$

which converges in a neighborhood of ∞. He obtained the following theorem. *Let in the above notation*

$$\Pi_\theta = \{z \in \mathbb{C} : Reze^{i\theta} > h(\theta;f)\}, \quad \theta \in (-\pi, \pi].$$

The system of Laplace integrals

$$G_\theta(z) = \int_0^{\infty(\arg t=\theta)} f(t)e^{-tz}dt, \quad z \in \Pi_\theta; \quad \theta \in (-\pi, \pi]$$

determines an analytic extension of the series G given by (6) to the set $D = \bigcup \Pi_\theta | \theta \in (-\pi, \pi]$. Moreover, $K = \mathbb{C} \setminus D = \{z \in \mathbb{C} : \bar{z} \in I_h\}$ (the conjugate diagram of f) is the smallest compact convex set in $\mathbb{C}$ outside of which the series F admits analytic continuation. Here I_h is the indicator diagram of the entire function f (see (5)).

These results by Borel and Polya as well as their one-dimensional and multi-dimensional analogues can be considered as a kind of a bridge between methods of the theory of entire functions and various other sections of complex analysis, in particular, summation theory of the Laurent series.

The paper by Maergoiz is devoted to investigations of the Laplace-Borel integral transformation of functions *equivalent to entire functions* of several variables.

5.2 Optimization Control Problems

Chapter 13 is Serovaisky's "Optimization Control Problems for Systems Described by Elliptic Variational Inequalities with State Constraints". The optimization control theory has many practical applications. Among other things, it is applicable for different mathematical models. One of the most important and difficult classes of mathematical models are the variational inequalities. Different optimization control problems of systems described by variational inequalities were considered by V. Barbu, D. Tiba, J. Bonnans, E. Casas, and other. The most difficult class of the optimization control problems are problems with state constraints. Special optimization control problems for systems described by variational inequalities have been analysed by He Zheng-Xu, G. Wang, Y. Zhao, W. Li, J. Bonnans, E. Casas. In his paper, Serovajsky considers an analogous optimization control problem with state constraint in the form of the general inclusion for the elliptic case. His analysis is based on the Warga's concept of the search of minimizing sequences, but not optimal controls. He proposes a double regularization of the optimization control problem. At first the variational inequality, which defines the state of the system, is approximated by a nonlinear equation by using the penalty method. Then he obtains an optimization control problem for a nonlinear elliptic equation with state constraints. This problem is approximated by a minimization problem for a penalty functional on the set of admissible "state-control" pairs. J.L. Lions used the analogous technique for distributed singular systems without state constraints. However, the system is regular, and state constraints are present at this case. Lions applied penalty method for obtaining necessary conditions of optimality for the initial solvable optimization problem. Nevertheless, this problem can be unsolvable. The author uses the idea of finding minimizing sequences but not optimal control here. Warga proposed this method for the extension of unsolvable optimization controls problems. However, the author proves the solvability of his problem, and this method is applied for finding an optimal control. Thus, an approximate solution of the initial optimization control problem is chosen as the optimal control for approximate problem for a large enough step of the algorithm. Then the necessary conditions of optimality for the approximate optimization control problem are obtained in the standard form.

5.3 Optimization Control Problems for Parabolic Equation

Chapter 14 is Ilyas Shakenov's "Two Approximation Methods of the Functional Gradient for a Distributed Optimization Control Problem". Practical solving of optimization control problems uses different approximation methods. Therefore, there arises a serious problem: One can use standard optimization methods (gradient methods, necessary conditions of optimality, etc.) with finite difference approximation of the state and adjoint systems. However, one can also first approximate the system and then solve the obtained discrete optimization problem. The question is: what will be more efficient, to use the approximation before or after the application of optimization methods?

A. Karchevsky, V. Zubov, B. Rysbaiuly considered this question for inverse problems of mathematical physics. Here the author considers corresponding practical computing aspects of the order of approximation and optimization for an optimization control problem. The author solves an optimization control problem for the system described by a one-dimensional parabolic equation. He uses two different methods of numerical analysis for finding an optimal control. He approximates functional gradient and calculates the gradient of functional approximation. He discusses the results of computer experiments for different values of the system parameters and compares the application of different forms of the approximation.

5.4 Numerical Modeling of the Linear Filtration

Chapter 15 is "Numerical Modeling of the Linear Relaxational Filtration by Monte Carlo Methods" by Kanat Shakenov.

The numerical methods for the relaxational filtration phenomenon is an important branch of mathematics directed at practical solutions of problems of mathematical physics. Problems of relaxational filtration have been considered by Yu.M. Molokovich, P.P. Osipov, A.V. Kazhihov, V.M. Monahov, etc. For example, in the book [23], a process of non stationary filtration flow of uniform droplet-compressible monophase fluid in isotropic weakly-deformable porous environment has been considered, etc.

A. Haji-Sheikh, E.M. Sparrow, S.M. Ermakov, V.V. Nekrutkin, G.A. Mihailov, and others solved a number of problems of mathematical physics by Monte Carlo methods. H. Kushner used probability difference methods for solving boundary problems for elliptic and parabolic systems.

The author previously considered a process of nonstationary filtration flow of uniform droplet-compressible monophase fluid in isotropic weakly-deformable porous environment. He analysed four models of linear “relaxational” filtration in a one-dimensional domain. However, these models are not applicable for the practical situations. Now the author considers these four models of linear “relaxational” filtration process in a three-dimensional domain. These are the model of classical elastic filtration, the simplest model of filtration with a constant speed of disturbance spread, the filtration model in relaxationaly-compressed porous environment realised by the linear Darcy law, and the model of filtration by the simplest unbalanced law in elastic porous environment. Dirichlet, Neumann and mixed problems for these models have been solved. The author approaches these problems by using “random walk on spheres”, “random walk on balls” and “random walk on lattices” algorithms by Monte Carlo methods and by probability difference methods. He obtains and discusses the results of the computer experiments.

Acknowledgements The first author “Michael Ruzhansky” was supported in parts by the EPSRC grant EP/K039407/1 and by the Leverhulme Grant RPG-2014-02. The second author “Sergey Tikhonov” was partially supported by MTM2014-59174-P, 2014 SGR 289, and RFFI 13-01-00043. We thank the authors for the assistance in writing this survey.

References

1. P. Ahern, W. Rudin, Bloch functions, BMO, and boundary zeros. Indiana Univ. Math. J. **36**(1), 131–148 (1987)
2. G.A. Akishev, Approximation of function classes in spaces with mixed norm. Sb. Math. **197**(7–8), 1121–1144 (2006)
3. K. Andersen, Inequalities with weights for discrete Hilbert transforms. Can. Math. Bull. **20**, 9–16 (1977)
4. P. Ausher, S. Hofmann, M. Lacey, A. McIntosh, Ph. Tchamitchian, The solution of the Kato square root problem for second order elliptic operators on R^n. Ann. Math. **2**, 633–654 (2002)
5. A. Axelson, S. Keith, A. McIntosh, Quadratic estimates and functional calculi of perturbated dirac operators. Invent. Math. **163**(3), 455–497 (2006)
6. E.E. Berdysheva, Two related extremal problems for entire functions of several variables. Math. Notes **66**(3), 271–282 (1999)
7. H. Bohman, Approximate Fourier analysis of distribution functions. Ark. Mat. **4**, 99–157 (1960)
8. E. Borel, *Leçons sur les séries divergentes* (Gaunter-Villars, Paris, 1901)
9. P.G. Casazza, M. Fickus, J. Kovacevic, M.T. Leon, J.C. Tremain, A physical interpretation of finite frames. Appl. Numer. Harmon. Anal. **2–3**, 51–76 (2006)
10. P. Casazza, G. Kutyniok, *Finite Frames: Theory and Applications* (Birkhauser, New York, 2013)
11. R.R. Coifman, Y. Meyer, *Wavelets, Calderon-Zygmund and Multilinear Operators* (Cambridge University Press, Cambridge, 1997)
12. P. De Nápoli, I. Drelichman, Weighted convolution inequalities for radial functions. Ann. Mat. Pura Appl. **194**, 167–181 (2015)
13. E. Doubtsov, Bloch-to-BMOA compositions on complex balls. Proc. Am. Math. Soc. **140**(12), 4217–4225 (2012)
14. W. Ehm, T. Gneiting, D. Richards, Convolution roots of radial positive definite functions with compact support. Trans. Am. Math. Soc. **356**, 4655–4685 (2004)

15. A.S. Kechris, A. Louveau, *Descriptive Set Theory and the Structure of Sets of Uniqueness*. London Mathematical Society lecture series, vol. 128 (Cambridge University Press, Cambridge, 1987)
16. E.G. Kwon, Hyperbolic mean growth of bounded holomorphic functions in the ball. Trans. Am. Math. Soc. **355**(3), 1269–1294 (2003)
17. J. Leray, *Hyperbolic Differential Equations* (Institute for Advanced Study, Princeton, 1953)
18. E. Liflyand, S. Tikhonov, Weighted Paley-Wiener theorem on the Hilbert transform. C.R. Acad. Sci. Paris, Ser. I **348**, 1253–1258 (2010)
19. E. Liflyand, S. Tikhonov, A concept of general monotonicity and applications. Math. Nachr. **284**, 1083–1098 (2011)
20. E. Liflyand, W. Trebels, On asymptotics for a class of radial Fourier transforms. Z. Anal. Anwen. **17**, 103–114 (1998)
21. P.L. Lions, Symétrie e compacité dans les espaces de Sobolev. J. Funct. Anal. **49**, 315–334 (1982)
22. B.F. Logan, Extremal problems for positive-definite bandlimited functions. I. Eventually positive functions with zero integral. SIAM J. Math. Anal. **14**(2), 249–252 (1983)
23. Y.M. Molokovich, P.P. Osipov, *Basics of Relaxation Filtration Theory* (Kazan University, Kazan, 1987)
24. W.M. Ni, A nonlinear Dirichlet problem on the unit ball and its applications. Indiana Univ. Math. J. **31**(6), 801–807 (1982)
25. E.D. Nursultanov, N.T. Tleukhanova, On the approximate computation of integrals for functions in the spaces $W_p^\alpha([0,1]^n)$. Russ. Math. Surv. **55**(6), 1165–1167 (2000)
26. M.G. Plotnikov, Quasi-measures, Hausdorff p-measures and Walsh and Haar Series. Izv. RAN: Ser. Mat. **74**, 157–188 (2010); Engl. Transl. Izvestia: Math. **74**, 819–848 (2010)
27. M. Plotnikov, V. Skvortsov, *On various types of continuity of multiple dyadic integrals*, Acta Math. Acad. Paedagog. Nyházi (N. S.) (2015, to appear)
28. A.N. Podkorytov, Linear means of spherical Fourier sums, in *Operator Theory and Function Theory*, ed. by M.Z. Solomyak, vol. 1 (Leningrad University, Leningrad) (1983), pp. 171–177 (Russian)
29. G. Polya, Untersuchungen über Lücken and Singularitäten von Potenzsrihen. Math. Zeits. Bd. **29**, 549–640 (1929)
30. M.K. Potapov, B.V. Simonov, S.Yu. Tikhonov, Mixed moduli of smoothness in L_p, $1 < p < \infty$: a survey. Surv. Approx. Theory **8**, 1–57 (2013)
31. M.K. Potapov, B.V. Simonov, S.Yu. Tikhonov, Relations between the mixed moduli of smoothness and embedding theorems for Nikol'skii classes, in *Proceeding of the Steklov Institute of Mathematics*, vol. 269 (2010), pp. 197–207; translation from Russian: Trudy Matem. Inst. V.A. Steklova **269**, 204–214 (2010)
32. W. Ramey, D. Ullrich, Bounded mean oscillation of Bloch pull-backs. Math. Ann. **291**(4), 591–606 (1991)
33. S.G. Samko, A.A. Kilbas, O.I. Marichev, *Fractional Integrals and Derivatives: Theory and Applications* (Gordon and Breach, New York, 1993)
34. J.H. Shapiro, *Composition Operators and Classical Function Theory*. Universitext: Tracts in Mathematics (Springer, New York, 1993)
35. B. Simonov, S. Tikhonov, Sharp Ul'yanov-type inequalities using fractional smoothness. J. Approx. Theory **162**, 1654–1684 (2010)
36. V.A. Skvortsov, Henstock-Kurzweil type integrals in $\mathcal{P}$-adic harmonic analysis. Acta Math. Acad. Paedagog. Nyházi (N. S.) **20**, 207–224 (2004)
37. W.A. Strauss, Existence of solitary waves in higher dimensions. Commun. Math. Phys. **55**, 149–162 (1977)
38. S. Tikhonov, Trigonometric series with general monotone coefficients. J. Math. Anal. Appl. **326**, 721–735 (2007)

39. S. Tikhonov, Best approximation and moduli of smoothness: computation and equivalence theorems. J. Approx. Theory **153**, 19–39 (2008)
40. S. Tikhonov, Weak type inequalities for moduli of smoothness: the case of limit value parameters. J. Fourier Anal. Appl. **16**(4), 590–608 (2010)
41. P. Villarroya, A characterization of compactness for singular integrals. J. Math. Pure Appl. (arXiv:1211.0672) (to appear)
42. S. Yamashita, Hyperbolic Hardy class H^1. Math. Scand. **45**(2), 261–266 (1979)

Part I
Fourier Analysis

Parseval Frames with $n + 1$ Vectors in $\mathbb{R}^n$

Laura De Carli and Zhongyuan Hu

Abstract We prove a uniqueness theorem for triangular Parseval frame with $n + 1$ vectors in $\mathbb{R}^n$. We also provide a characterization of unit-norm frames that can be scaled to Parseval frames.

Keywords Parseval frames • Scalable frames

Mathematics Subject Classification (2000). Primary 42C15, Secondary 46C99

1 Introduction

Let $\mathcal{B} = \{v_1, \dots, v_N\}$ be a set of vectors in $\mathbb{R}^n$. We say that $\mathcal{B}$ is a *frame* if it contains a basis of $\mathbb{R}^n$, or equivalently, if there exist constants $A,\ B > 0$ for which $A||v||^2 \le \sum_{j=1}^N < v, v_j >^2 \le B||v||^2$ for every $v \in \mathbb{R}^n$. Here and throughout the paper, $<\ ,\ >$ and $||\ ||$ are the usual scalar product and norm in $\mathbb{R}^n$. In general $A < B$, but we say that a frame is *tight* if $A = B$, and is *Parseval* if $A = B = 1$.

Parseval frames are nontrivial generalizations of orthonormal bases. Vectors in a Parseval frame are not necessarily orthogonal or linearly independent, and do not necessarily have the same length, but the *Parseval identities* $v = \sum_{j=1}^N < v, v_j > v_j$ and $||v||^2 = \sum_{j=1}^N < v, v_j >^2$ still hold. In the applications, frames are more useful than bases because they are resilient against the corruptions of additive noise and quantization, while providing numerically stable reconstructions [6, 7, 9]. Appropriate frame decomposition may reveal hidden signal characteristics, and have been employed as detection devices. Specific types of finite tight frames have been

L. De Carli (✉)
Department of Mathematics, Florida International University, Miami, FL 33199, USA
e-mail: decarlil@fiu.edu

Z. Hu
Department of Economics, Florida International University, Miami, FL 33199, USA
e-mail: zyhu999@gmail.com

M. Ruzhansky, S. Tikhonov (eds.), *Methods of Fourier Analysis and Approximation Theory*, Applied and Numerical Harmonic Analysis,
DOI 10.1007/978-3-319-27466-9_2

studied to solve problems in information theory. The references are too many to cite, but see [4], the recent book [2] and the references cited there.

In recent years, several inquiries about tight frames have been raised. In particular: how to characterize Parseval frames with N elements in $\mathbb{R}^n$ (or *Parseval N frames*), and whether it is possible to scale a given frame so that the resulting frame is Parseval.

Following [1, 11], we say that a frame $\mathcal{B} = \{v_1, \ldots, v_N\}$ is *scalable* if there exists positive constants $\ell_1, \ldots, \ell_N$ such that $\{\ell_1 v_1, \ldots, \ell_N v_N\}$ is a Parseval frame. Two Parseval $N-$frames are *equivalent* if one can be transformed into the other with a rotation of coordinates and the reflection of one or more vectors. A frame is *nontrivial* if no two vectors are parallel. In the rest of the paper, when we say "unique" we will always mean "unique up to an equivalence", and we will often assume without saying that frames are nontrivial.

It is well known that Parseval $n-$frames are orthonormal (see also Corollary 3.3). Consequently, for given unit vector w, there is a unique Parseval $n-$frame that contains w. If $||w|| \neq 1$, no Parseval $n-$frame contains w.

When $N > n$ and $||w|| \leq 1$, there are infinitely many non-equivalent Parseval N-frames that contain w.[1] By the main theorem in [3], it is possible to construct a Parseval frame $\{v_1, \ldots, v_N\}$ with vectors of prescribed lengths $0 < \ell_1, \ldots, \ell_N \leq 1$ that satisfy $\ell_1^1 + \ldots + \ell_N^2 = n$. We can let $\ell_N = ||w||$ and, after a rotation of coordinates, assume that $v_N = w$, thus proving that the Parseval frames that contain w are as many as the sets of constants $\ell_1, \ldots \ell_{N-1}$.

But when $N = n + 1$, there is a class of Parseval frames that can be uniquely constructed from a given vector: precisely, all *triangular* frames, that is, frames $\{v_1, \ldots, v_N\}$ such that the matrix $(v_1, \ldots v_n)$ whose columns are $v_1, \ldots, v_n$ is right triangular. We recall that a matrix $\{a_{i,j}\}_{1 \leq i,j \leq n}$ is *right-triangular* if $a_{i,j} = 0$ if $i > j$.

The following theorem will be proved in Sect. 3.

Theorem 1.1 *Let $\mathcal{B} = \{v_1, \ldots, v_n, w\}$ be a triangular Parseval frame, with $||w|| < 1$. Then $\mathcal{B}$ is unique, in the sense that if $\mathcal{B}' = \{v_1', \ldots, v_n', w\}$ is another triangular Parseval frame, then $v_j' = \pm v_j$.*

Every frame is equivalent, through a rotation of coordinates ρ, to a triangular frame, and so Theorem 1.1 implies that every Parseval $(n+1)$-frame that contains a given vector w is equivalent to one which is uniquely determined by $\rho(w)$. However, that does not imply that the frame itself is uniquely determined by w because the rotation ρ depends also on the other vectors of the frame.

We also study the problem of determining whether a given frame $\mathcal{B} = \{v_1, \ldots, v_n, v_{n+1}\} \subset \mathbb{R}^n$ is scalable or not. Assume $||v_j|| = 1$, and let $\theta_{i,j} \in [0, \pi)$ be the angle between v_i and v_j.

If $\mathcal{B}$ contains an orthonormal basis, then the problem has no solution, so we assume that this not the case. We prove the following

[1] We are indebted to P. Casazza for this remark.

Theorem 1.2 *$\mathcal{B}$ is scalable if and only there exist constants $\ell_1, \ldots, \ell_{n+1}$ such that for every $i \neq j$*

$$(1-\ell_i^2)(1-\ell_j^2) = \ell_i^2\ell_j^2 \cos\theta_{i,j}^2. \tag{1}$$

The identity (1) has several interesting consequences (see corollary 3.2). First of all, it shows that $\ell_j^2 \leq 1$; if $\ell_i = 1$ for some i, we also have $\cos\theta_{i,j}^2 = 0$ for every j, and so v_i is orthogonal to all other vectors. This interesting fact is also true for other Parseval frames, and is a consequence of the following.

Theorem 1.3 *Let $\mathcal{B} = \{v_1, \ldots, v_N\}$ be a Parseval frame. Let $||v_j|| = \ell_j$. Then*

$$\sum_{j=i}^{N} \ell_j^2 \cos^2\theta_{ij} = 1 \quad \sum_{j=i}^{N} \ell_j^2 \sin^2\theta_{ij} = n-1. \tag{2}$$

The identities (2) are probably known, but we did not find a reference in the literature. It is worthwhile to remark that from (2) follows that

$$\sum_{j=i}^{N} \ell_j^2 \cos^2\theta_{ij} - \sum_{j=i}^{N} \ell_j^2 \sin^2\theta_{ij} = \sum_{j=i}^{N} \ell_j^2 \cos(2\theta_{ij}) = 2-n.$$

When $n = 2$, this identity is proved in Proposition 2.1.

Another consequence of Theorem 1.1 is the following

Corollary 1.4 *If $\mathcal{B} = \{v_1, \ldots, v_n, v_{n+1}\}$ is a scalable frame, then, for every $1 \leq i \leq n+1$, and every $j \neq k \neq i$ and $k' \neq j' \neq i$,*

$$\frac{|\cos\theta_{k,j}|}{|\cos\theta_{k,j}| + |\cos\theta_{k,i}\cos\theta_{j,i}|} = \frac{|\cos\theta_{k',j'}|}{|\cos\theta_{k',j'}| + |\cos\theta_{k',i}\cos\theta_{j',i}|}. \tag{3}$$

We prove Theorems 1.1, 1.2 and 1.3 and their corollaries in Sect. 3. In Sect. 2 we prove some preliminary results and lemmas.

2 Preliminaries

We refer to [2] or to [10] for the definitions and basic properties of finite frames.

We recall that $\mathcal{B} = \{v_1, \ldots, v_N\}$ is a Parseval frames in $\mathbb{R}^n$ if and only if the rows of the matrix $(v_1, \ldots, v_N)$ are orthonormal. Consequently, $\sum_{i=1}^{N} ||v_i||^2 = n$. If the vectors in $\mathcal{B}$ have all the same length, then $||v_i|| = \sqrt{n/N}$. See e.g. [2].

We will often let $\vec{e}_1 = (1, 0, \ldots, 0), \ldots \vec{e}_n = (0, \ldots, 0, 1)$, and we will denote by $(v_1, \ldots, \hat{v}_k, \ldots v_N)$ the matrix with the column v_k removed.

To the best of our knowledge, the following proposition is due to P. Casazza (unpublished, 2000) but can also be found in [8] and in the recent preprint [5].

Proposition 2.1 *$\mathcal{B} = \{v_1, \dots, v_N\} \subset \mathbb{R}^2$ is a tight frame if and only if for some index $i \le N$,*

$$\sum_{j=1}^{N} ||v_j||^2 e^{2i\theta_{i,j}} = 0. \tag{4}$$

It is easy to verify that if (4) is valid for some index i, then it is valid for all other i's.

Proof Let $\ell_j = ||v_j||$. After a rotation, we can let $v_i = v_1 = (\ell_1, 0)$ and $\theta_{1,j} = \theta_j$, so that $v_j = (\ell_j \cos\theta_j, \ell_j \sin\theta_j)$.

$\mathcal{B}$ is a tight frame with frame constant A if and only if the rows of the matrix $(v_1, \dots, v_N)$ are orthogonal and have length A. That implies

$$\sum_{j=1}^{N} \ell_j^2 \cos^2\theta_j = \sum_{j=1}^{N} \ell_j^2 \sin^2\theta_j = A \tag{5}$$

and

$$\sum_{j=1}^{N} \ell_j^2 \cos\theta_j \sin\theta_j = 0. \tag{6}$$

From (5) follows that $\sum_{j=1}^{N} \ell_j^2(\cos^2\theta_j - \sin^2\theta_j) = \sum_{j=1}^{N} \ell_j^2 \cos(2\theta_j) = 0$, and from (6) that $\sum_{j=1}^{N} \ell_j^2 \sin(2\theta_j) = 0$, and so we have proved (4).

If (4) holds, then (5) and (6) hold as well, and from these identities follows that $\mathcal{B}$ is a tight frame. □

Corollary 2.2 *Let $\mathcal{B} = \{v_1, v_2, v_3\} \subset \mathbb{R}^2$ be a tight frame. Assume that the v_i's have all the same length. Then, $\theta_{1,2} = \pi/3$, and $\theta_{1,3} = 2\pi/3$.*

So, every such frame $\mathcal{B}$ is equivalent to a dilation of the "Mercedes-Benz frame" $\left\{(1,0), \left(-\frac{1}{2}, \frac{\sqrt{3}}{2}\right), \left(-\frac{1}{2}, -\frac{\sqrt{3}}{2}\right)\right\}$.

Proof Let $v_1 = (1,0)$, and $\theta_{1,i} = \theta_i$ for simplicity. By Proposition 2.1, $1 + \cos(2\theta_2) + \cos(2\theta_3) = 0$, and $\sin(2\theta_2) + \sin(2\theta_3) = 0$. It is easy to verify that these equations are satisfied only when $\theta_2 = \frac{\pi}{3}$ and $\theta_3 = \frac{2\pi}{3}$ or viceversa. □

The following simple proposition is a special case of Theorem 1.1, and will be a necessary step in the proof.

Lemma 2.3 *Let $w = (\alpha_1, \alpha_2)$ be given. Assume $||w|| < 1$. There exists a unique nontrivial Parseval frame $\{v_1, v_2, w\} \subset \mathbb{R}^2$, with $v_1 = (a_{11}, 0)$, $v_2 = (a_{1,2}, a_{2,2})$, and $a_{1,1}, a_{2,2} > 0$.*

Proof We find $a_{1,1}$, $a_{1,2}$ and $a_{2,2}$ so that the rows of the matrix $\begin{pmatrix} a_{11}, & a_{12}, & \alpha_1 \\ 0, & a_{22}, & \alpha_2 \end{pmatrix}$ are orthonormal. That is,

$$\alpha_1^2 + a_{1,1}^2 + a_{1,2}^2 = 1, \quad \alpha_2^2 + a_{2,2}^2 = 1, \quad \alpha_1\alpha_2 + a_{1,2}a_{2,2} = 0. \tag{7}$$

From the second equation, $a_{2,2} = \pm\sqrt{1-\alpha_2^2}$; if we can chose $a_{2,2} > 0$, from the third equation we obtain $a_{1,2} = -\frac{\alpha_1\alpha_2}{\sqrt{1-\alpha_2^2}}$ and from the first equation

$$a_{1,1}^2 = 1 - \alpha_1^2 - a_{12}^2 = 1 - \alpha_1^2 - \frac{\alpha_1^2\alpha_2^2}{1-\alpha_2^2} = \frac{1-\alpha_1^2-\alpha_2^2}{1-\alpha_2^2}.$$

Note that $a_{1,1}^2 > 0$ because $||w||^2 = \alpha_1^2 + \alpha_2^2 < 1$. We can chose then

$$a_{1,1} = \frac{\sqrt{1-\alpha_2^2-\alpha_1^2}}{\sqrt{1-\alpha_2^2}}.$$

Note also that v and v_2 cannot be parallel; otherwise, $\frac{a_{1,2}}{a_{2,2}} = -\frac{\alpha_1\alpha_2}{1-\alpha_2^2} = \frac{\alpha_1}{\alpha_2} \iff -\alpha_2^2 = 1 - \alpha_2^2$, which is not possible. □

Remark The proof shows that v_1 and v_2 are uniquely determined by w. It shows also that if $||w|| = 1$, then $a_{1,1} = 0$, and consequently $v_1 = 0$.

3 Proofs

In this section we prove Theorem 1.1 and some of its corollaries.

Proof of Theorem 1.1 Let $w = (\alpha_1, \ldots, \alpha_n)$. We construct a nontrivial Parseval frame $\mathcal{M} = \{v_1, \ldots, v_n, w\} \subset \mathbb{R}^n$ with the following properties: the matrix $(v_1, \ldots, v_n) = \{a_{i,j}\}_{1\le i,j\le n}$ is right triangular, and

$$a_{j,j} = \begin{cases} \sqrt{1-\alpha_n^2} & \text{if } j = n \\ \frac{\sqrt{1-\sum_{k=j}^n \alpha_k^2}}{\sqrt{1-\sum_{k=j+1}^n \alpha_k^2}} & \text{if } 1 \le j < n. \end{cases} \tag{8}$$

The proof will show that $\mathcal{M}$ is unique, and also that the assumption that $||w|| < 1$ is necessary in the proof.

To construct the vectors v_j we argue by induction on n. When $n = 2$ we have already proved the result in Lemma 2.3. We now assume that the lemma is valid in dimension $n-1$, and we show that it is valid also in dimension n.

Let $\tilde{w} = (\alpha_2, \ldots, \alpha_n)$. By assumptions, there exist vectors $\tilde{v}_2, \ldots, \tilde{v}_n$ such that the set $\widetilde{\mathcal{M}} = \{\tilde{v}_2, \ldots, \tilde{v}_n, \tilde{w}\} \subset \mathbb{R}^{n-1}$ is a Parseval frame, and the matrix $(\tilde{v}_2, \ldots, \tilde{v}_n, \tilde{w})$ is right triangular and invertible. If we assume that the elements of the diagonal are positive, the $\tilde{v}_j$'s are uniquely determined by w. We let $\tilde{v}_j = (a_{2,j}, \ldots, a_{n,j})$, with $a_{k,j} = 0$ if $k < j$ and $a_{j,j} > 0$.

We show that $\widetilde{\mathcal{M}}$ is the projection on $\mathbb{R}^{n-1}$ of a Parseval frame in $\mathbb{R}^n = \mathbb{R} \times \mathbb{R}^{n-1}$ that satisfies the assumption of the theorem. To this aim, we prove that there exist scalars $x_1, \ldots, x_n$ so that the vectors $\{v_1, \ldots, v_{n+1}\}$ which are defined by

$$v_1 = (x_1, 0, \ldots, 0), \quad v_j = (x_j, \tilde{v}_j) \text{ if } 2 \le j \le n, \; v_{n+1} = w \tag{9}$$

form a Parseval frame of $\mathbb{R}^n$. The proof is in various steps: first, we construct a unit vector $(y_2, \ldots, y_{n+1})$ which is orthogonal to the rows of the matrix $(\tilde{v}_2, \ldots, \tilde{v}_n, \tilde{w})$. Then, we show that there exists $-1 < \lambda < 1$ so that $\lambda y_{n+1} = \alpha_1$. Finally, we chose $x_1 = \sqrt{1-\lambda^2}$, $x_j = \lambda y_j$, and we prove that the vectors $v_1, \ldots, v_{n+1}$ defined in (9) form a Parseval frame that satisfies the assumption of the lemma.

First of all, we observe that $\{v_1, \ldots, v_{n+1}\}$ is a Parseval frame if and only if $\vec{x} = (x_1, x_2, \ldots, x_n, \alpha_1)$ is a unit vector that satisfies the orthogonality conditions:

$$(\tilde{v}_2, \ldots \tilde{v}_n, \tilde{w})\vec{x} = \begin{pmatrix} a_{22} & a_{23} & \ldots & a_{2,n} & \alpha_2 \\ 0 & a_{33} & \ldots & a_{3,n} & \alpha_3 \\ \vdots & \vdots & \vdots & \vdots & \vdots \\ 0 & 0 & \ldots & a_{n,n} & \alpha_n \end{pmatrix} \begin{pmatrix} x_2 \\ \vdots \\ x_n \\ \alpha_1 \end{pmatrix} = \begin{pmatrix} 0 \\ 0 \\ \vdots \\ 0 \end{pmatrix}. \tag{10}$$

By a well known formula of linear algebra, the vector

$$\vec{y} = y_2\vec{e}_2 + \ldots + \vec{e}_{n+1}y_{n+1} = \det \begin{pmatrix} \vec{e}_2 & \vec{e}_3 & \ldots & \vec{e}_n & \vec{e}_{n+1} \\ a_{2,2} & a_{2,3} & \ldots & a_{2,n} & \alpha_2 \\ 0 & a_{3,3} & \ldots & a_{3,n} & \alpha_3 \\ \vdots & \vdots & \vdots & \vdots & \vdots \\ 0 & 0 & \ldots & a_{n-1,n} & \alpha_{n-1} \\ 0 & 0 & \ldots & a_{n,n} & \alpha_n \end{pmatrix} \tag{11}$$

is orthogonal to the rows of the matrix in (10), and so it is a constant multiple of $\vec{x}$. That is, $\vec{x} = \lambda \vec{y}$ for some $\lambda \in \mathbb{R}$.

Let us prove that $||\vec{y}|| = 1$. The rows of the matrix $(\tilde{v}_2, \ldots, \tilde{v}_n, \tilde{w})$ are orthonormal, and so after a rotation

$$(\tilde{v}_2, \ldots, \tilde{v}_n, \tilde{w}) = \begin{pmatrix} 0 & 1 & \ldots & 0 & 0 \\ 0 & 0 & \ldots & 0 & 0 \\ \vdots & \vdots & \vdots & \vdots & \vdots \\ 0 & 0 & \ldots & 1 & 0 \\ 0 & 0 & \ldots & 0 & 1 \end{pmatrix}. \tag{12}$$

The formula in (11) applied with the matrix in (12) produces the vector $\vec{e}_1 = (1, 0, \ldots, 0)$. Thus, $\vec{y}$ in (11) is a rotation of $\vec{e}_1$, and so it is a unit vector as well.

We now prove that $|\lambda| < 1$. From $\vec{x} = (x_2, \ldots, x_n, \alpha_1) = \lambda(y_2, \ldots, y_n, y_{n+1})$, we obtain $\lambda = \alpha_1 / y_{n+1}$. By (11),

$$y_{n+1} = (-1)^{n+1}\det \begin{pmatrix} a_{2,2} & a_{2,3} & \ldots & a_{2,n-1} & a_{2,n} \\ 0 & a_{3,3} & \ldots & a_{3,n-1} & a_{3,n} \\ \vdots & \vdots & \vdots & \vdots & \vdots \\ 0 & 0 & \ldots & a_{n-1,n-1} & a_{n-1,n} \\ 0 & 0 & \ldots & 0 & a_{n,n} \end{pmatrix} = (-1)^{n+1} \prod_{j=2}^{n} a_{j,j}.$$

Recalling that by (8), $a_{j,j} = \frac{\sqrt{1-\sum_{k=j}^{n}\alpha_k^2}}{\sqrt{1-\sum_{k=j+1}^{n}\alpha_k^2}}$, we can see at once that

$$y_{n+1} = (-1)^{n+1} \prod_{j=2}^{n} a_{j,j} = (-1)^{n+1} \sqrt{1 - \alpha_2^2 - \ldots - \alpha_{n-1}^2 - \alpha_n^2}$$

$$= (-1)^{n+1} \sqrt{1 - ||w||^2 + \alpha_1^2}. \tag{13}$$

In view of $\lambda y_{n+1} = \alpha_1$, we obtain

$$\lambda = (-1)^{n+1} \frac{\alpha_1}{\sqrt{1 - ||w||^2 + \alpha_1^2}}.$$

Clearly, $|\lambda| < 1$ because $||w|| < 1$. We now let

$$x_1 = \sqrt{1 - \lambda^2} = \frac{\sqrt{1 - ||w||^2}}{\sqrt{1 - ||w||^2 + \alpha_1^2}}, \tag{14}$$

and we define the v_j's as in (9). The first rows of the matrix $(v_1, \ldots, v_{n+1})$ is $(\sqrt{1-\lambda^2}, \vec{x}) = (\sqrt{1-\lambda^2}, \lambda\vec{y})$, and so it is unitary and perpendicular to the other rows. Therefore, the $\{v_j\}$ form a tight frame that satisfies the assumption of the theorem. □

The proof of Theorem 1.1 shows the following interesting fact: By (13) and (14)

$$\det(v_1, \ldots, v_n) = \prod_{j=1}^{n} a_{jj} = x_1 \prod_{j=2}^{n} a_{jj} = \sqrt{1 - ||w||^2}.$$

This formula does not depend on the fact that $(v_1, \dots, v_n)$ is right triangular, because every $n \times n$ matrix can be reduced in this form with a rotation that does not alter its determinant and does not alter the norm of w. This observation proves the following

Corollary 3.1 *Let $\{w_1, \dots, w_{n+1}\}$ be a Parseval frame. Then,*

$$det(w_1, \dots, \hat{w}_j, \dots, w_{n+1}) = \pm\sqrt{1 - ||w_j||^2}.$$

Proof of Theorem 1.2 Let $\ell_j = ||v_j||$. If $\{v_1, \dots, v_{n+1}\}$ is a Parseval frame, then the rows of the matrix $B = (v_1, \dots, v_{n+1})$ are orthonormal. While proving Theorem 1.1, we have constructed a vector $\vec{x} = (x_1, \dots, x_{n+1})$, with $x_j = (-1)^{j+1}\lambda$ $\det(v_1, \dots, \hat{v}_j, \dots, v_{n+1})$, which is perpendicular to the rows of B. By Corollary 3.1, $x_j = \pm\sqrt{1 - \ell_j^2}$. Since $\mathcal{B}$ is a Parseval frame, $\ell_1^2 + \dots + \ell_{n+1}^2 = n$, and so

$$||\vec{x}||^2 = x_1^2 + \dots + x_{n+1}^2 = (1 - \ell_1^2) + \dots + (1 - \ell_{n+1}^2) = 1.$$

So, the $(n+1) \times (n+1)$ matrix $\tilde{B}$ which is obtained from B with the addition of the row $\vec{y}$, is unitary, and therefore also the columns of $\tilde{B}$ are orthonormal. For every $i, j \le n+1$,

$$\langle v_i, v_j \rangle \pm \sqrt{1 - \ell_i^2}\sqrt{1 - \ell_j^2} = \ell_i \ell_j \cos\theta_{ij} \pm \sqrt{1 - \ell_i^2}\sqrt{1 - \ell_j^2} = 0 \qquad (15)$$

which implies (1).

Conversely, suppose that (1) holds. By (15), the vectors $\tilde{v}_j = (\pm\sqrt{1 - \ell_j^2},\ v_j)$ are orthonormal for some choice of the sign $\pm$; therefore, the columns of the matrix $\tilde{B}$ are orthonormal, and so also the rows are orthonormal, and $\mathcal{B}$ is a Parseval frame.

□

Corollary 3.2 *Let $\mathcal{B} = \{v_1, \dots v_{n+1}\}$ be a nontrivial Parseval frame. Then, $\frac{1}{n+1} < \ell_j^2$ for every j. Moreover, for all j with the possible exception of one, $\frac{1}{2} < \ell_j^2$.*

Proof The identity (1) implies that, for $i \ne j$,

$$1 - \ell_j^2 - \ell_i^2 + \ell_i^2 \ell_j^2 \sin^2\theta_{ij} = 0. \qquad (16)$$

That implies $\ell_j^2 + \ell_i^2 \ge 1$ for every $i \ne j$, and so all ℓ_j^2's, with the possible exception of one, are $\ge \frac{1}{2}$. Recalling that $\sum_{i=1}^{n+1} \ell_i^2 = n$,

$$1 - (n+1)\ell_j^2 + \sum_{i=1}^{n+1} \ell_j^2 \ell_i^2 \sin^2\theta_{ij} = 0,$$

and so $\ell_j^2 > \frac{1}{n+1}$.

□

Proof of Theorem 1.3 After a rotation, we can assume $v_i = v_1 = (\ell_1, 0, \ldots, 0)$. We let $\theta_{1,j} = \theta_j$ for simplicity. With this rotation $v_j = (\ell_j \cos\theta_j,\ \ell_j \sin\theta_j w_j)$ where w_j is a unitary vector in $\mathbb{R}^{n-1}$. The rows of the matrix $(v_1, \ldots v_N)$ are orthonormal, and so the norm of the first row is

$$\sum_{j\geq 1} \ell_j^2 \cos^2\theta_j + \ell_1^2 = 1 \tag{17}$$

The projections of $v_2, \ldots.\ v_N$ over a hyperplane that is orthogonal to v_1 form a tight frame on this hyperplane. That is to say that $\{\ell_2 \sin\theta_2 w_2,\ \ldots,\ \ell_N \sin\theta_N w_N\}$ is a tight frame in $\mathbb{R}^{n-1}$, and so it satisfies

$$\begin{aligned} &\ell_2^2 \sin\theta_2^2 ||w_2||_2^2 + \ldots + \ell_N^2 \sin\theta_N^2 ||w_N||_2^2 \\ &= \ell_2^2 \sin\theta_2^2 + \ldots + \ell_N^2 \sin\theta_N^2 = n - 1. \end{aligned} \tag{18}$$

□

Corollary 3.3 *$\{v_1, v_2, \ldots, v_n\}$ is a Parseval frame in $\mathbb{R}^n$ if and only if the v_i's are orthonormal.*

Proof By (17), all vectors in a Parseval frame have length ≤ 1. By (18)

$$\sum_{j=1}^{n} \ell_j^2 \sin^2\theta_{i,j} = n - 1$$

which implies that $\ell_j = 1$ and $\sin\theta_{ij} = 1$ for every $j \neq i$, and so all vectors are orthonormal. □

Proof of Corollary 1.4 Assume that $\mathcal{B}$ is scalable; fix $i < n+1$, and chose $j \neq k \neq i$. By 1

$$\begin{aligned} (1 - \ell_i^2)(1 - \ell_j^2) &= \ell_i^2 \ell_j^2 \cos\theta_{i,j}^2, \\ (1 - \ell_i^2)(1 - \ell_k^2) &= \ell_i^2 \ell_k^2 \cos\theta_{i,k}^2, \\ (1 - \ell_k^2)(1 - \ell_j^2) &= \ell_k^2 \ell_j^2 \cos\theta_{k,j}^2. \end{aligned}$$

These equations are easily solvable for ℓ_1^2, ℓ_j^2 and ℓ_k^2; we obtain

$$\ell_i^2 = \frac{|\cos\theta_{k,j}|}{|\cos\theta_{k,j}| + |\cos\theta_{k,i}\cos\theta_{j,i}|}.$$

This expression for ℓ_i must be independent of the choice of j and k, and so (3) is proved. □

Acknowledgements We wish to thank Prof. P. Casazza and Dr. J. Cahill for stimulating conversations.

References

1. J. Cahill, X. Chen, A note on scalable frames. ArXiv:1301.7292v1 (2013)
2. P. Casazza, G. Kutyniok, *Finite Frames: Theory and Applications* (Birkhauser, New York, 2013)
3. P. Casazza, M.T. Leon, Existence and construction of finite tight frames. J. Concr. Appl. Math. **4**(2), 277–289 (2006)
4. P.G. Casazza, M. Fickus, J. Kovacevic, M.T. Leon, J.C. Tremain, A physical interpretation of finite frames. Appl. Numer. Harmon. Anal. **2**(3), 51–76 (2006)
5. M. Copenhaver, Y. Kim, C. Logan, K. Mayfield, S.K. Narayan, M. Petro, J. Sheperd, Diagram vectors and tight frame scaling in finite dimensions. Oper. Matrices **8**(1), 73–88 (2014)
6. Z. Cvetkovic, Resilience properties of redundant expansions under additive noise and quantization. IEEE Trans. Inf. Thesis (2002)
7. I. Daubechies, *Ten Lectures on Wavelets* (SIAM, Philadelphia, 1992)
8. V.K. Goyal, J. Kovacevic, Quantized frames expansion with erasures. Appl. Comput. Harmon. Anal. **10**, 203–233 (2001)
9. V.K Goyal, M. Vetterli, N.T. Thao, Quantized overcomplete expansions in $\mathbb{R}^N$: analysis, synthesis, and algorithms. IEEE Trans. Inf. Theory **44**(1), 16–31 (1998)
10. D. Han, K. Kornelson, D. Larson, E. Weber, *Frames for Undergraduates*. Student Mathematical Library Series (American Mathematical Society, Providence, 2007)
11. G. Kutyniok, K. Okoudjou, F. Philipp, E. Tuley, Scalable frames. Linear Algebra Appl. **438**(5), 2225–2238 (2013)

Hyperbolic Hardy Classes and Logarithmic Bloch Spaces

Evgueni Doubtsov

Abstract Let φ be a holomorphic mapping between complex unit balls. We use the composition operators $C_\varphi : f \mapsto f \circ \varphi$ to relate the hyperbolic Hardy classes and the logarithmic Bloch spaces.

Keywords Hardy space • Logarithmic Bloch space

Mathematics Subject Classification (2000). Primary 32A35, Secondary 32A18, 47B33

1 Introduction

1.1 Hyperbolic Hardy Classes

Let $H(B_n)$ denote the space of holomorphic functions on the unit ball B_n of $\mathbb{C}^n$, $n \in \mathbb{N}$. For $0 < p < \infty$, the Hardy space $H^p(B_n)$ consists of $f \in H(B_n)$ such that

$$\|f\|^p_{H^p(B_n)} = \sup_{0 \le r < 1} \int_{\partial B_n} |f(r\zeta)|^p \, d\sigma_n(\zeta) < \infty,$$

where σ_n denotes the normalized Lebesgue measure on the unit sphere ∂B_n.

For $m, n \in \mathbb{N}$ and $p > 0$, the hyperbolic Hardy class $H^p_h = H^p_h(B_n, B_m)$ consists of those holomorphic mappings $\varphi : B_n \to B_m$ for which

$$\sup_{0 \le r < 1} \int_{\partial B_n} \beta^p_m(\varphi(r\zeta), 0) \, d\sigma_n(\zeta) < \infty,$$

E. Doubtsov (✉)
St. Petersburg Department of V.A. Steklov Mathematical Institute, Fontanka 27, St. Petersburg 191023, Russia
e-mail: dubtsov@pdmi.ras.ru

M. Ruzhansky, S. Tikhonov (eds.), *Methods of Fourier Analysis and Approximation Theory*, Applied and Numerical Harmonic Analysis,
DOI 10.1007/978-3-319-27466-9_3

where β_m denotes the Bergman metric on B_m. Yamashita [11] introduced the classes $H^p_h(B_1, B_1)$, $p > 0$, using the following formal rule: the definition of $H^p_h(B_1, B_1)$ reduces to that of $H^p(B_1)$ when φ is replaced by $f \in H(B_1)$ and β_1 is replaced by the Euclidean metric.

One has

$$\beta_m(w, 0) = \frac{1}{2} \log \frac{1 + |w|}{1 - |w|}, \qquad w \in B_m.$$

Thus, $\varphi \in H^p_h(B_n, B_m)$ if and only if

$$\sup_{0 \le r < 1} \int_{\partial B_n} \left(\log \frac{e}{1 - |\varphi(r\zeta)|^2} \right)^p d\sigma_n(\zeta) < \infty.$$

1.2 Bloch-to-Hardy Composition Operators

The Bloch space $\mathcal{B}(B_m)$ consists of those functions $f \in H(B_m)$ for which

$$\|f\|_{\mathcal{B}(B_m)} = |f(0)| + \sup_{w \in B_m} |\mathcal{R}f(w)|(1 - |w|^2) < \infty,$$

where

$$\mathcal{R}f(w) = \sum_{j=1}^{m} w_j \frac{\partial f}{\partial w_j}(w)$$

is the radial derivative of f.

Given a holomorphic mapping $\varphi : B_n \to B_m$, $m, n \in \mathbb{N}$, the composition operator $C_\varphi : H(B_m) \to H(B_n)$ is defined as

$$(C_\varphi f)(z) = f(\varphi(z)), \quad f \in H(B_m), \ z \in B_n.$$

Kwon [8] argued that various hyperbolic classes naturally arise in the studies of the composition operators C_φ defined on the Bloch space. In particular, the following result is known in the setting of the hyperbolic Hardy classes $H^p_h(B_n, B_1)$:

Theorem 1.1 (Kwon [7, 8]) *Let $0 < p < \infty$ and let $\varphi : B_n \to B_1$ be a holomorphic mapping. Then the following properties are equivalent:*

$C_\varphi : \mathcal{B}(B_1) \to H^{2p}(B_n)$ *is a bounded operator;*

$\varphi \in H^p_h(B_n, B_1)$;

$$\int_0^1 \frac{|\mathcal{R}\varphi(r\zeta)|^2}{(1 - |\varphi(r\zeta)|^2)^2} (1 - r)\, dr \in L^p(\partial B_n).$$

1.3 Hyperbolic Hardy Classes and Logarithmic Bloch Spaces

For $\alpha \in \mathbb{R}$, the logarithmic Bloch space $L^\alpha\mathcal{B}(B_m)$ consists of $f \in H(B_m)$ such that

$$\|f\|_{L^\alpha\mathcal{B}(B_m)} = |f(0)| + \sup_{w\in B_m} |\mathcal{R}f(w)|(1-|w|^2)\left(\log\frac{e}{1-|w|^2}\right)^\alpha < \infty. \tag{1}$$

In the present paper, we show that the Bloch space $\mathcal{B}(B_m) = L^0\mathcal{B}(B_m)$ is not an exception in the scale $L^\alpha\mathcal{B}(B_m)$. Namely, we prove an analog of Theorem 1.1 for $L^\alpha\mathcal{B}(B_m)$, $\alpha < \frac{1}{2}$ and $0 < p \le 1$, assuming that $m, n \in \mathbb{N}$ are arbitrary and $\varphi : B_n \to B_m$ is a *regular* mapping. By definition, a holomorphic mapping $\varphi : B_n \to B_m$ is called regular if the following property holds:

$$\begin{gathered}\text{There exist constants } s \in (0,1) \text{ and } \tau > 0 \text{ such that}\\ |\langle \mathcal{R}\varphi(z), \varphi(z)\rangle| \ge \tau|\mathcal{R}\varphi(z)||\varphi(z)| \quad \text{when } s < |\varphi(z)| < 1.\end{gathered} \tag{2}$$

The above definition was used in [6] to study composition operators on $\mathcal{B}(B_m)$; a very close notion was earlier introduced in the Bloch-to-BMOA setting (see [3]). In fact, we show that the scheme applied in [6] is adaptable to the logarithmic Bloch spaces.

The main result of the present paper is the following theorem.

Theorem 1.2 *Let $\alpha < \frac{1}{2}$, $0 < p \le 1$ and let $\varphi : B_n \to B_m$ be a regular holomorphic mapping. Then the following properties are equivalent:*

$$C_\varphi : L^\alpha\mathcal{B}(B_m) \to H^{2p}(B_n) \text{ is a bounded operator;} \tag{3}$$

$$\varphi \in H_h^{(1-2\alpha)p}(B_n, B_m); \tag{4}$$

$$\int_0^1 \frac{|\mathcal{R}\varphi(r\zeta)|^2}{(1-|\varphi(r\zeta)|^2)^2}\left(\log\frac{e}{(1-|\varphi(r\zeta)|^2}\right)^{-2\alpha}(1-r)\,dr \in L^p(\partial B_n). \tag{5}$$

Since every holomorphic mapping $\varphi : B_n \to B_1$ is regular, we obtain, as a corollary, the following direct extension of Theorem 1.1 for $0 < p \le 1$:

Corollary 1.3 *Let $\alpha < \frac{1}{2}$, $0 < p \le 1$ and let $\varphi : B_n \to B_1$ be a holomorphic mapping. Then the following properties are equivalent:*

$$C_\varphi : L^\alpha\mathcal{B}(B_1) \to H^{2p}(B_n) \text{ is a bounded operator;} \tag{6}$$

$$\varphi \in H_h^{(1-2\alpha)p}(B_n, B_1); \tag{7}$$

$$\int_0^1 \frac{|\mathcal{R}\varphi(r\zeta)|^2}{(1-|\varphi(r\zeta)|^2)^2}\left(\log\frac{e}{(1-|\varphi(r\zeta)|^2}\right)^{-2\alpha}(1-r)\,dr \in L^p(\partial B_n). \tag{8}$$

1.4 Comments

1.4.1 Property (7) indicates that the parameter $\alpha = \frac{1}{2}$ is critical. In fact, if $\alpha > \frac{1}{2}$, then $L^\alpha \mathcal{B}(B_1) \subset H^q(B_1)$ for all $q > 0$. In particular, property (6) with $\alpha > \frac{1}{2}$ holds for all holomorphic mappings $\varphi : B_n \to B_1$. The case $\alpha = \frac{1}{2}$ is related to the hyperbolic Nevanlinna class (cf. [4]).

1.4.2 Let $\alpha < \frac{1}{2}$ and let $0 < p \le 1$. If the product $(1 - 2\alpha)p$ is fixed, then Corollary 1.3 guarantees that the operators $C_\varphi : L^\alpha \mathcal{B}(B_1) \to H^{2p}(B_n)$ are bounded for the same collection of mappings φ. Remark that the property $\varphi \in H_h^{(1-2\alpha)p}(B_n, B_1)$ also characterizes the bounded composition operators C_φ between appropriate pairs of growth and Hardy spaces (see [1]).

1.5 Notation

We denote by C an absolute positive constant whose value may change from line to line. The notation $C(p, q, \dots)$ indicates that the constant depends on the parameters $p, q, \dots$. The symbol C_φ is reserved for a composition operator.

1.6 Organization of the Paper

Section 2 contains preliminary technical results related to reverse estimates in the logarithmic Bloch spaces, properties of regular mappings and applications of Green's formula. Theorem 1.2 is proved in Sect. 3.

2 Auxiliary Results

2.1 Reverse Estimates

The key technical tool of the present paper is the following lemma which provides reverse estimates, that is, integral estimates from below for appropriate test functions.

Lemma 2.1 ([5, 9]) *Let $m \in \mathbb{N}$, $\alpha < \frac{1}{2}$ and let $0 < q < \infty$. Then there exist functions $F_x \in L^\alpha \mathcal{B}(B_m)$, $0 \le x \le 1$, such that $\|F_x\|_{L^\alpha \mathcal{B}(B_m)} \le 1$ and*

$$\int_0^1 |F_x(w)|^q \, dx \ge \kappa \left(\log \frac{1}{1 - |w|^2} \right)^{\left(\frac{1}{2}-\alpha\right)q}, \quad w \in B_m, \tag{9}$$

for a constant $\kappa = \kappa(m, \alpha, q) > 0$.

It is shown in [5] that the exponent $\left(\frac{1}{2} - \alpha\right) q$ in (9) is sharp. Corollary 1.3 also indicates that Lemma 2.1 is sharp for $0 < q \le 2$, since estimate (9) guarantees that (6) implies (7); see Sect. 3.

2.2 Hardy Spaces

The classical Littlewood–Paley characterization of $H^p(B_1)$ extends to $H^p(B_n)$ for all $n \in \mathbb{N}$.

Theorem 2.2 (cf. [2, Theorem 3.1]) *Let* $q > 0$ *and let* $n \in \mathbb{N}$. *Then there exist constants* $C_1, C_2 > 0$ *such that*

$$C_1 \|f\|^q_{H^q(B_n)} \le |f(0)|^q + \int_{\partial B_n} \left(\int_0^1 |\mathcal{R}f(r\zeta)|^2 (1-r)\, dr \right)^{\frac{q}{2}} d\sigma_n(\zeta) \le C_2 \|f\|^q_{H^q(B_n)}$$

for all $f \in H(B_n)$.

2.3 Estimates for Regular Holomorphic Mappings

Given a holomorphic mapping $\varphi : B_n \to B_m$, put

$$\Phi(z) = \log \frac{e}{1 - |\varphi(z)|^2}, \qquad z \in B_n.$$

Lemma 2.3 *Let* $0 < q < \infty$ *and let* $\varphi : B_n \to B_m$ *be a regular holomorphic mapping. Then*

$$\Delta \Phi^q \ge C(s, \tau, q) \left(\log \frac{e}{1 - |\varphi|^2} \right)^{q-1} \frac{|\mathcal{R}\varphi|^2}{(1 - |\varphi|^2)^2}, \tag{10}$$

where $s \in (0, 1)$ *and* $\tau > 0$ *are the constants provided by property* (2).

Proof For $k = 1, \dots, n$, we have

$$\begin{aligned}\frac{1}{q}\frac{\partial^2 \Phi^q}{\partial \bar{z}_k \partial z_k} &= (q-1)\left(\log\frac{e}{1-|\varphi|^2}\right)^{q-2} \frac{\left|\langle \frac{\partial \varphi}{\partial z_k}, \varphi\rangle\right|^2}{(1-|\varphi|^2)^2} \\ &\quad + \left(\log\frac{e}{1-|\varphi|^2}\right)^{q-1} \frac{\left|\frac{\partial \varphi}{\partial z_k}\right|^2 (1-|\varphi|^2) + \left|\langle \frac{\partial \varphi}{\partial z_k}, \varphi\rangle\right|^2}{(1-|\varphi|^2)^2} \\ &= S_{k,1} + S_{k,2}.\end{aligned}$$

Also, observe that

$$\sum_{k=1}^{n} \left|\frac{\partial \varphi}{\partial z_k}\right|^2 \geq |\mathcal{R}\varphi|^2; \tag{11}$$

$$\sum_{k=1}^{n} \left|\langle \frac{\partial \varphi}{\partial z_k}, \varphi\rangle\right|^2 \geq \sum_{k=1}^{n} \left|\langle z_k \frac{\partial \varphi}{\partial z_k}, \varphi\rangle\right|^2 \geq \frac{|\langle \mathcal{R}\varphi, \varphi\rangle|^2}{n}. \tag{12}$$

First, assume that $1 \leq q < \infty$. Let the constants $s \in (0, 1)$ and $\tau > 0$ be those provided by property (2). To estimate $\Delta\Phi^q$ from below, we replace $S_{k,1} + S_{k,2}$ by $S_{k,2}$. So, if $s < |\varphi(z)| < 1$, then (2), (11) and (12) imply (10). Clearly, estimate (10) holds when $|\varphi(z)| \leq s$. Hence, the proof of the lemma is finished for $1 \leq q < \infty$.

Second, assume that $0 < q < 1$. To estimate $\Delta\Phi^q$ from below, we replace $S_{k,1}$ by

$$(q-1)\left(\log\frac{e}{1-|\varphi|^2}\right)^{q-1} \frac{\left|\langle \frac{\partial \varphi}{\partial z_k}, \varphi\rangle\right|^2}{(1-|\varphi|^2)^2}.$$

Therefore,

$$\frac{\partial^2 \Phi^q}{\partial \bar{z}_k \partial z_k} \geq q\left(\log\frac{e}{1-|\varphi|^2}\right)^{q-1} \frac{\left|\frac{\partial \varphi}{\partial z_k}\right|^2 (1-|\varphi|^2) + q\left|\langle \frac{\partial \varphi}{\partial z_k}, \varphi\rangle\right|^2}{(1-|\varphi|^2)^2}.$$

Hence, repeating the arguments used in the case $1 \leq q < \infty$, we obtain (10). □

2.4 A Maximal Function Estimate

Lemma 2.3 guarantees that Φ^q is a subharmonic function for any $q > 0$. So, we have the following lemma:

Lemma 2.4 ([6, Lemma 4.7]) *Let $0 < q < \infty$. For $0 < \rho < 1$ and a holomorphic mapping $\varphi : B_n \to B_m$, consider the radial maximal function*

$$M_{\Phi_\rho}(\zeta) = \sup_{0 \le r < 1} \Phi_\rho(r\zeta) = \sup_{0 \le r < 1} \Phi(\rho r\zeta), \qquad \zeta \in \partial B_n.$$

Then

$$\int_{\partial B_n} M^q_{\Phi_\rho}(\zeta)\, d\sigma_n(\zeta) \le C \int_{\partial B_n} \Phi^q_\rho(\zeta)\, d\sigma_n(\zeta), \qquad 0 < \rho < 1.$$

2.5 Applications of Green's Formula

Given $u \in C^2(B_n)$, put $u_\rho(z) = u(\rho z)$, $0 < \rho < 1$. Green's formula guarantees that

$$\int_{\partial B_n} u(\rho\zeta)\, d\sigma_n(\zeta) - u(0) = C \int_{B_n} G(z)\Delta u_\rho(z)\, d\nu_n(z), \qquad 0 < \rho < 1, \tag{13}$$

where ν_n is the normalized Lebesgue measure on B_n,

$$\Delta = 4 \sum_{k=1}^{n} \frac{\partial^2}{\partial z_k \partial \overline{z}_k}$$

is the real Laplacian and $G(z)$ is the Green function for the ball B_n, that is,

$$G(z) = \log \frac{1}{|z|}, \qquad n = 1; \qquad G(z) = |z|^{2-2n} - 1, \qquad n \ge 2.$$

Lemma 2.5 *Let $\alpha < \frac{1}{2}$, $0 < p \le 1$ and let $\varphi : B_n \to B_m$ be a regular holomorphic mapping. Then*

$$\int_{\partial B_n} \left[\int_0^1 \frac{|\mathcal{R}\varphi_\rho(r\zeta)|^2}{(1 - |\varphi_\rho(r\zeta)|^2)^2} \left(\log \frac{e}{1 - |\varphi_\rho(r\zeta)|^2} \right)^{-2\alpha} (1 - r) r\, dr \right]^p d\sigma_n(\zeta)$$
$$\le C(s, \tau, \alpha, p) \int_{\partial B_n} \left(\log \frac{e}{1 - |\varphi_\rho(\zeta)|^2} \right)^{(1-2\alpha)p} d\sigma_n(\zeta)$$

for all $0 < \rho < 1$.

Proof Given $\alpha < \frac{1}{2}$ and $0 < p \le 1$, put $\beta = (1 - 2\alpha)p$. By Lemma 2.3, we have

$$\Phi^{1-\beta}_\rho(z)\Delta\Phi^\beta_\rho(z)\Phi^{-2\alpha}_\rho(z) \ge C(s, \tau, \beta) \frac{|\mathcal{R}\varphi_\rho(z)|^2}{(1 - |\varphi_\rho(z)^2|)^2} \Phi^{-2\alpha}_\rho(z) \tag{14}$$

for $z \in B_n$ and $0 < \rho < 1$.

Observe that $1-\beta-2\alpha = (1-2\alpha)(1-p)$. Hence, applying estimate (14), the definition of M_{Φ_ρ}, Hölder's inequality and Lemma 2.4, we obtain the following chain of estimates:

$$
\begin{aligned}
&C(s,\tau,\beta)\int_{\partial B_n}\left(\int_0^1 \frac{|\mathcal{R}\varphi_\rho(r\zeta)|^2}{(1-|\varphi_\rho(r\zeta)|^2)^2}\Phi_\rho^{-2\alpha}(r\zeta)(1-r)r\,dr\right)^p d\sigma_n(\zeta)\\
&\le \int_{\partial B_n}\left(\int_0^1 \Phi_\rho^{(1-2\alpha)(1-p)}(r\zeta)\Delta\Phi_\rho^\beta(r\zeta)(1-r)r\,dr\right)^p d\sigma_n(\zeta)\\
&\le \int_{\partial B_n} M_{\Phi_\rho}^{(1-2\alpha)(1-p)p}(\zeta)\left(\int_0^1 \Delta\Phi_\rho^\beta(r\zeta)(1-r)r\,dr\right)^p d\sigma_n(\zeta)\\
&\le \left(\int_{\partial B_n} M_{\Phi_\rho}^{(1-2\alpha)p}(\zeta)\,d\sigma_n(\zeta)\right)^{1-p}\left(\int_{\partial B_n}\int_0^1 \Delta\Phi_\rho^\beta(r\zeta)(1-r)r\,dr\,d\sigma_n(\zeta)\right)^p\\
&\le C\left(\int_{\partial B_n}\Phi_\rho^\beta(\zeta)\,d\sigma_n(\zeta)\right)^{1-p}\left(\int_{B_n}\frac{1-|z|}{|z|^{2n-2}}\Delta\Phi_\rho^\beta(z)\,d\nu_n(z)\right)^p.
\end{aligned}
$$

We claim that the latter product is estimated by

$$\int_{\partial B_n}\Phi_\rho^\beta(\zeta)\,d\sigma_n(\zeta).$$

Indeed, applying Green's formula (13) with $u=\Phi^\beta$, we have

$$
\begin{aligned}
\int_{\partial B_n}\Phi_\rho^\beta(\zeta)\,d\sigma_n(\zeta) &\ge C\int_{B_n} G(z)\Delta\Phi_\rho^\beta(z)\,d\nu_n(z)\\
&\ge C\int_{B_n}\frac{1-|z|}{|z|^{2n-2}}\Delta\Phi_\rho^\beta(z)\,d\nu_n(z).
\end{aligned}
$$

So, the proof of the lemma is finished. □

3 Proof of Theorem 1.2

(3)⇒(4) Let the constant $\kappa=\kappa(m,\alpha,q)>0$ and the functions F_x, $0\le x\le 1$, be those provided by Lemma 2.1 for $q=2p$. We have $\|F_x\|_{L^\alpha\mathcal{B}(B_m)}\le 1$, therefore,

$$\int_{\partial B_n}|F_x\circ\varphi(r\zeta)|^{2p}\,d\sigma_n(\zeta)\le C \quad \text{for all } 0\le r<1$$

by (3). Thus, by Fubini's theorem and Lemma 2.1,

$$C \geq \int_0^1 \int_{\partial B_n} |F_x \circ \varphi(r\zeta)|^{2p} \, d\sigma_n(\zeta) \, dx \geq \kappa \int_{\partial B_n} \left(\log \frac{1}{1 - |\varphi(r\zeta)|^2} \right)^{(1-2\alpha)p} d\sigma_n(\zeta)$$

for all $0 \leq r < 1$. So, (3) implies (4).

(4)⇒(5) If (4) holds, then Lemma 2.5 and Fatou's lemma guarantee that

$$\int_{\partial B_n} \left[\int_0^1 \frac{|\mathcal{R}\varphi(r\zeta)|^2}{(1 - |\varphi(r\zeta)|^2)^2} \left(\log \frac{e}{(1 - |\varphi(r\zeta)|^2} \right)^{-2\alpha} (1 - r) r \, dr \right]^p d\sigma_n(\zeta) < \infty.$$

Hence, (5) holds.

(5)⇒(3) Assume that $f \in L^\alpha \mathcal{B}(B_m)$. As shown in [10], we obtain an equivalent norm on $L^\alpha \mathcal{B}(B_m)$ if the radial derivative in (1) is replaced by the complex gradient

$$\nabla f = \left(\frac{\partial f}{\partial w_1}, \ldots, \frac{\partial f}{\partial w_m} \right).$$

Thus, we have

$$|\nabla f(w)|(1 - |w|^2) \left(\log \frac{e}{1 - |w|^2} \right)^\alpha \leq C(f), \quad w \in B_m.$$

Hence,

$$\begin{aligned} |\mathcal{R}(f \circ \varphi)(z)| &\leq |\nabla f(\varphi(z))| |\mathcal{R}\varphi(z)| \\ &\leq \frac{C(f) |\mathcal{R}\varphi(z)|}{1 - |\varphi(z)|^2} \left(\log \frac{e}{(1 - |\varphi(r\zeta)|^2} \right)^{-\alpha}, \quad z \in B_n. \end{aligned}$$

So, $f \circ \varphi \in H^{2p}(B_n)$ by (5) and Theorem 2.2. Therefore, (3) holds by the closed graph theorem. The proof of Theorem 1.2 is finished.

Acknowledgements The author "Evgueni Doubtsov" is grateful to the anonymous referee for helpful comments and suggestions. The author was partially supported by RFBR (grant No. 14-01-00198-a).

References

1. E. Abakumov, E. Doubtsov, Reverse estimates in growth spaces. Math. Z. **271**(1–2), 399–413 (2012)
2. P. Ahern, J. Bruna, Maximal and area integral characterizations of Hardy-Sobolev spaces in the unit ball of $\mathbf{C}^n$. Rev. Mat. Iberoam. **4**(1), 123–153 (1988)

3. O. Blasco, M. Lindström, J. Taskinen, Bloch-to-BMOA compositions in several complex variables. Complex Var. Theory Appl. **50**(14), 1061–1080 (2005)
4. E. Doubtsov, Characterizations of the hyperbolic Nevanlinna class in the ball. Complex Var. Ellipt. Equat. **54**(2), 119–124 (2009)
5. E. Doubtsov, Weighted Bloch spaces and quadratic integrals. J. Math. Anal. Appl. **412**(1), 269–276 (2014)
6. E. Doubtsov, A.N. Petrov, Bloch-to-Hardy composition operators. Cent. Eur. J. Math. **11**(6), 985–1003 (2013)
7. E.G. Kwon, Composition of Blochs with bounded analytic functions. Proc. Am. Math. Soc. **124**(5), 1473–1480 (1996)
8. E.G. Kwon, Hyperbolic mean growth of bounded holomorphic functions in the ball. Trans. Am. Math. Soc. **355**(3), 1269–1294 (2003)
9. A.N. Petrov, Reverse estimates in logarithmic Bloch spaces. Arch. Math. (Basel) **100**(6), 551–560 (2013)
10. X. Tang, Extended Cesàro operators between Bloch-type spaces in the unit ball of $\mathbb{C}^n$. J. Math. Anal. Appl. **326**(2), 1199–1211 (2007)
11. S. Yamashita, Hyperbolic Hardy class H^1. Math. Scand. **45**(2), 261–266 (1979)

Multidimensional Extremal Logan's and Bohman's Problems

D.V. Gorbachev

Abstract Multidimensional extremal Logan's and Bohman's problems for entire functions of exponential type are considered. They have applications in metric geometry (estimations of parameters of n-dimensional lattices), approximation theory (sharp Jackson's inequalities in the space $L^2(\mathbb{R}^n)$) and spatial statistics (minimum variance problem over the class of characteristic functions on $\mathbb{R}^n$).

Keywords Bohman's problem • Covering radius • Fourier transform • Lattice • Logan's extremal problem • Packing radius • The Laplacian • Yudin's function

Mathematics Subject Classification (2000). Primary 99Z99; Secondary 00A00

1 Introduction

For $n \in \mathbb{N}$, let $\mathbb{R}^n$ denote n-dimensional Euclidean space. For $x \in \mathbb{R}^n$, we write $x = (x_1, \dots, x_n)$. The inner product of $x, y \in \mathbb{R}^n$ is denoted by $xy = x_1y_1 + \cdots + x_ny_n$, and the Euclidean norm of x is denoted by $|x| = (xx)^{1/2}$. The unit Euclidean sphere S^{n-1} and the unit Euclidean ball B^n of $\mathbb{R}^n$ are defined by

$$S^{n-1} = \{x \in \mathbb{R}^n \colon |x| = 1\} \quad \text{and} \quad B^n = \{x \in \mathbb{R}^n \colon |x| \le 1\}.$$

Let $|x|_p = (|x_1|^p + \cdots + |x_n|^p)^{1/p}$, where $1 \le p < \infty$, and let $|x|_\infty = \max\{|x_1|, \dots, |x_n|\}$. We call $|x|_p$ the l_p^n-norm of $x \in \mathbb{R}^n$. For $1 \le p < \infty$, the unit l_p^n-ball B_p^n is defined by $B_p^n = \{x \in \mathbb{R}^n \colon |x|_p \le 1\}$. The set $B_\infty^n = [-1, 1]^n$ is called the n-dimensional cube. Note that $|x| = |x|_2$ and $B^n = B_2^n$.

The set of closed convex 0-centrally symmetric bodies $C \subset \mathbb{R}^n$ with piecewise smooth boundary ∂C is denoted by $\mathcal{C}_n$. Let $\|x\|_C = \inf\{r > 0 \colon x \in rC\}$ be the

D.V. Gorbachev (✉)
Lenina pr. 92, Tula State University, 300012 Tula, Russia
e-mail: dvgmail@mail.ru

M. Ruzhansky, S. Tikhonov (eds.), *Methods of Fourier Analysis and Approximation Theory*, Applied and Numerical Harmonic Analysis,
DOI 10.1007/978-3-319-27466-9_4

Minkowski functional, $rB^n_C = rC$ a ball of radius $r > 0$ in the C-norm,

$$C^\circ = \{y \in \mathbb{R}^n : xy \le 1\ \forall\, x \in C\} \in \mathcal{C}_n$$

the polar set of C with $\|y\|_{C^\circ} = \sup\{|xy| : x \in C\}$ [25, Chap. III]. If $C = B^n_p$, then $\|x\|_{B^n_p} = |x|_p$, $(B^n_p)^\circ = B^n_{p'}$, where $p' = p/(p-1)$ is the exponent conjugate to p. Note that $\|rx\|_C = r\|x\|_C = \|x\|_{r^{-1}C}$, $r > 0$.

On $\mathbb{R}^n$, all C-norms are equivalent. In particular, $\|x\|_{C'} \le M_n\|x\|_C$, where the constant $M_n > 0$ does not depend on $x \in \mathbb{R}^n$. Set

$$M_n(C')_C = \sup\{\|x\|_{C'} : \|x\|_C = 1\}.$$

Notation $M_n(C')$ without the lower index, as well as all similar notations below, refer to the B^n_2-case. Let $M_n(C')_p = M_n(C')_{B^n_p}$. We have $M_n(B^n_p) = 1, p > 2$,

$$M_n(B^n_p) = \sup_{x_1^2+\cdots+x_n^2=1} (|x_1|^p + \cdots + |x_n|^p)^{1/p} = n^{1/p-1/2}, \quad 1 \le p \le 2.$$

Suppose that $e(u) = e^{2\pi i u}$,

$$\hat{f}(s) = \int_{\mathbb{R}^n} f(x)e(-sx)\,dx, \quad s \in \mathbb{R}^n,$$

is the Fourier transform of a function f [25, Chap. I], $\operatorname{supp} f$ is the support of f (the closure of $\{x \in \mathbb{R}^n : f(x) \ne 0\}$), and $\chi_A(x)$ is the characteristic function of a set A ($\chi_A(x) = 1$ if $x \in A$, $\chi_A(x) = 0$ if $x \notin A$).

Introduce the class $\mathcal{F}_n(C)$ consisting of real-valued functions $f \in C(\mathbb{R}^n) \cap L^1(\mathbb{R}^n)$ such that $\operatorname{supp} \hat{f} \subset C$. It follows from the theory of the Fourier transform [25, Chap. I] that $f \in L^2(\mathbb{R}^n)$, $\hat{f} \in C(\mathbb{R}^n) \cap L^1(\mathbb{R}^n)$ and

$$f(x) = \int_C \hat{f}(s)e(sx)\,ds, \quad x \in \mathbb{R}^n. \tag{1}$$

Using (1), we can extend the function f to an entire function of exponential type C° [25, Chap. III], that is,

$$\forall\, \varepsilon > 0 \quad \exists A_\varepsilon > 0 : \quad |f(z)| \le A_\varepsilon e^{2\pi(1+\varepsilon)\|z\|_{C^\circ}} \quad \forall\, z \in \mathbb{C}^n.$$

On the other hand, if the restriction of an entire function of exponential type C° to $\mathbb{R}^n$ is a real-valued function belonging to $L^1(\mathbb{R}^n)$, then $f \in \mathcal{F}_n(C)$. This inverse result follows from [25, Theorem 4.9].

Functions of type τB^n with $\tau > 0$ are also called entire functions of exponential spherical type $2\pi\tau$ [23]. One example of such functions is the normalized Bessel function [5, Chap. VII]

$$j_\alpha(2\pi\tau|x|) = \frac{1}{\omega_{n-1}} \int_{S^{n-1}} e(\tau xs)\, ds, \quad x \in \mathbb{R}^n, \qquad \alpha = n/2 - 1,$$

where $\omega_{n-1} = 2\pi^{n/2}/\Gamma(n/2)$ is the surface area of S^{n-1},

$$j_\alpha(u) = \Gamma(\alpha + 1)(2/u)^\alpha J_\alpha(u).$$

We denote by q_α the smallest positive zero of $j_\alpha(u)$. For $n = 1$, we have $\alpha = -1/2$, $j_{-1/2}(u) = \cos u$ and $q_{-1/2} = \pi/2$.

1. Let us formulate extremal Logan's problem. Suppose $U, V \in \mathcal{C}_n$.
Let us introduce the quantity

$$\mathrm{l}(f)_U = \sup\{\|x\|_U : f(x) > 0\}.$$

This quantity is equal to the radius of the smallest ball in the norm of the body U outside which the function f is nonpositive. For brevity, define $\mathrm{l}(f)_p = \mathrm{l}(f)_{B_p^n}$, $\mathrm{l}(f) = \mathrm{l}(f)_2$.

Consider the following classes of functions:

$$\mathcal{L}_n(V) = \{f \in \mathcal{F}_n(V) : f(0) = 1,\ \hat{f}(0) \geq 0\},$$

$$\mathcal{L}_n^+(V) = \{f \in \mathcal{L}_n(V) : \hat{f} \geq 0\} \subset \mathcal{L}_n(V).$$

From the definition of the class $\mathcal{L}_n^+(V)$ it follows that it coincides with the class of positive-definite entire functions $f \in L^1(\mathbb{R}^n)$ of exponential type V° such that $\operatorname{supp}\hat{f} \subset V$, $\int_{\mathbb{R}^n} f(x)\, dx \geq 0$ and $\int_V \hat{f}(s)\, ds = 1$.

The class $\mathcal{L}_n^+(V)$ is nonempty and contains functions f for which $\mathrm{l}(f)_U < \infty$ (see, for example, Yudin's function (14)). Extremal Logan's problem consists in the calculation of the quantity

$$\mathrm{L}_n^+(V)_U = \inf\{\mathrm{l}(f)_U : f \in \mathcal{L}_n^+(V)\}.$$

Similarly, we can define the quantity $\mathrm{L}_n(V)_U$. Obviously,

$$\mathrm{L}_n(V)_U \leq \mathrm{L}_n^+(V)_U. \tag{2}$$

The quantities $\mathrm{L}_n(V)_U$ and $\mathrm{L}_n^+(V)_U$ are homogeneous. Indeed, if $f_\tau(x) = f(\tau x)$, $\tau > 0$, then $\hat{f}_\tau(s) = \tau^{-n}\hat{f}(\tau^{-1}s)$ [25, Chap. I],

$$\mathrm{l}(f_\tau)_U = \tau^{-1}\mathrm{l}(f)_U = \mathrm{l}(f)_{\tau^{-1}U}, \quad \operatorname{supp}\hat{f}_\tau = \tau \operatorname{supp}\hat{f}.$$

Thus, for example,

$$\mathrm{L}_n(V)_U = \tau^{-1}\mathrm{L}_n(\tau V)_U = \mathrm{L}_n(\tau V)_{\tau^{-1}U}. \tag{3}$$

Let $f(x) \le 0$ for $x \notin rB^n_C$. If $\|x\|_C \le r$, then $\|x\|_{C'} \le M_n\|x\|_C \le rM_n$, where $M_n = M_n(C')_C$. So $rB^n_C \subset rM_nB^n_{C'}$ and $f(x) \le 0$ for $x \notin rM_nB^n_{C'}$. Consequently, $\mathrm{l}(f)_{C'} \le M_n\mathrm{l}(f)_C$ and, for example,

$$\mathrm{L}_n^+(V)_{C'} \le M_n(C')_C\mathrm{L}_n^+(V)_C. \tag{4}$$

For $n = 1$, Logan [22] proved that

$$\mathrm{L}_1^+([-\tau,\tau]) = \mathrm{l}(f^*) = \frac{1}{2\tau}, \quad \tau > 0,$$

where

$$f^*(x) = \frac{\cos^2(\pi\tau x)}{1-(2\tau x)^2} \in \mathcal{L}_1^+([-\tau,\tau]), \quad \hat{f^*}(s) = \frac{\pi}{4\tau}\sin\Big(\frac{\pi|s|}{\tau}\Big)\chi_{[-\tau,\tau]}(s). \tag{5}$$

This result also follows from [3, 8].

In the multidimensional case, Logan's problem was studied by E. Berdysheva and the author. Let $n \in \mathbb{N}$, $\tau > 0$.

Theorem 1.1 ([6]) *Let $1 \le p \le 2$. If $V = \tau B^n_\infty$ and $U = B^n_p$, then*

$$\mathrm{L}_n(\tau B^n_\infty)_p = \mathrm{L}_n^+(\tau B^n_\infty)_p = \mathrm{l}(F_{\tau B^n_\infty}) = \frac{\sqrt{n}}{2\tau},$$

where

$$F_{\tau B^n_\infty}(x) = \Big(1 - \frac{2\tau^2|x|^2}{n}\Big)\prod_{j=1}^n\Big(\frac{\cos(\pi\tau x_j)}{1-(2\tau x_j)^2}\Big)^2 \in \mathcal{L}_n^+(\tau B^n_\infty).$$

Theorem 1.2 ([13]) *If $V = \tau B^n$ and $U = B^n$, then*

$$\mathrm{L}_n(\tau B^n) = \mathrm{L}_n^+(\tau B^n) = \mathrm{l}(F_{\tau B^n}) = \frac{q_\alpha}{\pi\tau},$$

where

$$F_{\tau B^n}(x) = \frac{j_\alpha^2(\pi\tau|x|)}{1-(\pi\tau|x|/q_\alpha)^2} \in \mathcal{L}_n^+(\tau B^n).$$

In both cases, the extremal function is Yudin's function $F_C \in \mathcal{L}_n^+(C)$ (see Sect. 3, (14)). Note that $\mathrm{l}(F_C) = \sigma_1(C)/\pi$, where $\sigma_1(C)$ is the first eigenvalue of the eigenvalue problem [11]

$$-\Delta u(x) = \sigma^2 u(x), \quad x \in C, \qquad u|_{\partial C} = 0. \tag{6}$$

Here $\Delta = \partial^2/\partial x_1^2 + \cdots + \partial^2/\partial x_n^2$ is the Laplacian.

To prove the desired lower estimates, the authors use the multidimensional Poisson summation formula in the case of Theorem 1.1 and the one-dimensional Gaussian quadrature formula based on the zeros of the Bessel function in the case of Theorem 1.2.

We shall discuss the following result of Yudin [28]:

$$2R(L^*)\rho(L) \le \frac{q_{n/2-1}}{\pi} = \frac{n}{2\pi}\bigl(1 + O(n^{-2/3})\bigr), \quad n \to \infty, \tag{7}$$

where $L \subset \mathbb{R}^n$ is a lattice of rank n, $\rho(L)$ is the packing radius of L, $R(L^*)$ is the covering radius of the dual lattice L^*. In particular, we shall show the connection between inequalities of type (7) and Logan's problem.

Logan's problem allows us to get sharp Jackson's inequality in the space $L^2(\mathbb{R}^n)$ [27]. By the best approximation of a function $f \in L^2(\mathbb{R}^n)$ we mean the quantity

$$E_U(f) = \inf\bigl\{\|f - g\|_{L^2(\mathbb{R}^n)} \colon \forall\, g \in L^2(\mathbb{R}^n),\ \operatorname{supp}\hat{g} \subset B_U^n\bigr\}.$$

The modulus of continuity of a function $f \in L^2(\mathbb{R}^n)$ is defined by the equality

$$\omega(V, f) = \sup\bigl\{\|f(x+t) - f(x)\|_{L^2(\mathbb{R}^n)} \colon t \in V\bigr\}.$$

It follows from [6] that

$$E_U(f) \le 2^{-1/2}\omega(\mathrm{L}_n^+(V)_U, f) \quad \forall f \in L^2(\mathbb{R}^n),$$

where $2^{-1/2}$ is a sharp constant and argument in the modulus of continuity is optimal. Thus, in two specific situations Yudin's problem [27] to determine the optimal argument in the sharp Jackson inequality in the space $L^2(\mathbb{R}^n)$ can be solved by Theorems 1.1 and 1.2.

Logan's problem is closely related to the Turán, Delsarte, and other extremal problems [1, 2, 9, 12–15, 18–20, 24].

2. To solve extremal Bohman's problem, we can apply the methods used to solve Logan's problem. Let us formulate the Bohman's problem. Let $V \in \mathcal{C}_n$,

$$\mathcal{B}_n(V) = \{f \in \mathcal{F}_n(V) \colon |x|^2 f \in L^1(\mathbb{R}^n),\ \hat{f}(0) = 1,\ f \ge 0\},$$

$$\mathcal{B}_n^+(V) = \{f \in \mathcal{B}_n(V) \colon \hat{f} \ge 0\} \subset \mathcal{B}_n(V).$$

In terms of probability theory, the class $\mathcal{B}_n(V)$ coincides with the set of probability density functions f such that $\varphi = \hat{f} \in C^2(\mathbb{R}^n)$ and $\operatorname{supp}\varphi \subset V$. Recall that φ is called a characteristic function [12].

The quantity

$$\mathrm{b}(f) = \int_{\mathbb{R}^n} |x|^2 f(x)\,dx = -(2\pi)^{-2}\Delta\hat{f}(0)$$

is called the second moment of probability density function f [12].

Bohman's problem consists in the determination of the quantity

$$\mathrm{B}_n(V) = \inf\{\mathrm{b}(f){:}f \in \mathcal{B}_n(V)\}.$$

Similarly, we can define the quantity $\mathrm{B}_n^+(V)$. Obviously, $\mathrm{B}_n(V) \le \mathrm{B}_n^+(V)$.

Setting $f_\tau(x) = \tau^n f(\tau x)$, we have $\hat{f_\tau}(s) = \hat{f}(\tau^{-1}s)$, $\mathrm{b}(f_\tau) = \tau^{-2}\mathrm{b}(f)$, and, for example,

$$\mathrm{B}_n(\tau V) = \tau^{-2}\mathrm{B}_n(V). \tag{8}$$

For $n = 1$, Bohman [7] proved that

$$\mathrm{B}_1([-\tau,\tau]) = \mathrm{b}(g^*) = \Big(\frac{1}{2\tau}\Big)^2, \qquad \tau > 0,$$

where

$$\begin{aligned} g^*(x) &= \frac{8\tau}{\pi^2}\Big(\frac{\cos(\pi\tau x)}{1-(2\tau x)^2}\Big)^2 \in \mathcal{B}_1^+([-\tau,\tau]),\\ \hat{g^*}(s) &= \Big\{\Big(1-\frac{|s|}{\tau}\Big)\cos\Big(\frac{\pi|s|}{\tau}\Big) + \frac{1}{\pi}\sin\Big(\frac{\pi|s|}{\tau}\Big)\Big\}\chi_{[-\tau,\tau]}(s). \end{aligned} \tag{9}$$

It follows that $\mathrm{B}_1([-\tau,\tau]) = \mathrm{B}_1^+([-\tau,\tau])$.

W. Ehm, T. Gneiting and D. Richards solved multidimensional Bohman's problem for the Euclidean ball B^n.

Theorem 1.3 ([12]) *For $n \in \mathbb{N}$ and $\tau > 0$,*

$$\mathrm{B}_n(\tau B^n) = \mathrm{B}_n^+(\tau B^n) = \mathrm{b}(G_{\tau B^n}) = \Big(\frac{q_\alpha}{\pi\tau}\Big)^2,$$

where the extremal function is of the form

$$G_{\tau B^n}(x) = \frac{\omega_{n-1}\tau^n}{2^{n-1}q_\alpha^2}\Big(\frac{j_\alpha(\pi\tau|x|)}{1-(\pi\tau|x|/q_\alpha)^2}\Big)^2 \in \mathcal{B}_n^+(\tau B^n).$$

We shall prove that

$$\mathrm{B}_n(\tau B^n_\infty) = \mathrm{B}^+_n(\tau B^n_\infty) = \mathrm{b}(G_{\tau B^n_\infty}) = \frac{n}{4\tau^2}, \quad \tau > 0,$$

where G_C is a function such that $G_C \in \mathcal{B}^+_n(C)$ (16), $\mathrm{l}(G_C) = (\sigma_1(C)/\pi)^2$. The function G_C is closely related to the function F_C (see (17)). For any $C \subset \mathcal{C}_n$, the functions G_C and F_C were constructed by Yudin in [26, 27], respectively. Note that the function $G_{\tau B^n}$ is used in the proof of sharp Jackson's inequalities in the spaces $L^p(\mathbb{T}^n)$ for $1 \leq p < 2$ [17].

2 Main Results

Suppose Λ_n is the set of lattices $L \subset \mathbb{R}^n$ of rank n [10],

$$L = \{l = \upsilon_1\nu_1 + \cdots + \upsilon_n\nu_n = M\nu : \nu \in \mathbb{Z}^n\},$$

where $\{\upsilon_1, \ldots, \upsilon_n\} \subset \mathbb{R}^n$ is a basis of L, $M = (\upsilon_1 \ \ldots \ \upsilon_n) \in \mathbb{R}^{n\times n}$ is a generator matrix of L, $\det L = (\det M)^2$ is the lattice determinant.

The quantity $\lambda_1(L) = \inf\{|l| : l \in L,\ l \neq 0\}$ is called the first successive minimum of L [21]. We have $\lambda_1(L) = 2\rho(L)$, where $\rho(L)$ is the packing radius of L. Recall that the packing radius of L is defined by Conway and Sloane [10]

$$\rho(L) = \sup\{r > 0 : (l + rB^n) \cap (l' + rB^n) = \emptyset \ \forall\, l, l' \in L,\ l \neq l'\}.$$

Suppose $R(L) = \sup_{x\in\mathbb{R}^n} \inf_{l\in L} |x - l|$ is the covering radius of L. A vector $x_0 \in \mathbb{R}^n$ such that $|x_0| = R(L)$ is called a deep hole of L.

For $C \in \mathcal{C}_n$, we define [21]

$$\lambda_1(L)_C = \inf\{\|l\|_C : l \in L,\ l \neq 0\} = \inf\{r > 0 : (L \setminus \{0\}) \cap rC \neq \emptyset\},$$

$$R(L)_C = \sup_{x\in\mathbb{R}^n} \inf_{l\in L} \|x - l\|_C = \inf\{r > 0 : L + rC = \mathbb{R}^n\}.$$

We denote by L^* the dual lattice to L:

$$L^* = \{l^* \in \mathbb{R}^n : l^* l \in \mathbb{Z} \ \forall\, l \in L\} = M^*\mathbb{Z}^n,$$

where $M^* = (M^{-1})^T$ is a generator matrix of L^*, $\det L^* = (\det L)^{-1}$ is the determinant of L^*.

The problem of estimate of $R(L)_U \lambda_1(L^*)_V$, $U, V \in \mathcal{C}_n$, is studied in lattice theory, especially in the case $U = V = B^n$, $V = U^\circ$ [4, 21]. Using the following theorem, we can obtain some useful estimates.

Theorem 2.1 *If $n \in \mathbb{N}$, $L \in \Lambda_n$, $U, V \in \mathcal{C}_n$, and $\lambda_1(L^*)_V = 1$, then*

$$R(L)_U \le \mathrm{L}_n(V)_U \le \mathrm{L}_n^+(V)_U \le M_n(U)\mathrm{L}_n^+(V) \le M_n(U)\mathrm{l}(F_V) = \frac{M_n(U)\sigma_1(V)}{\pi},$$

where $F_V \in \mathcal{L}_n^+(V)$ is the function (14).

The stated theorem is one of the main results of the paper. In the case when $L = \mathbb{Z}^n$, $U = B_p^n$, $1 \le p \le 2$, and $V = B_\infty^n$, this can be deduced using the results of [6]. In this situation, we have $\lambda_1(B_\infty^n) = 1$, $\sigma_1(B_\infty^n) = \pi\sqrt{n}/2$, $R(\mathbb{Z}^n)_p = |x_0|_p = n^{1/p}/2$, where $x_0 = (1/2, \dots, 1/2) \in \mathbb{R}^n$ is a deep hole of $\mathbb{Z}^n$ (for any $p \ge 1$), $M_n(B_p^n) = n^{1/p-1/2}$. Thus,

$$\frac{n^{1/p}}{2} \le \mathrm{L}_n(B_\infty^n)_p \le \mathrm{L}_n^+(B_\infty^n)_p \le \frac{n^{1/p-1/2}\pi\sqrt{n}/2}{\pi} = \frac{n^{1/p}}{2}.$$

Hence, using the relation (3), we obtain Theorem 1.1.

Theorem 1.1 does not hold for $p > 2$, since $M_n(B_p^n) = 1$. For such p, the determination of the quantity $\mathrm{L}_n^+(B_\infty^n)_p$ is an open problem.

Let $L \in \Lambda_n$, $U = B^n$, $V = \lambda_1(L^*)B^n$. Then $\lambda_1(L^*)_V = 1$ and it follows from Theorems 2.1, 1.2 that

$$R(L) \le \mathrm{L}_n(\lambda_1(L^*)B^n) = \frac{q_\alpha}{\pi\lambda_1(L^*)}.$$

Hence $R(L)\lambda_1(L^*) \le q_{n/2-1}/\pi$. Thus, Yudin's inequality (7) is established. In [21] the upper estimate of $R(L)\lambda_1(L^*)$ is equal to $n^{3/2}/2$. It was known [21] that there exist self-dual lattices $L_n \in \Lambda_n$ such that

$$R(L_n)\lambda_1(L_n^*) \ge \frac{n}{4\pi e}(1 + o(1)), \quad n \to \infty.$$

The inequality $R(L)\lambda_1(L^*) \le n/2$ was proved in [4]. Also, it was mentioned without a detailed proof that

$$R(L)\lambda_1(L^*) \le \frac{n}{2\pi}\big(1 + O(n^{-1/2})\big).$$

The method of proving Theorem 2.1 can be used to obtain some new case of Bohman's problem.

Theorem 2.2 *If $n \in \mathbb{N}$, $L \in \Lambda_n$, $V \in \mathcal{C}_n$, and $\lambda_1(L^*)_V = 1$, then*

$$R^2(L) \le \mathrm{B}_n(V) \le \mathrm{B}_n^+(V) \le \mathrm{b}(G_V) = \Big(\frac{\sigma_1(V)}{\pi}\Big)^2,$$

where $G_V \in \mathcal{B}_n^+(V)$ is the function (16).

Suppose $L = \mathbb{Z}^n$, $V = B^n_\infty$; then

$$\frac{n}{4} \le \mathrm{B}_n(B^n_\infty) \le \mathrm{B}^+_n(B^n_\infty) \le \Big(\frac{\pi\sqrt{n}/2}{\pi}\Big)^2 = \frac{n}{4}.$$

Consequently, by (8) and (18), we obtain

$$\mathrm{B}_n(\tau B^n_\infty) = \mathrm{B}^+_n(\tau B^n_\infty) = \mathrm{b}(G_{\tau B^n_\infty}) = \frac{n}{4\tau^2},$$

where

$$G_{\tau B^n_\infty}(x) = \Big(\frac{8\tau}{\pi^2}\Big)^n \prod_{j=1}^{n}\Big(\frac{\cos(\pi\tau x_j)}{1-(2\tau x_j)^2}\Big)^2, \quad \tau > 0.$$

3 Proof of Theorem 2.1

1. The inequalities

$$\mathrm{L}_n(V)_U \le \mathrm{L}^+_n(V)_U \le M_n(U)\mathrm{L}^+_n(V)$$

follows from (2) and (4).

Our first objective is the construction of Yudin's function $F_C \in \mathcal{L}^+_n(C)$ [27] such that $\mathrm{l}(F_C) = \sigma_1/\pi$, where $\sigma_1 = \sigma_1(C)$. Let $n \ge 2$ (for $n = 1$, see (5)).

We denote by u_1 the first eigenfunction of the problem (6) with $\sigma = \sigma_1$. So $\Delta u_1 = -\sigma_1^2 u_1$. It is known [11, Chap. VI] that the function u_1 is even and

$$u_1\big|_{C\setminus\partial C} > 0, \qquad \frac{\partial u_1}{\partial n}\Big|_{\partial C} \le 0. \tag{10}$$

Put

$$v(x) = \begin{cases} u_1(x), & x \in C, \\ 0, & x \notin C. \end{cases} \tag{11}$$

The function v is continuous, even, nonnegative with support C. Find the Fourier transform $\hat{v}(s)$, $s \in \mathbb{R}^n$. By Green's identity,

$$\int_C \big(\Delta v(x)\, e(-sx) - v(x)\Delta e(-sx)\big)\, dx = \int_{\partial C}\Big(\frac{\partial v(x)}{\partial n}\, e(-sx) - v(x)\,\frac{\partial e(-sx)}{\partial n}\Big)\, dx.$$

Therefore,

$$\widehat{\Delta v} + |2\pi s|^2 \hat{v} = -\sigma_1^2 \hat{v} + |2\pi s|^2 \hat{v} = \hat{v}_n, \quad \hat{v} = \frac{-\hat{v}_n}{\sigma_1^2 - |2\pi s|^2}, \tag{12}$$

where

$$\hat{v}_n(s) = \int_{\partial C} \frac{\partial v(x)}{\partial n}\, e(-sx)\, dx.$$

Introduce the following continuous, even, nonnegative function:

$$w(x) = \int_{\partial C} v(x-y)\Big(-\frac{\partial v(y)}{\partial n}\Big)\, dy.$$

By (10) and (12), we have $\operatorname{supp} w \subset 2C$, $w(0) = 0$,

$$\hat{w} = -\hat{v}\,\hat{v}_n = \frac{(\hat{v}_n)^2}{\sigma_1^2 - |2\pi s|^2}, \quad \hat{w}(0) > 0, \qquad \hat{w}(s) \le 0, \quad |2\pi s| \ge \sigma_1.$$

Then it follows from the results of [25, Chap. I] that

$$0 = w(0) = \lim_{\varepsilon\to 0} \int_{\mathbb{R}^n} \hat{w}(s) e^{-2\pi\varepsilon|s|}\, ds = \int_{|2\pi s|<\sigma_1} \hat{w}(s)\, ds - \int_{|2\pi s|\ge\sigma_1} |\hat{w}(s)|\, ds.$$

Thus,

$$\int_{\mathbb{R}^n} |\hat{w}(s)|\, ds = \int_{|2\pi s|<\sigma_1} (|\hat{w}(s)| + \hat{w}(s))\, ds \le 2\int_{|2\pi s|<\sigma_1} |\hat{w}(s)|\, ds < \infty. \tag{13}$$

Let

$$w_a(x) = \frac{a^{-n} w(a^{-1}x)}{\hat{w}(0)}, \quad a > 0.$$

Then $\operatorname{supp} w_a \subset 2aC$,

$$\hat{w}_a(s) = \frac{\hat{w}(as)}{\hat{w}(0)}, \quad \hat{w}_a(0) = 1, \qquad \hat{w}_a(s) \le 0, \quad |2\pi a s| \ge \sigma_1.$$

Finally, put

$$F_C(x) = \hat{w}_{1/2}(x) = \frac{\{\hat{v}_n(x/2)/\hat{v}_n(0)\}^2}{1 - (\pi|x|/\sigma_1)^2}, \quad x \in \mathbb{R}^n. \tag{14}$$

Then $F_C \in C(\mathbb{R}^n) \cap L^1(\mathbb{R}^n)$, F_C is an even function, $F_C(0) = 1$,

$$\operatorname{supp} \widehat{F}_C \subset C, \quad \widehat{F}_C \geq 0, \quad \widehat{F}_C(0) = 0.$$

Thus, $F_C \in \mathcal{L}_n^+(C)$, $\mathrm{l}(F_C) = \sigma_1/\pi$.

In the case of the Euclidean ball $C = B^n$ [27], we have $u_1(x) = j_\alpha(q_\alpha|x|)$, $\sigma_1(B^n) = q_\alpha$,

$$\frac{\partial v(x)}{\partial n} = -\frac{q_\alpha^2 j_{\alpha+1}(q_\alpha)}{n} = k < 0, \quad x \in S^{n-1},$$

$$\hat{v}_n(s) = k \int_{S^{n-1}} e(-sx)\,dx = k\omega_{n-1} j_\alpha(2\pi|s|),$$

and

$$F_{\tau B^n}(x) = \frac{j_\alpha^2(\pi\tau|x|)}{1 - (\pi\tau|x|/q_\alpha)^2}, \quad \tau > 0.$$

For the cube $C = B_\infty^n$ [6], we get

$$u_1(x) = \prod_{j=1}^n \cos\Bigl(\frac{\pi x_j}{2}\Bigr), \quad \sigma_1(B_\infty^n) = \frac{\pi\sqrt{n}}{2},$$

$$F_{\tau B_\infty^n}(x) = \Bigl(1 - \frac{2\tau^2|x|^2}{n}\Bigr) \prod_{j=1}^n \Bigl(\frac{\cos(\pi\tau x_j)}{1 - (2\tau x_j)^2}\Bigr)^2, \quad \tau > 0.$$

2. This part of the proof is based on an idea of Logan [22].

Let us prove that $R(L)_U \leq \mathrm{L}_n(V)_U$, where $U, V \in \mathcal{C}_n$, $\lambda_1(L^*)_V = 1$, $L \in \Lambda_n$ is the lattice with a generator matrix M.

The Poisson summation formula [10]

$$\sum_{l \in L} f(x - l) = \frac{1}{\det M} \sum_{l^* \in L^*} \hat{f}(l^*) e(l^* x), \tag{15}$$

holds for functions f such that $|f(x)| + |\hat{f}(x)| = O(1 + |x|^{-n-\varepsilon})$, $\varepsilon > 0$. Also, it holds for functions $f \in \mathcal{F}_n(V)$ such that $\mathrm{l}(f)_U < \infty$. This fact can be shown by applying the properties $\operatorname{supp}\hat{f}(x) \subset V$ and $|f(x)| = -f(x)$, $\|x\|_U \geq \mathrm{l}(f)_U$, in an obvious manner (see (13)).

Let $f \in \mathcal{L}_n(V)$, $\mathrm{l}(f)_U < \infty$, $\lambda_1(L^*)_V = 1$. Then f is an analytic function, $\operatorname{supp}\hat{f} \subset V = B_V^n$, $\|l^*\|_V \geq 1\ \forall\, l^* \in L^*$, $l^* \neq 0$. Therefore, $\hat{f}(l^*) = 0\ \forall\, l^* \in L^* \setminus \{0\}$. By (15), we get

$$\sum_{l \in L} f(x-l) = (\det M)^{-1}\hat{f}(0) \geq 0 \quad \forall\, x \in \mathbb{R}^n.$$

Suppose $R(L)_U = \|x_0\|_U$, where x_0 is a deep hole in the U-norm of the lattice L. Then $\|x_0 - l\|_U \geq \|x_0\|_U\ \forall\, l \in L$. It follows that if $\mathrm{l}(f)_U = \|x_0\|_U - \varepsilon$, $\varepsilon > 0$, then there exists a neighborhood N_ε of x_0 such that $\|x - l\|_U \geq \mathrm{l}(f)_U$ and $f(x-l) \leq 0$ $\forall\, l \in L$, $x \in N_\varepsilon$. Thus, we have

$$0 \geq \sum_{l \in L} f(x-l) \geq 0 \quad \forall\, x \in N_\varepsilon,$$

which contradicts the analyticity of f. This proves that $\mathrm{l}(f)_U \geq \|x_0\|_U$ and $\mathrm{L}_n(V)_U \geq R(L)_U$. The theorem is proved.

Theorem 2.1 allows us to prove Theorem 1.1. In the case of Theorem 1.2, we need to use the function averaging and the Bessel quadrature formulas [13].

We see that Theorem 2.1 leads to the eigenvalue problem for the Laplacian Δ [16]. Consider one example. For $n = 2$, let $L = A_2$ be the hexagonal lattice with $M = \begin{pmatrix} 1 & 1/2 \\ 0 & \sqrt{3}/2 \end{pmatrix}$. Then $\lambda_1(A_2) = 1$. The lattice A_2 is self-dual. We have $M^* = \begin{pmatrix} 1 & 0 \\ -1/\sqrt{3} & 2/\sqrt{3} \end{pmatrix}$ and $R(A_2^*) = 2/3$.

Suppose $V_0 = MB_\infty^2$, $V_1 = 2C$, where C is the Voronoi cell of L around the origin; then $\lambda_1(V_i) = 1$, $\operatorname{mes}(V_i) = 2\sqrt{3}$. By Theorem 2.1, $2/3 \leq \mathrm{L}_2(V_i) \leq \sigma_1(V_i)/\pi$.

On the other hand, from the Faber–Krahn inequality [16, Chap. 3]

$$\sigma_1^2(C) \geq \frac{\pi q_0^2}{\operatorname{mes}(C)}$$

it follows that $\pi^{-1}\sigma_1(V_i) \geq 0.728 > 2/3$.

4 Proof of Theorem 2.2

1. Let $f \in \mathcal{B}_n(V)$ and $\lambda_1(L^*)_V = 1$. Then $g(x) = |x|^2 f(x) \in L^1(\mathbb{R}^n)$,

$$\hat{g}(s) = -(2\pi)^{-2}\Delta\hat{f}(s) \in C(\mathbb{R}^n), \quad \hat{g}(0) = \mathrm{b}(f), \qquad \hat{g}(s) = 0, \quad \|s\|_V > 1.$$

Therefore, $\hat{g}(l^*) = 0 \;\forall\, l^* \in L^* \setminus \{0\}$. By (15),

$$\sum_{l\in L} g(x-l) = (\det M)^{-1}\hat{g}(0) \quad \forall\, x \in \mathbb{R}^n.$$

Similarly, we have

$$\sum_{l\in L} f(x-l) = (\det M)^{-1}\hat{f}(0) = (\det M)^{-1}.$$

Suppose x_0 is a deep hole of the lattice L; then $|x_0 - l| \geq R(L) \;\forall\, l \in L$. Consequently, using $f \geq 0$, we conclude that

$$\sum_{l\in L} g(x_0-l) = \sum_{l\in L} |x_0-l|^2 f(x_0-l) \geq R^2(L) \sum_{l\in L} f(x_0-l) = (\det M)^{-1}R^2(L).$$

Thus, $(\det M)^{-1}\hat{g}(0) \geq (\det M)^{-1}R^2(L)$ and $\mathrm{b}(f) \geq R^2(L)$.

2. Construct a function $G_C \in \mathcal{B}_n^+(C)$ such that $\mathrm{b}(G_C) = (\sigma_1/\pi)^2$ [26]. Let $n \geq 2$ (for $n = 1$, see (9)).

Let us consider the convolution

$$z(x) = (v * v)(x) = \int_{\mathbb{R}^n} v(x-y)v(y)\,dy = \int_C v(x-y)v(y)\,dy,$$

where the function v is defined in (11). This convolution is a continuous, even, nonnegative function with support in $2C$. By (12),

$$\hat{z}(s) = (\hat{v}(s))^2 = \Big(\frac{\hat{v}_n(s)}{\sigma_1^2 - |2\pi s|^2}\Big)^2 \geq 0 \quad \forall\, s \in \mathbb{R}^n.$$

Let $z_a(x) = z(a^{-1}x)/z(0)$, $a > 0$. Then

$$\operatorname{supp} z_a \subset 2aC, \quad z_a(0) = 1, \quad \hat{z}_a(s) = \frac{a^n\hat{z}(as)}{z(0)}.$$

Finally, for $x \in \mathbb{R}^n$, we set

$$G_C(x) = \hat{z}_{1/2}(x) = A_C\Big(\frac{\hat{v}_n(x/2)/\hat{v}_n(0)}{1-(\pi|x|/\sigma_1)^2}\Big)^2, \quad A_C = G_C(0) = \frac{(\hat{v}_n(0))^2}{2^n\sigma_1^4 z(0)}. \tag{16}$$

Then the function G_C is continuous, even, nonnegative, $\operatorname{supp}\widehat{G}_C \subset C$, $\widehat{G}_C \geq 0$, $\widehat{G}_C(0) = 1$.

Comparing (14) with (16) yields

$$F_C(x) = A_C^{-1}\Big(1 - \frac{\pi^2|x|^2}{\sigma_1^2}\Big)G_C(x). \tag{17}$$

It follows that $|x|^2 G_C \in L^1(\mathbb{R}^n)$ and

$$0 = \widehat{F}_C(0) = A_C^{-1}\int_{\mathbb{R}^n}\Big(1 - \frac{\pi^2|x|^2}{\sigma_1^2}\Big)G_C(x)\,dx = A_C^{-1}\Big(1 - \frac{\pi^2}{\sigma_1^2}\,\mathrm{b}(G_C)\Big).$$

Consequently, $G_C \in \mathcal{B}_n^+(V)$ and $\mathrm{b}(G_C) = (\sigma_1/\pi)^2$. The theorem is proved.

In the case $C = B^n$, we have $\hat{v}_n(0) = k\omega_{n-1}$, $\sigma_1 = q_\alpha$,

$$z(0) = \int_{|x|\le 1} j_\alpha^2(q_\alpha|x|)\,dx = \omega_{n-1}\int_0^1 j_\alpha^2(q_\alpha r)r^{n-1}\,dr = \frac{\omega_{n-1}2^{2\alpha}\Gamma^2(\alpha+1)}{q_\alpha^{2\alpha}}\,I,$$

where [5, Chap. VII]

$$I = \int_0^1 J_\alpha^2(q_\alpha u)u\,du = \frac{J_{\alpha+1}^2(q_\alpha)}{2} = \frac{q_\alpha^{2\alpha+2} j_{\alpha+1}^2(q_\alpha)}{2^{2\alpha+3}\Gamma^2(\alpha+2)}.$$

It follows that

$$z(0) = \frac{\omega_{n-1}2^{2\alpha}\Gamma^2(\alpha+1)}{q_\alpha^{2\alpha}}\,\frac{q_\alpha^{2\alpha+2} j_{\alpha+1}^2(q_\alpha)}{2^{2\alpha+3}\Gamma^2(\alpha+2)} = \frac{\omega_{n-1}q_\alpha^2 j_{\alpha+1}^2(q_\alpha)}{2^3(\alpha+1)^2}.$$

Thus,

$$A_2 = \frac{(\hat{v}_n(0))^2}{2^n\sigma_1^4 z(0)} = \frac{1}{2^n q_\alpha^4}\Big(\frac{q_\alpha^2 j_{\alpha+1}(q_\alpha)\omega_{n-1}}{n}\Big)^2\,\frac{2^3(\alpha+1)^2}{\omega_{n-1}q_\alpha^2 j_{\alpha+1}^2(q_\alpha)} = \frac{\omega_{n-1}}{2^{n-1}q_\alpha^2}$$

with

$$\hat{v}_n(0) = k\omega_{n-1} = -\frac{q_\alpha^2 j_{\alpha+1}(q_\alpha)\omega_{n-1}}{n}, \qquad \sigma_1 = q_\alpha.$$

For $\tau = 1$, using (17), we can construct the function $G_{\tau B^n}$ from Theorem 1.3. For $C = B_\infty^n$, we obtain

$$G_{B_\infty^n}(x) = A_\infty\prod_{j=1}^n\Big(\frac{\cos(\pi x_j)}{1-(2x_j)^2}\Big)^2, \quad A_\infty = \Big(\frac{8}{\pi^2}\Big)^n. \tag{18}$$

Acknowledgements The author "D.V. Gorbachev" thanks the referee and R. Veprintsev for their careful reading and valuable comments. This research was supported by the RFFI (no. 13-01-00045), the Ministry of Education and Science of the Russian Federation (no. 5414GZ), and Dmitry Zimin's Foundation "Dynasty".

References

1. V.V. Arestov, E.E. Berdysheva, Turán's problem for positive definite functions with supports in a hexagon. Proc. Steklov Inst. Math. (Suppl.), **suppl. 1**, S20–S29 (2001)
2. V.V. Arestov, E.E. Berdysheva, The Turán problem for a class of polytopes. East J. Approx. **8**(3), 381–388 (2002)
3. V.V. Arestov, N.I. Chernykh, On the L_2-approximation of periodic functions by trigonometric polynomials, in *Approximation and Function Spaces: Conference Proceedings, Gdansk (1979)* (North-Holland, Amsterdam, 1981), pp. 25–43
4. W. Banaszczyk, New bounds in some transference theorems in the geometry of numbers. Math. Ann. **296**(4), 625–635 (1993)
5. G. Bateman, A. Erdélyi, et al., *Higher Transcendental Functions, II* (McGraw Hill, New York, 1953)
6. E.E. Berdysheva, Two related extremal problems for entire functions of several variables. Math. Notes **66**(3), 271–282 (1999)
7. H. Bohman, Approximate Fourier analysis of distribution functions. Ark. Mat. **4**, 99–157 (1960)
8. N.I. Chernykh, On best approximation of periodic functions by trigonometric polynomials in L_2. Mat. Zametki **2**(5), 513–522 (1967)
9. H. Cohn, New upper bounds on sphere packings II. Geom. Topol. **6**, 329–353 (2002)
10. J.H. Conway, N.J.A. Sloane, *Sphere Packings, Lattices and Groups*, 3rd edn. (Springer, New York, 1999)
11. R. Courant, D. Hilbert, *Methods of Mathematical Physics I* (Interscience, New York, 1953)
12. W. Ehm, T. Gneiting, D. Richards, Convolution roots of radial positive definite functions with compact support. Trans. Am. Math. Soc. **356**, 4655–4685 (2004)
13. D.V. Gorbachev, Extremal problems for entire functions of exponential spherical type. Math. Notes **68**(1–2), 159–166 (2000)
14. D.V. Gorbachev, An extremal problem for entire functions of exponential spherical type, which is connected with the Levenshtein bound for the density of a packing of $\mathbb{R}^n$ by balls (Russian). Izv. Tul. Gos. Univ. Ser. Mat. Mekh. Inf. **6**(1), 71–78 (2000)
15. D.V. Gorbachev, An extremal problem for periodic functions with support in a ball. Math. Notes **69**(3–4), 313–319 (2001)
16. A. Henrot, *Extremum Problems for Eigenvalues of Elliptic Operators*. Frontiers in Mathematics (Birkhäuser, Basel, 2006)
17. V.I. Ivanov, Approximation of functions in spaces L_p. Math. Notes **56**(2), 770–789 (1994)
18. A.V. Ivanov, V.I. Ivanov, Jackson's theorem in the space $L_2(\mathbb{R}^d)$ with Power Weight. Math. Notes **88**(1), 140–143 (2010)
19. A.V. Ivanov, V.I. Ivanov, Optimal arguments in Jackson's inequality in the power-weighted space $L_2(\mathbb{R}^d)$. Math. Notes **94**(3), 320–329 (2013)
20. M.N. Kolountzakis, Sz.Gy. Révész, On pointwise estimates of positive definite functions with given support. Can. J. Math. **58**(2), 401–418 (2006)
21. J.C. Lagarias, H.W. Lenstra Jr., C.P. Schnorr, Korkin–Zolotarev bases and successive minima of a lattice and its reciprocal lattice. Combinatorica **10**(4), 333–348 (1990)
22. B.F. Logan, Extremal problems for positive-definite bandlimited functions. II. Eventually negative functions. SIAM J. Math. Anal. **14**(2), 253–257 (1983)

23. S.M. Nikolskii, *Approximation of Functions of Several Variables and Imbedding Theorems* (Springer, Berlin/Heidelberg/New York, 1975)
24. Sz.Gy. Révész, Turan's extremal problem on locally compact abelian groups. Anal. Math. **37**(1), 15–50 (2011)
25. E.M. Stein, G. Weiss, *Introduction to Fourier Analysis on Euclidean Spaces* (Princeton University Press, Princeton, 1971)
26. V.A. Yudin, The multidimensional Jackson theorem. Math. Notes **20**(3), 801–804 (1976)
27. V.A. Yudin, Multidimensional Jackson theorem in L_2. Math. Notes **29**(2), 158–162 (1981)
28. V.A. Yudin, Two external problems for trigonometric polynomials. Sb. Math. **187**(11), 1721–1736 (1996)

Weighted Estimates for the Discrete Hilbert Transform

E. Liflyand

Abstract The Paley-Wiener theorem states that the Hilbert transform of an integrable odd function, which is monotone on $\mathbb{R}_+$, is integrable. In this paper we prove weighted analogs of this theorem for sequences and their discrete Hilbert transforms under the assumption of general monotonicity for an even/odd sequence.

Keywords Discrete Hilbert transform • General monotone functions and sequences • Hilbert transform • Weighted integrability

Mathematics Subject Classification (2000). Primary 42A50, Secondary 40A99, 26A48, 44A15

1 Introduction

Weighted estimates for the Hilbert transform is one of the most popular and important topics in today harmonic analysis. Being somewhat apart of the mainstream, the Paley-Wiener theorem (see [13]) asserts that for an odd and monotone decreasing on $\mathbb{R}_+$ function $g \in L^1$ its Hilbert transform is also integrable, i.e., g is in the (real) Hardy space $H^1(\mathbb{R})$ (for alternative proofs, see [21] and [15, Chap. IV, 6.2]). The oddness of g is essential, since by Kober's result [9], if $g \in H^1(\mathbb{R})$, then the cancelation property holds

$$\int_{\mathbb{R}} g(t)\, dt = 0. \tag{1}$$

In [12], the monotonicity assumption has been relaxed in the Paley-Wiener theorem, and in [11] weighted versions have been obtained for both the sine and

E. Liflyand (✉)
Department of Mathematics, Bar-Ilan University, 52900 Ramat-Gan, Israel
e-mail: liflyand@math.biu.ac.il

M. Ruzhansky, S. Tikhonov (eds.), *Methods of Fourier Analysis and Approximation Theory*, Applied and Numerical Harmonic Analysis,
DOI 10.1007/978-3-319-27466-9_5

cosine Fourier transforms of functions more general than monotone ones. The periodic case has also been covered in [11] in the same manner. One can find historical background relevant to these problems in [11].

However, for sequences rather than functions such problems are of similar interest and importance as well. The goal of this note is to prove the weighted analogues of the Paley-Wiener theorem for odd and even sequences, that is, for the discrete Hilbert transform. More precisely, the estimates will be given in the weighted ℓ^1 spaces $\ell(w)$, the space of sequences $a = \{a_k\}_{k=0}^{\infty}$ endowed with the norm

$$\|a\|_{\ell(w)} = \sum_{k=0}^{\infty} |a_k| w_k < \infty,$$

where $w = \{w_k\}$, $k = 0, 1, 2, \ldots$, is a non-negative sequence, or weight (if the sequence is two-sided, that is, defined (and summed) for $-\infty < k < \infty$, then we assume w to be an even sequence; we get a usual ℓ^1 sequence if the weight is constant). In particular, we show that for a weight $w_k = k^\alpha$, the discrete Hilbert transform is bounded in $\ell(w)$, when $-1 < \alpha < 1$ provided that a is an odd and monotone on $\mathbb{Z}_+$ sequence, or, when $-2 < \alpha < 0$ provided that a is even and monotone on $\mathbb{R}_+$. Thus assuming monotonicity or general monotonicity of a allows us to extend the results of [11] to sequences; recall that the estimates in [11] were, in turn, generalizations of Flett's [4], Hardy-Littlewood's [5], and Andersen's [1] results to the case $p = 1$. Our results can also be considered as generalizations of the known results for sequences in [2]. We mention that for $p = 1$ not only that two-weighted estimates (for monotone sequences and power weights) are obtained in [16] but also their sharpness is proved there (for more details for functions, see also [17]).

The outline of the paper is as follows. In the next section, we start with the needed prerequisites and then formulate the mentioned results for discrete Hilbert transforms. The proofs in the third section go along the same lines as those in [11], roughly speaking, the integrals and functions are replaced with sums and sequences. In the last section, we present an application of the obtained analog of the Paley-Wiener theorem to an old problem on the absolute convergence of Fourier series.

To fix certain notation, let $\lesssim$ mean $\le C$, with some absolute constant C, while $\gtrsim$ similarly means $\ge C$. Notation $\asymp$ is used when simultaneously $\lesssim$ and $\gtrsim$ are valid, of course, with different constants. We denote by $[y]$ the integer part of y.

2 Definitions and Statement of Results

We first give some prerequisites for discrete Hilbert transforms, then we formulate the above mentioned results.

2.1 Discrete Hilbert Transforms

For the sequence $a = \{a_k\} \in \ell^1$, the discrete Hilbert transform is defined for $n \in \mathbb{Z}$ as (see, e.g., [8, (13.127)])

$$\hbar a(n) = \sum_{\substack{k=-\infty \\ k \neq n}}^{\infty} \frac{a_k}{n-k}. \tag{2}$$

If the sequence a is either even or odd, the corresponding Hilbert transforms $\hbar_e$ and $\hbar_o$ may be expressed in a special form (see, e.g., [2] or [8, (13.130) and (13.131)]). More precisely, if a is even, with $a_0 = 0$, we have $\hbar_e(0) = 0$ and for $n = 1, 2, \ldots$

$$\hbar_e a(n) = \sum_{\substack{k=1 \\ k \neq n}}^{\infty} \frac{2na_k}{n^2 - k^2} + \frac{a_n}{2n}. \tag{3}$$

If a is odd, with $a_0 = 0$, we have for $n = 0, 1, 2, \ldots$

$$\hbar_o a(n) = \sum_{\substack{k=1 \\ k \neq n}}^{\infty} \frac{2ka_k}{n^2 - k^2} - \frac{a_n}{2n}. \tag{4}$$

Of course, $\frac{a_0}{0}$ is considered to be zero.

2.2 Estimates for the Discrete Hilbert Transforms

A null sequence a, that is, vanishing at infinity, is said to be general monotone (see [11, 12, 18]), or $a \in GMS$, if it satisfies the conditions

$$\sum_{k=n}^{2n} |\Delta a_k| \leq C \sum_{k=[n/c]}^{[cn]} \frac{|a_k|}{k}, \qquad n = 1, 2, \ldots, \tag{5}$$

and

$$\sum_{k=2n}^{n} |\Delta a_k| \leq C \sum_{k=[cn]}^{[n/c]} \frac{|a_k|}{k}, \qquad n = -1, -2, \ldots, \tag{6}$$

with $\Delta a_k = a_k - a_{k+1}$, where $C > 1$ and $c > 1$ are independent of n. If a is even or odd, then both conditions are the same. Note that any monotone or quasi-monotone sequence a is general monotone.

Definition 2.1 Let a non-negative sequence, or weight, $w = \{w_k\}$, $k = 1, 2, \ldots,$ belong to the ω class, written $w \in \omega$, if there exists $\varepsilon > 0$ such that

$$w_k\, k^{1-\varepsilon} \uparrow \qquad \text{for all} \qquad k, \tag{7}$$

$$w_k\, k^{\varepsilon-1} \downarrow \qquad \text{for all} \qquad k, \tag{8}$$

where $\uparrow$ and $\downarrow$ mean almost increase and almost decrease.

Recall that sequence d is called almost increasing (respectively, decreasing) if $d_m \le C d_n$ or, equivalently, $d_m \lesssim d_n$ when $m < n$ (respectively, $m > n$).

Theorem 2.2 *Let a be an odd sequence from $\ell(w)$, with $a_0 = 0$ and weight w. If $a \in GMS$ and $w \in \omega$, then*

$$\|\hbar a\|_{\ell(w)} \lesssim \|a\|_{\ell(w)}. \tag{9}$$

A counterpart of Theorem 2.2 for even sequences reads as follows.

Definition 2.3 Let a weight w belong to the ω^* class, written $w \in \omega^*$, if there exists $\varepsilon > 0$ such that

$$w_k\, k^{2-\varepsilon} \uparrow \qquad \text{for all} \qquad k, \tag{10}$$

$$w_k\, k^{\varepsilon} \downarrow \qquad \text{for all} \qquad k. \tag{11}$$

Theorem 2.4 *Let a be an even sequence from $\ell(w)$, with $a_0 = 0$ and weight w. If $a \in GMS$ and $w \in \omega^*$, then (9) holds.*

Not posing assumptions of evenness or oddness, we obtain a weighted estimate for the discrete Hilbert transform of a sequence summable on the whole $\mathbb{Z}$.

Corollary 2.5 *Let $w \in \omega \cap \omega^*$, that is, w satisfies (7) and (11). If $a = \{a_k\}_{k=-\infty}^{\infty} \in GMS$, then (9) holds.*

In particular, for the weight $w_k = |k|^\alpha$, $-1 < \alpha < 0$, the discrete Hilbert transform $\hbar a$ is in $\ell(w)$ provided $a \in \ell(w) \cap GMS$.

3 Proofs

In this section we present the proofs of the above formulated results. Observe that $\sum\limits_{n=1}^{\infty} w_n \frac{|a_n|}{n}$ is trivially dominated by $\|a\|_{\ell(w)}$, therefore we will not discuss the terms $\pm\frac{a_n}{2n}$ in (3) and (4).

Proof of Theorem 2.2 Let $n > 0$, calculations for $n < 0$ are the same. Since a is odd, we obtain

$$\sum_{n=1}^{\infty} w_n \left| \left(\sum_{k=[3n/2]}^{\infty} + \sum_{k=-\infty}^{-[3n/2]} \right) \frac{a_k}{n-k} \right| \leq \sum_{n=1}^{\infty} w_n \sum_{k=[3n/2]}^{\infty} |a_k| \frac{2k}{k^2 - n^2}$$

$$\lesssim \sum_{k=1}^{\infty} k|a_k| \sum_{n=1}^{[2k/3]+1} \frac{w_n}{k^2 - n^2}. \tag{12}$$

Applying (7), we get

$$\sum_{n=1}^{[2k/3]+1} \frac{w_n}{k^2 - n^2} \lesssim \frac{w_k k^{1-\varepsilon}}{k^2} \sum_{n=1}^{[2k/3]+1} n^{\varepsilon - 1} \lesssim \frac{w_k}{k}.$$

By this, the right-hand side of (12) is dominated by $\|a\|_{\ell(w)}$.

Similarly, but making use of (8), we get

$$\sum_{n=1}^{\infty} w_n \left| \left(\sum_{k=1}^{[n/2]} + \sum_{k=-[n/2]}^{1} \right) \frac{a_k}{n-k} \right| \lesssim \sum_{k=1}^{\infty} k|a_k| \sum_{n=2k}^{\infty} \frac{w_n}{n^2 - k^2}$$

$$\lesssim \sum_{k=1}^{\infty} k|a_k| \sum_{n=2k}^{\infty} \frac{w_n n^{\varepsilon-1}}{n^{2+\varepsilon-1}} \lesssim \sum_{k=1}^{\infty} w_k |a_k|.$$

We remark that (7) and (8) imply

$$w_n \asymp w_k, \qquad \beta n \leq k \leq \gamma n, \qquad 0 < \beta < \gamma, \tag{13}$$

i.e., $C_1 w_n \leq w_k \leq C_2 w_n$, where $C_1, C_2 > 0$ are independent of k and n. Then

$$\sum_{n=1}^{\infty} w_n \left| \sum_{k=-[3n/2]+1}^{-[n/2]-1} \frac{a_k}{n-k} \right| \lesssim \sum_{k=1}^{\infty} |a_k| \sum_{n=[2k/3]}^{2k+1} \frac{w_n}{n+k} \lesssim \sum_{k=1}^{\infty} w_k |a_k|.$$

Collecting estimates from above, we obtain

$$\sum_{n=1}^{\infty} w_n \left| \sum_{\substack{k=-\infty \\ k \neq n}}^{\infty} \frac{a_k}{n-k} \right| \lesssim \sum_{n=1}^{\infty} \left| w_n \sum_{\substack{k=[n/2]+1 \\ k \neq n}}^{[3n/2]-1} \frac{a_k}{n-k} \right| + \sum_{k=1}^{\infty} w_k |a_k|$$

$$\lesssim I + \sum_{k=1}^{\infty} w_k |a_k|,$$

where

$$I=\sum_{n=1}^{\infty}w_n\left|\sum_{k=1}^{[n/2]}\frac{a_{n-k}-a_{n+k}}{k}\right|.$$

We then have

$$\begin{aligned}I&\le\sum_{n=1}^{\infty}w_n\sum_{k=1}^{[n/2]}\Big(\sum_{s=n-k}^{n+k}\frac{|\Delta a_s|}{k}\Big)\\&\le\sum_{k=1}^{\infty}\sum_{n=2k}^{\infty}w_n\Big(\sum_{s=n-k}^{n+k}\frac{|\Delta a_s|}{k}\Big)\\&=\sum_{k=1}^{\infty}\Big[\sum_{s=k}^{3k}|\Delta a_s|\sum_{n=2k}^{s+k}w_n+\sum_{s=3k+1}^{\infty}|\Delta a_s|\sum_{n=s-k}^{s+k}w_n\Big]\frac{1}{k}=:I_1+I_2.\end{aligned}$$

By (13),

$$\frac{1}{k}\sum_{n=2k}^{s+k}w_n\le\frac{1}{k}\sum_{n=2k}^{4k}w_n\asymp w_k,\qquad k\le s\le 3k$$

and

$$\frac{1}{k}\sum_{n=s-k}^{s+k}w_n\asymp w_s,\qquad s\ge 3k.$$

Hence, since $a\in GMS$,

$$\begin{aligned}I_1&\lesssim\sum_{k=1}^{\infty}w_k\Big(\sum_{s=k}^{3k}|\Delta a_s|\Big)\lesssim\sum_{k=1}^{\infty}w_k\Big(\sum_{s=[k/c]}^{[ck]}\frac{|a_s|}{s}\Big)\\&\lesssim\sum_{s=1}^{\infty}|a_s|\Big(\sum_{k=[s/c]}^{[c(s+1)]}\frac{w_k}{k}\Big)\asymp\sum_{s=1}^{\infty}|a_s|w_s.\end{aligned}$$

Changing the order of summation yields

$$I_2\lesssim\sum_{k=1}^{\infty}\Big(\sum_{s=3k}^{\infty}w_s|\Delta a_s|\Big)\lesssim\sum_{s=1}^{\infty}sw_s|\Delta a_s|\lesssim\sum_{k=1}^{\infty}\frac{1}{k}\sum_{s=k}^{2k}sw_s|\Delta a_s|.$$

Using (13) and general monotonicity of a, we get, as above,

$$I_2 \lesssim \sum_{k=1}^{\infty} w_k \sum_{s=k}^{2k} |\Delta a_s| \lesssim \sum_{k=1}^{\infty} |a_k| w_k,$$

which completes the proof. □

Proof of Theorem 2.4 The proof goes along the same lines as that of Theorem 2.2. Using the evenness of a, we obtain

$$\sum_{n=1}^{\infty} w_n \left| \left(\sum_{k=[3n/2]}^{\infty} + \sum_{k=-\infty}^{-[3n/2]} \right) \frac{a_k}{n-k} \right| du \lesssim \sum_{k=1}^{\infty} |a_k| \sum_{n=1}^{[2k/3]+1} \frac{n\, w_n}{k^2 - n^2}$$
$$\lesssim \sum_{k=1}^{\infty} w_k |a_k|,$$

since, by (10), we have

$$\sum_{n=1}^{[2k/3]+1} \frac{n\, w_n}{k^2 - n^2} \lesssim \frac{w_k k^{2-\varepsilon}}{k^2} \sum_{n=1}^{[2k/3]+1} n^{\varepsilon - 1} \lesssim w_k.$$

Taking into account (11), we get

$$\sum_{n=1}^{\infty} w_n \left| \left(\sum_{k=1}^{[n/2]} + \sum_{k=-[n/2]}^{1} \right) \frac{a_k}{n-k} \right| \lesssim \sum_{k=1}^{\infty} |a_k| \sum_{n=2k}^{\infty} \frac{n\, w_n}{n^2 - k^2}$$
$$\lesssim \sum_{k=1}^{\infty} |a_k| \sum_{n=2k}^{\infty} \frac{w_n n^{\varepsilon}}{n^{1+\varepsilon}} \lesssim \sum_{k=1}^{\infty} w_k |a_k|.$$

Finally, we note that (10) and (11) also imply (13) and we can repeat the rest of the proof of Theorem 2.2. □

Proof of Corollary 2.5 Representing a in a standard way as the sum of its even and odd parts

$$a_k = \frac{a_k + a_{-k}}{2} + \frac{a_k - a_{-k}}{2},$$

we apply the same calculations as in the proof of Theorem 2.2 to the odd part and of Theorem 2.4 to the even part. Using then $|a_k \pm a_{-k}| \le |a_k| + |a_{-k}|$ and $w \in \omega \cap \omega^*$, we obtain the required estimate. □

4 Absolutely Convergent Fourier Series

In 1950s (see, e.g., [6] or in more detail [7, Chaps. II and VI]), the following problem in Fourier Analysis attracted much attention:

> Let $\{a_k\}_{k=0}^{\infty}$ be the sequence of the Fourier coefficients of the absolutely convergent sine (cosine) Fourier series of a function $f : \mathbb{T} = [-\pi, \pi) \to \mathbb{C}$, that is $\sum |a_k| < \infty$. Under which conditions on $\{a_k\}$ the re-expansion of $f(t)$ ($f(t) - f(0)$, respectively) in the cosine (sine) Fourier series will also be absolutely convergent?

The obtained condition is quite simple and is the same in both cases:

$$\sum_{k=1}^{\infty} |a_k| \ln(k+1) < \infty. \tag{14}$$

Analyzing the proof, say, in [6], one can see that in fact more general results are hidden in the proofs. They can be given in terms of the (discrete) Hilbert transform.

Theorem 4.1 *In order than the re-expansion $\sum b_k \sin kt$ of $f(t) - f(0)$ with the absolutely convergent cosine Fourier series be absolutely convergent, it is necessary and sufficient that the discrete Hilbert transform $\hbar a$ of the sequence a of the cosine Fourier coefficients of f be summable.*

Similarly, in order than the re-expansion $\sum b_k \cos kt$ of f with the absolutely convergent sine Fourier series be absolutely convergent, it is necessary and sufficient that the discrete Hilbert transform $\hbar a$ of the sequence a of the sine Fourier coefficients of f be summable.

What, in fact, is proven in the mentioned papers, for $b = \{b_k\}_{k=0}^{\infty}$,

$$b = \hbar_e a \tag{15}$$

in the first part of Theorem 4.1, and

$$b = \hbar_o a \tag{16}$$

in the second one.

In this case (14) is only a sufficient condition for the summability of the discrete Hilbert transform, though sharp on the whole class. We will use the obtained results to show that (14) is not necessary in the second part of Theorem 4.1. Indeed, Theorem 2.2 includes the non-weighted case, that is, when $w_n = 1$ for all n. This is a direct analog of the Paley-Wiener theorem for monotone sequences. It suffices to take any ℓ^1 monotone sequence a such that

$$\sum_{k=1}^{\infty} |a_k| \ln(k+1) = \infty.$$

This does not work for the cosine Fourier series, since the non-weighted case is excluded. An appropriate counter-example can be built, of course, but it is not related to the present work.

An interesting open problem appears in a different topic related both to the absolute convergence of Fourier series and to properties of the discrete Hilbert transform. It is proven by Wiener [20, Sect. 11] that if f is supported on $[-\pi, \pi - \varepsilon]$, $0 < \varepsilon < \pi$, then $\hat{f} \in L^1(\mathbb{R})$ if and only if the 2π-periodic extension of f has absolutely convergent Fourier series. This is no more true, in general, if $\varepsilon = 0$ (see, e.g., [19, 6.1.1]). Here, two independent conditions are necessary and sufficient for the integrability of the Fourier transform of a function f supported on $[-\pi, \pi]$: not only the 2π-periodic extension of f has absolutely convergent Fourier series but also $f_1(t) = tf(t)$ has.

Let us show how this follows from [3, Thm. 6]. Let ℓ^p, $0 < p < \infty$, be the space of sequences $\{d_j\}$ satisfying

$$\|\{d_j\}\|_{\ell^p} = \left(\sum_{j=1}^{\infty} |d_j|^p\right)^{\frac{1}{p}}.$$

We denote by h^p the class of ℓ^p sequences whose discrete Hilbert transform is also in ℓ^p. Denoting by E^p the class of L^p functions whose (inverse) Fourier transform is supported in $[-\pi, \pi]$, we have Theorem 6 from [3] in the following form.

Theorem 4.2 *Let $0 < p \le 1$. If g belongs to E^p, then $\{(-1)^k g(k)\}$ belongs to h^p. Conversely, if $\{a_k\}$ belongs to h^p, there is a unique $g \in E^p$ such that $g(k) = (-1)^k a_k$.*

In this terminology, we consider $\hat{f} \in E^1$. If $c_k(f) = \frac{1}{2\pi}\hat{f}(k)$, then for the n-th Fourier coefficient of f_1, written $c_n(f_1)$, we have

$$\frac{1}{2\pi}\int_{-\pi}^{\pi} f(t)te^{-int}dt = \frac{1}{2\pi}\int_{-\pi}^{\pi}\sum_{k=-\infty}^{\infty} c_k e^{ikt}te^{-int}dt = \sum_{\substack{k=-\infty \\ k\ne n}}^{\infty} c_k \frac{1}{2\pi}\int_{-\pi}^{\pi} te^{i(k-n)t}dt.$$

Let us calculate the last integral. Integrating by parts, we obtain

$$\begin{aligned} &\frac{1}{i(k-n)} te^{i(k-n)t}\Big|_{-\pi}^{\pi} - \frac{1}{i(k-n)}\int_{-\pi}^{\pi} e^{i(k-n)t}dt \\ &= \frac{2\pi\cos(k-n)\pi}{i(k-n)} = 2\pi\frac{(-1)^{k-n}}{i(k-n)}. \end{aligned}$$

Therefore, with $f(t) = \sum\limits_{k=-\infty}^{\infty} c_k e^{ikt}$,

$$f_1(t) = \frac{1}{i} \sum_{n=-\infty}^{\infty} (-1)^n \sum_{\substack{k=-\infty \\ k\neq n}}^{\infty} \frac{(-1)^k c_k}{k-n} e^{int},$$

and hence

$$c_n(f_1) = \frac{(-1)^n}{i} \sum_{\substack{k=-\infty \\ k\neq n}}^{\infty} \frac{(-1)^k c_k}{k-n} = \frac{(-1)^n}{i} \hbar\tilde{c}(n),$$

where $c = \{c_k\}_{k=-\infty}^{\infty}$, $\tilde{c} = \{(-1)^k c_k\}_{k=-\infty}^{\infty}$ and $\hbar\tilde{c}(n) = \sum\limits_{\substack{k=-\infty \\ k\neq n}}^{\infty} \frac{(-1)^k c_k}{k-n}$ is the n-th element of the discrete Hilbert transform of $\tilde{c}$. Belonging of $\tilde{c}$ to h^1 is just equivalent to the absolute convergence of the Fourier series of f and f_1 and, by Theorem 4.2, is equivalent to the integrability of $\hat{f}$. □

Since the same argument goes through for any $0 < p < 1$, we in fact have the following theorem for f supported in $[-\pi, \pi]$ (see [19, 6.1.1 b)]).

Theorem 4.3 *Let* $0 < p \leq 1$. *We have* $\hat{f} \in L^p(\mathbb{R})$ *if and only if both* f *and* f_1 *after* 2π*-periodic extension have the sequences of Fourier coefficients from* ℓ^p.

Recall (see [14]) that if $1 < p < \infty$ then the Fourier transform of a compactly supported function is in L^p if and only if this function, being extended periodically, has the sequence of Fourier coefficients from ℓ^p.

Back to Wiener's result, one may ask whether it is possible that there exists a set, containing both π and $-\pi$ and maybe rather thin near one or both endpoints, for which every function with absolutely convergent Fourier series (after 2π-periodic extension) supported on that set has integrable Fourier transform. The author failed to prove in [10] that no such set can exist, or, in other words, that Wiener's result is sharp. Therefore, the problem of sharpness of Wiener's result remains open.

References

1. K. Andersen, Weighted norm inequalities for Hilbert transforms and conjugate functions of even and odd functions. Proc. Am. Math. Soc. **56**(1), 99–107 (1976)
2. K. Andersen, Inequalities with weights for discrete Hilbert transforms. Can. Math. Bull. **20**, 9–16 (1977)
3. C. Eoff, The discrete nature of the Paley-Wiener spaces. Proc. Am. Math. Soc. **123**, 505–512 (1995)
4. T.M. Flett, Some theorems on odd and even functions. Proc. Lond. Math. Soc. (3) **8**, 135–148 (1958)

5. G.H. Hardy, J.E. Littlewood, Some more theorems concerning Fourier series and Fourier power series. Duke Math. J. **2**, 354–382 (1936)
6. S.I. Izumi, T. Tsuchikura, Absolute convergence of trigonometric expansions. Tôhoku Math. J. **7**, 243–251 (1955)
7. J.-P. Kahane, *Séries de Fourier Absolument Convergentes* (Springer, Berlin, 1970)
8. F.W. King, *Hilbert Transforms*. Encyclopedia of Mathematics and Its Applications, vol. 1 (Cambridge University Press, Cambridge, 2009)
9. H. Kober, A note on Hilbert's operator. Bull. Am. Math. Soc. **48**(1), 421–426 (1942)
10. E. Liflyand, Sharpness of Wiener's theorem on the Fourier transform of a compactly supported function, in *Theory of Mappings and Approximation of Functions* (Naukova dumka, Kiev, 1989), pp. 87–95 (Russian)
11. E. Liflyand, S. Tikhonov, Weighted Paley-Wiener theorem on the Hilbert transform. C.R. Acad. Sci. Paris, Ser. I **348**, 1253–1258 (2010)
12. E. Liflyand, S. Tikhonov, A concept of general monotonicity and applications. Math. Nachr. **284**, 1083–1098 (2011)
13. R.E.A.C. Paley, N. Wiener, Notes on the theory and application of Fourier transform, note II. Trans. Am. Math. Soc. **35**, 354–355 (1933)
14. M. Plancherel, G. Pólya, Fonctions entières et intégrales de Fourier multiples. Commun. Math. Helvetici I **9**, 224–248 (1937); II **10**, 110–163 (1938)
15. E.M. Stein, *Harmonic Analysis: Real-Variable Methods, Orthogonality, and Oscillatory Integrals* (Princeton University Press, Princeton, 1993)
16. V. Stepanov, S. Tikhonov, Two-weight inequalities for the Hilbert transform on monotone functions. Doklady RAN **437**, 606–608 (2011) (Russian). Engl. Transl. Dokl. Math. **83**, 241–242 (2011)
17. V. Stepanov, S. Tikhonov, Two power-weight inequalities for the Hilbert transform on the cones of monotone functions. Complex Var. Ellipt. Equat. **56**, 1039–1047 (2011)
18. S. Tikhonov, Trigonometric series with general monotone coefficients. J. Math. Anal. Appl. **326**, 721–735 (2007)
19. R.M. Trigub, E.S. Belinsky, *Fourier Analysis and Approximation of Functions* (Kluwer, London, 2004)
20. N. Wiener, *The Fourier Integral and Certain of Its Applications* (Dover, New York, 1932)
21. A. Zygmund, Some points in the theory of trigonometric and power series. Trans. Am. Math. Soc. **36**, 586–617 (1934)

Q-Measures on the Dyadic Group and Uniqueness Sets for Haar Series

Mikhail G. Plotnikov

Abstract The aim of article is to describe, in terms of the Q-measures, uniqueness sets for Haar series with non-decreasing partial sum's major sequences.

Keywords Dyadic group • Haar series • Hausdorff measures • Partial sum's major sequence • Q-measures • Uniqueness sets

Mathematics Subject Classification (2000). Primary 42C10, Secondary 28A12

1 Introduction

In 1870 G. Cantor proved the following theorem (see [1, Chap. 1], [15, vol. 1, Chap. 9]): *if a trigonometric series*

$$\frac{a_0}{2} + \sum_{n=1}^{\infty} a_n \cos(nx) + b_n \sin(nx)$$

converges to zero everywhere on $[0, 2\pi)$ *except possibly on a finite set, then this series is identically zero, that is, all its coefficients are equal to zero*. The branched theory of uniqueness of representation of functions by orthogonal series originates with the Cantor Theorem.

Definition 1.1 Let $\{f_n\}$ be a system of functions on some set X, and let

$$\sum_n c_n f_n(x) \qquad (1)$$

M.G. Plotnikov (✉)
Vologda Vereshchagin Academy, Schmidt Street 2, 160555 Molochnoe, Vologda, Russia
e-mail: mgplotnikov@gmail.com

M. Ruzhansky, S. Tikhonov (eds.), *Methods of Fourier Analysis and Approximation Theory*, Applied and Numerical Harmonic Analysis,
DOI 10.1007/978-3-319-27466-9_6

be a series with respect to this system. A set $A \subset X$ is called *uniqueness set* (or *U-set*) for series (1) if only identically zero series of the form (1) converges to zero on $X \setminus A$.

From the viewpoint of the last definition the Cantor Theorem means that every finite subset of $[0, 2\pi)$ is a uniqueness set for trigonometric series. The description of more complicated uniqueness sets for trigonometric series requires of involvement of arithmetic characteristics of sets, not only of metric ones [1, Chap. 14].

In present work we study uniqueness problems for Haar series. A. Haar proved (1910) that the empty set is a U-set for Haar series

$$\sum_{n=0}^{\infty} a_n H_n(x),$$

but the proof contained an error. A correct proof was obtained (1964), independently, by F.G. Arutunyan, and by F.G. Arutunyan & A.A. Talalyan, and by M.B. Petrovskaya, and by V.A. Skvortsov. In other side, every one-point set is not a U-set for Haar series (G. Faber, 1910, for the set $\{1/2\}$; J. McLaughlin & J. Price, 1969, for an arbitrary one-point set $A \subset [0, 1]$). Thus, in contrast of the case of trigonometric series, only $\emptyset$ is a uniqueness set for Haar ones. See [3] about mentioned results.

In several works uniqueness sets for some subclasses of Haar series (*conditional uniqueness sets*) are studied. Mushegyan proved [4] that *a Borel set A is a U-set for Haar series whose coefficients are satisfy the Arutyunyan–Talalyan condition*

$$a_n H_n(x) = o_x(n),$$

if and only if A is at most countable. W.R. Wade explored [12] properties of uniqueness sets for Haar series such that

$$a_n = o\,(n^{p-1/2}), \quad 0 \le p \le 1.$$

In [6] an attempt was made to describe U-sets for Haar series from Wade classes, with a help of some dimension characteristics. In [7] the following theorem was established: *a set $A \in [0, 1]$ is a U-set for Haar series such that*

$$a_n = O\,(n^{p-1/2}), \quad 0 < p \le 1,$$

if and only if A contains no perfect subset of positive Hausdorff $(1-p)$*-measure*. In [7] a similar result for multidimensional Haar series on the group $\mathbb{G}^m$ also was proved.

The aim of this work is to describe, in terms of the Q-measures, uniqueness sets for Haar series with non-decreasing partial sum's majorants.

2 The Group $\mathbb{G}$: Main

We shall denote the set of *positive integers by* $\mathbb{P}$, the set of *non-negative integers* by $\mathbb{N}$, the set of *real numbers* by $\mathbb{R}$. We use the convenient notation [8] $x = (x_n,\ n \in \mathbb{N})$ for the sequence $x = (x_0,\ x_1,\ x_2,\ \ldots)$.

The set of all 0–1 sequences $x = (x_j,\ j \in \mathbb{N})$ is called the *dyadic group* $\mathbb{G}$. The zero element of $\mathbb{G}$ is the sequence $0 := (x_j = 0,\ j \in \mathbb{N})$ and the group operation $\oplus$ is given by

$$x \oplus y = (|x_j - y_j|,\ j \in \mathbb{N})$$

for every $x = (x_j,\ j \in \mathbb{N}) \in \mathbb{G},\ y = (y_j,\ j \in \mathbb{N}) \in \mathbb{G}$.

Let $k,\ p \in \mathbb{N}$,

$$p = \sum_{j=0}^{k-1} p_j 2^j, \qquad (p_j \in \{0, 1\}, \quad 0 \le j \le k-1);$$

then the set

$$\Delta_k^p := \{x = (x_j, j \in \mathbb{N}) \in \mathbb{G} :\ x_j = p_{k-1-j},\ j = 0, \ldots, k-1\} \tag{2}$$

is called the *dyadic interval of rank k* on $\mathbb{G}$. If $t \in \mathbb{G}$, then we shall write $\Delta_{k,t}$ for the uniquely determined dyadic interval of rank k containing the point t.

Let Δ_k^p be an arbitrary dyadic interval of the form (2) with $k \geq 1$. Denote by $\Delta_k^{\overline{p}}$ the uniquely determined dyadic interval of rank k such that

$$\Delta_k^p \sqcup \Delta_k^{\overline{p}} = \Delta_{k-1}^q \tag{3}$$

for some $q \in \{0, \ldots, 2^{k-1} - 1\}$.

The *topology* on $\mathbb{G}$ is generated by the collection of all dyadic intervals. Obviously, each dyadic interval simultaneously open and closed in this topology. The dyadic group is metrizable (we omit details, see [8, Introduction]).

The dyadic group $(\mathbb{G},\ \oplus)$ is a compact Abelian group. Denote by μ the normed Haar measure on $\mathbb{G}$. Thus μ is a translation invariant Borel measure on $\mathbb{G}$ such that $\mu(\mathbb{G}) = 1$. It is well-known [8, Introduction] that

$$\mu(\Delta_k) = 2^{-k} \tag{4}$$

for each dyadic interval Δ_k of rank k.

Lemma 2.1 *Let* $\Phi :\ [1,\ +\infty) \to [1,\ +\infty)$ *be a function such that*

$$\Phi(qx) \le q\,\Phi(x) \tag{5}$$

for all $q > 1$ *and* $x \geq 1$. *Then the function* $Q : (0, 1] \to (0, 1]$ *defined by*

$$Q(x) = x\,\Phi\left(\frac{1}{x}\right) \tag{6}$$

is non-decreasing.

Proof Choose $x, y \in (0, 1]$, $x < y$. Then $x = y/q$ for some $q > 1$. We have

$$Q(x) \overset{(6)}{=} x\,\Phi\left(\frac{1}{x}\right) = \frac{y}{q}\,\Phi\left(\frac{q}{y}\right) \overset{(5)}{\leq} \frac{q \cdot y}{q}\,\Phi\left(\frac{1}{y}\right) \overset{(6)}{=} Q(y). \tag{7}$$

The statement of Lemma is a corollary of (7). □

3 Haar Series on the Group $\mathbb{G}$: Main

Haar functions H_n $(n \in \mathbb{N})$ on the group $\mathbb{G}$ are defined by the following way. $H_0 \equiv 1$ on $\mathbb{G}$. If $n = 2^k + p$ $(k = 0, 1, \ldots, p = 0, \ldots, 2^k - 1)$, then

$$H_n(t) = \begin{cases} 2^{k/2} & \text{for } t \in \Delta_{k+1}^{2p}, \\ -2^{k/2} & \text{for } t \in \Delta_{k+1}^{2p+1}, \\ 0 & \text{for } t \in \mathbb{G} \setminus \Delta_k^p. \end{cases}$$

A *Haar series* on $\mathbb{G}$ is defined as

$$\sum_{n=0}^{\infty} a_n H_n(t), \qquad t \in \mathbb{G}. \tag{8}$$

Let $N \in \mathbb{P}$. Then the *Nth rectangular partial sum* $S_N(t)$ of the series (8) at the point t is given by $S_N(t) := \sum_{n=0}^{N-1} a_n H_n(t)$.

4 Quasi-Measures on the Group $\mathbb{G}$: Main

Let $\mathcal{B}$ be the family of all dyadic intervals (2). Functions $\tau : \mathcal{B} \to \mathbb{R}$ are called *$\mathcal{B}$-functions.* A $\mathcal{B}$-function τ is said to be *$\mathcal{B}$-superadditive* (*$\mathcal{B}$-subadditive*) [5] if for any disjunct collection $\{\Delta_i\}_{i=1}^{p}$ of dyadic intervals such that $\bigsqcup_{i=1}^{p} \Delta_i \in \mathcal{B}$ we have the inequality

$$\sum_{i=1}^{p} \tau(\Delta_i) \leq \tau\left(\bigsqcup_{i=1}^{p} \Delta_i\right) \quad \left(\sum_{i=1}^{p} \tau(\Delta_i) \geq \tau\left(\bigsqcup_{i=1}^{p} \Delta_i\right)\right).$$

We denote by the symbol $\overline{\mathcal{A}}_{\mathcal{B}}$ (respectively, $\underline{\mathcal{A}}_{\mathcal{B}}$) the set of all $\mathcal{B}$-superadditive (respectively, $\mathcal{B}$-subadditive) functions. A $\mathcal{B}$-function $\tau \in \mathcal{A} := \overline{\mathcal{A}}_{\mathcal{B}} \cap \underline{\mathcal{A}}_{\mathcal{B}}$ is called *$\mathcal{B}$-additive*. Additive $\mathcal{B}$-functions are called in other words *quasi-measures*, [8, Chap. 7], [13].

The *upper derivative* of a $\mathcal{B}$-function τ at a point $t \in \mathbb{G}$ with respect to the family $\mathcal{B}$ is defined as

$$\overline{D}_{\mathcal{B}}\tau(t) = \overline{\lim} \frac{\tau(\Delta)}{\mu(\Delta)}, \quad \mu(\Delta) \to 0, \quad \Delta \in \mathcal{B}, \quad t \in \Delta,$$

see [11]. Similarly, the *lower derivative* is defined as

$$\underline{D}_{\mathcal{B}}\tau(t) = \underline{\lim} \frac{\tau(\Delta)}{\mu(\Delta)}, \quad \mu(\Delta) \to 0, \quad \Delta \in \mathcal{B}, \quad t \in \Delta.$$

If $\overline{D}_{\mathcal{B}}\tau(t) = \underline{D}_{\mathcal{B}}\tau(t) \neq \pm\infty$, we say that τ is *$\mathcal{B}$-differentiable* at t and denote the *$\mathcal{B}$-derivative* by $D_{\mathcal{B}}\tau(t)$.

Let $Q : (0, 1] \to (0, 1]$ be a non-decreasing function. We write H^{Q} for the class of all $\mathcal{B}$-functions τ such that the following analogue of Hölder's condition holds for some $M > 0$:

$$|\tau(\Delta)| \leq M \cdot Q(\mu(\Delta)), \quad \Delta \in \mathcal{B}. \tag{9}$$

Let $\mathrm{H}^{Q,*}$ be the class of all $\mathcal{B}$-functions τ satisfying the following condition everywhere on $\mathbb{G}$ for some $M_t > 0$:

$$|\tau(\Delta_{k,t})| \leq M_t \cdot Q(\mu(\Delta_{k,t})), \quad \Delta_{k,t} \in \mathcal{B}. \tag{10}$$

Lemma 4.1 *Let $F \subset \mathbb{G}$ be a closed set. Suppose that τ is a non-negative and subadditive $\mathcal{B}$-function satisfying*

$$\tau(\mathbb{G}) > 0. \tag{11}$$

Then there is a non-trivial quasi-measure ψ such that

$$0 \leq \psi(\Delta) \leq \tau(\Delta) \quad \textit{for each } \Delta \in \mathcal{B}. \tag{12}$$

Proof Consider an arbitrary dyadic interval Δ_k^p of the form (2) with $k \geq 1$. Then there exists the uniquely determined collection $I_k^p = \{\Delta_s^{p_s}\}_{s=0}^{k-1}$ of dyadic intervals of the form (2), such that

$$\Delta_s^{p_s} \supset \Delta_k^p \quad \text{for every } s = 0, \ldots, k-1.$$

Easily,

$$I_k^p = I_k^{\overline{p}}. \tag{13}$$

We set

$$\psi(\Delta_k^p) := \tau(\Delta_k^p) \cdot \prod_{s=1}^{k} \frac{\tau(\Delta_{s-1}^{p_{s-1}})}{\tau(\Delta_s^{p_s}) + \tau(\Delta_s^{\overline{p}_s})}, \tag{14}$$

where $\prod_{s=1}^{0} * := 1$. It follows from (3), (13), and (14), that

$$\psi(\Delta_k^{\overline{p}}) = \tau(\Delta_k^{\overline{p}}) \cdot \prod_{s=1}^{k} \frac{\tau(\Delta_{s-1}^{p_{s-1}})}{\tau(\Delta_s^{p_s}) + \tau(\Delta_s^{\overline{p}_s})}. \tag{15}$$

Combining (11) and (14), we get $\psi(\mathbb{G}) = \tau(\mathbb{G}) > 0$. Consequently, the $\mathcal{B}$-function ψ is non-trivial. Since τ is non-negative, we see from (14) that ψ is non-negative, too. Since τ is subadditive, every multiplier at product in (14) no more than 1. Therefore

$$\prod_{s=1}^{k} \frac{\tau(\Delta_{s-1}^{p_{s-1}})}{\tau(\Delta_s^{p_s}) + \tau(\Delta_s^{\overline{p}_s})} \leq 1$$

for all s, and $\psi(\Delta_k^p) \leq \tau(\Delta_k^p)$ for all $k \geq 1, p = 0, \ldots, 2^k - 1$.

Finally, we establish that ψ is a quasi-measure. It is sufficient to prove that

$$\psi(\Delta_k^p) + \psi(\Delta_k^{\overline{p}}) = \psi(\Delta_{k-1}^q) \tag{16}$$

for all $k \geq 1, p = 0, \ldots, 2^k - 1$, where the dyadic interval Δ_{k-1}^q has introduced by (3). We have

$$\psi(\Delta_k^p) + \psi(\Delta_k^{\overline{p}}) \overset{(14),(15)}{=} \left(\tau(\Delta_k^p) + \tau(\Delta_k^{\overline{p}})\right) \cdot \prod_{s=1}^{k} \frac{\tau(\Delta_{s-1}^{p_{s-1}})}{\tau(\Delta_s^{p_s}) + \tau(\Delta_s^{\overline{p}_s})}$$

$$= \left(\tau(\Delta_k^p) + \tau(\Delta_k^{\overline{p}})\right) \cdot \frac{\tau(\Delta_{k-1}^{p_{k-1}})}{\tau(\Delta_k^{p_k}) + \tau(\Delta_k^{\overline{p}_k})} \cdot \prod_{s=1}^{k-1} \frac{\tau(\Delta_{s-1}^{p_{s-1}})}{\tau(\Delta_s^{p_s}) + \tau(\Delta_s^{\overline{p}_s})}$$

$$= \tau(\Delta_{k-1}^{p_{k-1}}) \cdot \prod_{s=1}^{k-1} \frac{\tau(\Delta_{s-1}^{p_{s-1}})}{\tau(\Delta_s^{p_s}) + \tau(\Delta_s^{\overline{p}_s})} \overset{(3),(14)}{=} \psi(\Delta_{k-1}^q).$$

Thus, ψ is a quasi-measure. The lemma is proved. □

Remark 4.2 There exists another way to prove Lemma 4.1, using the Doob expansion, the well-known in the theory of martingales result [9], [14]. We see the similar way would be long in the format of our article.

5 Haar Series and Quasi-Measures

It is well-known (see, for example, [11]) that the study of general Haar series is equivalent in some sense to the study of quasi-measures.

For every series (8) we define a $\mathcal{B}$-function ψ by

$$\psi(\Delta_{k,t}) = S_{2^k}(t)\,\mu(\Delta_{k,t}). \tag{17}$$

The value of $\psi(\Delta_{k,t})$ defined by (17) is independent of the choice of $t \in \Delta$. It is well known [10, 11] that ψ is a quasi-measure and that the correspondence between the set of all series (8) and the set of all quasi-measures, defined by (17), becomes an isomorphism if we endow each of these sets with the natural structure of a vector space. In what follows, when speaking of the quasi-measure isomorphic to a series (8), we mean the quasi-measure defined by (17).

Lemma 5.1 (See [10, 11]) *Suppose we are given $t \in \mathbb{G}$ and a series (S) of the form* (8). *Consider the quasi-measure ψ isomorphic to (S). Then (S) converges to a finite sum A at t if and only if $D_{\mathcal{B}}\tau(t) = A$.*

Lemma 5.2 *Let (S) be a series of the form* (8), *ψ be the quasi-measure isomorphic to the series (S), $t \in \mathbb{G}$. Consider an arbitrary non-decreasing function $\Phi : [1, +\infty) \to [1, +\infty)$ satisfying* (5), *and the function $Q : (0, 1] \to (0, 1]$ defined by* (6). *Assume that for some $M_t > 0$ the partial sums $S_N(t)$ satisfy*

$$|S_N(t)| \le M_t \cdot \Phi(N), \qquad N = 1, 2, \ldots. \tag{18}$$

Then we have the inequality (10) *for the quasi-measure ψ.*

Proof Choose an arbitrary dyadic interval $\Delta_{k,t} \in \mathcal{B}$. We have

$$\begin{aligned} |\psi\left(\Delta_{k,t}\right)| &\overset{(4),\,(17)}{=} 2^{-k}\left|S_{2^k}(t)\right| \overset{(18)}{\le} M_t \cdot 2^{-k}\Phi\left(2^k\right) \\ &\overset{(4)}{=} M_t \cdot \mu\left(\Delta_{k,t}\right) \cdot \Phi\left(\frac{1}{\mu\left(\Delta_{k,t}\right)}\right) \overset{(6)}{=} M_t \cdot Q\left(\mu\left(\Delta_{k,t}\right)\right), \end{aligned} \tag{19}$$

and the formula (10) follows from (19). The lemma is proved. □

Lemma 5.3 *Let (S) be a series of the form* (8), *ψ be the quasi-measure isomorphic to the series (S). Consider an arbitrary non-decreasing function $\Phi : [1, +\infty) \to [1, +\infty)$ satisfying* (5), *and the function $Q : (0, 1] \to (0, 1]$ defined by* (6). *Suppose that for some $M > 0$ the quasi-measure ψ satisfies the inequality* (9). *Then the Nth partial sums of the series (S) satisfy*

$$|S_N(t)| \le 2M \cdot \Phi(N), \qquad N = 1, 2, \ldots, \quad t \in \mathbb{G}. \tag{20}$$

Proof Choose an arbitrary $N \in \mathbb{P}$ and $t \in \mathbb{G}$. It follows from the structure of Haar series that $S_N(t) = S_{2^k}(t)$ for some $k \in \mathbb{N}$ satisfying

$$2^k \leq 2 \cdot N. \tag{21}$$

We have

$$\begin{aligned} |S_N(t)| = |S_{2^k}(t)| &\overset{(17)}{=} \frac{|\psi\left(\Delta_{k,t}\right)|}{\mu\left(\Delta_{k,t}\right)} \overset{(9)}{\leq} \frac{M}{\mu\left(\Delta_{k,t}\right)} \cdot Q\left(\mu\left(\Delta_{k,t}\right)\right) \\ &\overset{(6)}{=} \frac{M}{\mu\left(\Delta_{k,t}\right)} \cdot \mu\left(\Delta_{k,t}\right) \cdot \Phi\left(\frac{1}{\mu\left(\Delta_{k,t}\right)}\right) \\ &\overset{(4)}{=} M \cdot \Phi\left(2^k\right) \overset{(5),\,(21)}{\leq} 2M \cdot \Phi\left(N\right), \end{aligned} \tag{22}$$

and the formula (20) follows from (22). The lemma is proved. □

6 The Q-Measures of Sets in $\mathbb{G}$

A *covering* of a set $E \subset \mathbb{G}$ is a family Ω of sets in $\mathbb{G}$ such that

$$E \subset \bigcup_{J \in \Omega} J.$$

Let $Q : (0, 1] \to (0, +\infty)$ be a non-decreasing function. Then the *Q-measure* of a set $E \subset \mathbb{G}$ is defined as

$$\operatorname{mes}_Q E := \lim_{\delta \to 0+0} \operatorname{mes}_Q^{\delta} E, \quad \operatorname{mes}_Q^{\delta} E := \inf \sum Q\left(\mu(J)\right), \tag{23}$$

where the infimum runs over all at most countable coverings $\{J\}$ of E by dyadic intervals J with $\mu(J) \leq \delta$. The set function Q is a Borel measure on $\mathbb{G}$. If $Q(x) \equiv x^q$ (q-const), then the x^q-measure is identical to the Hausdorff q-measure.

We need the next definition. A non-trivial $\mathcal{B}$-function τ is said to be *supported on a set* $E \subset \mathbb{G}$, if $\tau(\Delta) = 0$ for all $\Delta \in \mathcal{B}$ such that $\Delta \cap E = \emptyset$.

Theorem 6.1 *Suppose we are given a non-decreasing function* $Q : (0, 1] \to (0, 1]$, *and a closed set* $F \subset \mathbb{G}$ *with* $mes_Q F > 0$. *Then there is a non-trivial quasi-measure* $\psi \in H^Q$ *supported on the set* F.

Proof We introduce a $\mathcal{B}$-function τ by

$$\tau(\Delta) := \operatorname{mes}_Q^{\delta}(\Delta \cap F), \qquad \Delta \in \mathcal{B}. \tag{24}$$

Clearly, $\tau \in \underline{\mathcal{A}}_{\mathcal{B}}$. Further, $\text{mes}_Q F > 0$ implies that $\tau(\mathbb{G}) > 0$, and that the $\mathcal{B}$-function τ is non-trivial.

Let Δ be an arbitrary dyadic interval. If $\mu(\Delta) \leq \delta$, then the family containing only one set Δ is a covering of the set $\Delta \cap F$. Therefore

$$0 \leq \tau(\Delta) \overset{(23),\,(24)}{\leq} Q(\mu(\Delta)). \tag{25}$$

Note that there are finitely many dyadic intervals whose diameter exceed δ. This fact along with (25) implies

$$0 \leq \tau(\Delta) \leq M \cdot Q(\mu(\Delta)) \quad \text{for each } \Delta \in \mathcal{B}, \tag{26}$$

with some constant $M > 0$ independent of Δ.

Using Lemma 4.1, we find a non-trivial quasi-measure ψ satisfying the formula (12). It follows from (12) and (26) that $\psi \in \mathrm{H}^Q$. Also we see from (12), (24), and (26), that the quasi-measure ψ supported on the set F. The theorem is proved. □

7 The Monotonicity Theorem for $\mathcal{B}$-Functions

Here we prove a monotonicity theorem $\mathcal{B}$-functions. Below we apply this theorem to solve the uniqueness sets problem.

Theorem 7.1 *Suppose we are given a non-decreasing function* $Q : (0, 1] \to (0, 1]$ *satisfying*

$$\underline{\lim}_{x \to 0+} \frac{Q(x)}{x} > 0, \tag{27}$$

and a set $E \subset \mathbb{G}$, *and a* $\mathcal{B}$*-subadditive* $\mathcal{B}$*-function* $\tau \in H^{Q,*}$ *such that*

$$\underline{D}_{\mathcal{B}}\tau(t) \leq 0 \tag{28}$$

at every point $t \in \mathbb{G} \setminus E$. *If* $\tau(J) > 0$ *for some* $J \in \mathcal{B}$, *then* E *contains a closed subset* F *with* $\text{mes}_Q F > 0$.

Proof This proof repeat in general the proof of the Theorem 2 in [7]. Divide the interval J into finitely many dyadic intervals. Since $\tau \in \underline{\mathcal{A}}_{\mathcal{B}}$, we have $\tau(J_1) > 0$ for one of these intervals J_1. We introduce a $\mathcal{B}$-function $\tau_1 : \Delta \in \mathcal{B} \mapsto \tau(\Delta) - \varepsilon_1 \cdot \mu(\Delta)$, where $\varepsilon_1 > 0$ is so small that $\tau_1(J_1) > 0$. Evidently, $\varepsilon_1 \cdot \mu(\Delta) \in \underline{\mathcal{A}}_{\mathcal{B}}$. It follows from (10) and (27) that $\varepsilon_1 \cdot \mu(\Delta) \in \mathrm{H}^{Q,*}$. Therefore $\tau_1 \in \underline{\mathcal{A}}_{\mathcal{B}} \cap \mathrm{H}^{Q,*}$.

Let k_0 be the rank of the J_1 and consider the set

$$F = \bigcap_{k=k_0}^{\infty} \bigcup \Delta,$$

where the union runs over all dyadic intervals $\Delta \subseteq J_1$ of rank k such that $\tau_1(\Delta) > 0$. Being an intersection of closed sets, F is closed. The Lemmas 7.2 and 7.3 have proved in [7] (see the proof of the Theorem 2 in mentioned work).

Lemma 7.2 $F \neq \emptyset, \quad F \subset E.$

Lemma 7.3 *If $t \in F$, then $\tau_1(\Delta) > 0$ for every dyadic interval Δ of rank $k \geq k_0$ containing t.*

We return to the proof of the theorem. For every $M \in \mathbb{P}$ let F_M be the set of points $t \in F$ such that the inequality

$$\tau_1(\Delta) \leq M \cdot Q(\mu(\Delta)) \tag{29}$$

holds for all $\Delta = \Delta_{k,t}$. Since $\tau \in \mathrm{H}^{Q,*}$, we see that $F = \bigcup_{M=1}^{\infty} F_M$. Since F is closed, it follows from the Baire Category Theorem [2] that there is $M \in \mathbb{P}$ such that F_M is dense on a non-empty portion of F. Hence, there is a dyadic interval $J_2 \subset J_1$ with $\overline{F_M \cap J_2} = \widetilde{F}$, where $\widetilde{F} := F \cap J_2 \neq \emptyset$. By Lemma 7.3 we have $\tau_1(J_2) = C > 0$.

Consider an arbitrary covering $\{X_i\}$ of the closed set $\widetilde{F}$ by dyadic intervals. It was shown in [7] (see the proof of the Theorem 2 in the mentioned work) that there is a finite subcovering $\{Y_j\} \subset \{X_i\}$ of the set $\widetilde{F}$ by disjoint dyadic intervals, and a finite family $\{Z_l\}$ of disjoint dyadic intervals, such that

$$J_2 = \left(\bigsqcup_j Y_j\right) \bigsqcup_l \left(\bigsqcup_l Z_l\right);$$

$$\sum_j \tau_1(Y_j) + \sum_l \tau_1(Z_l) \geq \tau_1(J_2); \tag{30}$$

$$\tau_1(Z_l) \leq 0 \qquad \text{for all } l. \tag{31}$$

We have

$$\begin{aligned} \sum_i Q(\mu(X_i)) &\geq \sum_j Q(\mu(Y_j)) \overset{(29)}{\geq} \frac{1}{M} \sum_j \tau(Y_j) \\ &\overset{(30)}{\geq} \frac{1}{M}\left(\tau_1(J_2) - \sum_l \tau_1(Z_l)\right) \overset{(31)}{\geq} \frac{\tau_1(J_2)}{M} = \frac{C}{M}. \end{aligned} \tag{32}$$

Since $\{X_i\}$ is an arbitrary covering of $\widetilde{F}$ by dyadic intervals, the formula (32) yield that $\mathrm{mes}_Q \widetilde{F} > 0$. Since $F \supset \widetilde{F}$, we get $\mathrm{mes}_Q F > 0$. The theorem is proved. □

Theorem 7.4 *Suppose we are given a non-decreasing function $Q : (0, 1] \to (0, 1]$ satisfying the condition (27), and a set $E \subset \mathbb{G}$, and a quasi-measure $\tau \in H^{Q,*}$ such* *that*

$$D_{\mathcal{B}}\tau(t) = 0 \tag{33}$$

at all points $t \in \mathbb{G} \setminus E$. *If* $\tau(J) \neq 0$ *for some* $J \in \mathcal{B}$, *then* E *contains a closed subset* F *with* $mes_Q F > 0$.

Proof Replacing, if necessary, the $\mathcal{B}$-function τ by the $\mathcal{B}$-function $-\tau$, we obtain that $\tau(J) > 0$. All hypotheses of Theorem 7.1 are hold and, therefore, there is a closed set $F \subset E$ with $\text{mes}_Q F > 0$. The theorem is proved. □

8 Uniqueness Sets for Haar Series

Here we present new results for Haar series, concerning the uniqueness sets problem. Having a non-decreasing function $\Phi : [1, +\infty) \to [1, +\infty)$, we denote by $\mathcal{U}_\Phi$ the family of uniqueness sets for the series (8) whose Nth partial sums satisfy (20). Also we write $\mathcal{U}^*_\Phi$ for the class of uniqueness sets for the series (8) whose partial sums $S_N(t)$ satisfy (18) everywhere on $\mathbb{G}$. Clearly,

$$\mathcal{U}^*_\Phi \subseteq \mathcal{U}_\Phi \tag{34}$$

for all admissible Φ.

Theorem 8.1 *Consider an arbitrary non-decreasing function* $\Phi : [1, +\infty) \to [1, +\infty)$ *satisfying* (5), *and the function* $Q : (0, 1] \to (0, 1]$ *defined by* (6). *If* $E \notin \mathcal{U}^*_\Phi$, *then* E *contains a closed subset* F *with* $mes_Q F > 0$.

Proof Since $E \notin \mathcal{U}^*_\Phi$, there is a non-trivial series (S) of the form (8) such that (S) satisfies the condition (18) everywhere on $\mathbb{G}$, and (S) converges to zero on $\mathbb{G} \setminus E$. Let ψ be the quasi-measure isomorphic to this series. Since the series (S) satisfies the condition (18) everywhere on $\mathbb{G}$, Lemma 5.2 yields that the quasi-measure ψ satisfies the condition (10) everywhere on $\mathbb{G}$. Since (S) converges to zero for all $t \in \mathbb{G} \setminus E$, we have $D_{\mathcal{B}}\tau(t) = 0$ on $\mathbb{G} \setminus E$ by Lemma 5.1. Finally,

$$\frac{Q(x)}{x} \overset{(6)}{=} \Phi\left(\frac{1}{x}\right) \geq 1$$

for every $x > 0$. Consequently, the function Q satisfies (27). Thus all hypotheses of Theorem 7.4 hold and, therefore, there is a closed subset F with $\text{mes}_Q F > 0$. The theorem is proved. □

Theorem 8.2 *Consider an arbitrary non-decreasing function* $\Phi : [1, +\infty) \to [1, +\infty)$ *satisfying* (5), *and the function* $Q : (0, 1] \to (0, 1]$ *defined by* (6). *If* E *contains a closed subset* F *with* $mes_Q F > 0$, *then* $E \notin \mathcal{U}_\Phi$.

Proof By Theorem 6.1 there exists a non-trivial quasi-measure $\psi \in \mathrm{H}^Q$ supported on the set F. We consider the series (S) of the form (8) which is isomorphic to ψ. The quasi-measure ψ is non-trivial, hence the series (S) is non-trivial, too.

Further, $\psi \in \mathrm{H}^Q$ and, therefore, (S) satisfies the condition (20) by Lemma 5.3.

Choose an arbitrary point $t \in \mathbb{G} \setminus E$. Then $t \in \mathbb{G} \setminus F$. Since the set $\mathbb{G} \setminus F$ is open, there exists $\Delta_0 \in \mathcal{B}$ such that $\Delta_0 \ni t$ and $\Delta_0 \subset \mathbb{G} \setminus F$. Since the quasi-measure $\psi \in \mathrm{H}^Q$ is supported on the set F, $\psi(\Delta) = 0$ for each dyadic interval $\Delta \subset \Delta_0$. Hence, by Lemma 5.1 the series (S) converges to zero at the point t.

Thus there exists a non-trivial series (S) of the form (8) satisfying (20) and converging to zero on $\mathbb{G} \setminus E$. Therefore, $E \notin \mathcal{U}_\Phi$. The theorem is proved. □

Along with the formula (34), Theorems 8.1 and 8.2 immediately imply the next result.

Theorem 8.3 (Main Theorem) *Consider an arbitrary non-decreasing function* $\Phi : [1, +\infty) \to [1, +\infty)$ *satisfying* (5), *and the function* $Q : (0, 1] \to (0, 1]$ *defined by* (6). *Then* $E \in \mathcal{U}_\Phi \cap \mathcal{U}_\Phi^*$ *if and only if* E *contains no closed subsets* F *with* $mes_Q F > 0$.

Corollary 8.4 $\mathcal{U}_\Phi = \mathcal{U}_\Phi^*$ *for every non-decreasing function* $\Phi : [1, +\infty) \to [1, +\infty)$ *satisfying* (5).

Acknowledgements This work was completed with the support of the Russian Foundation for Basic Research (grant no. 14-01-00417), and of the program "Leading science schools" (grant no. NSh-3682.2014.1), and of the grant VGMHA-2014.

References

1. N.K. Bary (Bari), *A Treatise on Trigonometric Series* (Fizmatgiz, Moscow, 1961); Engl. transl., Pergamon Press, 1964
2. N. Bourbaki, *General Topology: Chapters 5–10. Elements of Mathematics* (Springer, Berlin, 1998)
3. B.I. Golubov, Series with respect to the Haar system. Itogi Nauki Ser. Mat. Anal. 109–146 (1971); Engl. transl., J. Soviet Math. **1**(6), 704–726 (1973)
4. G.M. Mushegyan, Uniqueness sets for the Haar system. Izv. Akad. Nauk Armen. SSR Ser. Mat. **2**(6), 350–361 (1967) (in Russian)
5. K.M. Ostaszewski, Henstock integration in the plane. Mem. Am. Math. Soc. **63** (1986), 1–106
6. M.G. Plotnikov, Uniqueness questions for some classes of Haar series. Mat. Zametki **75**(3), 392–404 (2004); Engl. transl., Math. Notes **75**(3), 360–371
7. M.G. Plotnikov, Quasi-measures, Hausdorff p-measures and Walsh and Haar series. Izv. RAN Ser. Mat. **74**(4), 157–188 (2010); Engl. transl., Izvestia Math. **74**(4), 819–848 (2010)
8. F. Schipp, W.R. Wade, P. Simon, *Walsh Series. An Introduction to Dyadic Harmonic Analysis* (Adam Hilger, Bristol/New York, 1990)
9. A.N. Shiryaev, *Probability* (Springer Science+Business Media New York, 1996)
10. V.A. Skvortsov, Calculation of the coefficients of an everywhere convergent Haar series. Mat. Sb. **75**(117):3, 349–360 (1968); Engl. transl., Math. USSR Sb. **4**(3), 317–327 (1968)
11. V.A. Skvortsov, Henstock-Kurzweil type integrals in $\mathcal{P}$-adic harmonic analysis. Acta Math. Acad. Paedagog. Nyházi (N.S.) **20**(2), 207–224 (2004)

12. W.R. Wade, Sets of uniqueness for Haar series. Acta Math. Acad. Sci. Hungaricae **30**(3–4), 265–281 (1977)
13. W.R. Wade, K. Yoneda, Uniqueness and quasi-measures on the group of integers of p-series field. Proc. Am. Math. Soc. **84**(2), 202–206 (1982)
14. D. Williams, *Probability with Martingales* (Cambridge University Press, Cambridge, 1991)
15. A. Zygmund, *Trigonometric Series*, vol. I, II (Cambridge University Press, Cambridge, 1959)

Off-Diagonal and Pointwise Estimates for Compact Calderón-Zygmund Operators

Paco Villarroya

Abstract We prove several off-diagonal and pointwise estimates for singular integral operators that extend compactly on $L^p(\mathbb{R}^n)$.

Keywords Calderón-Zygmund operator • Compact operator • Off-diagonal estimates, Singular integral

Mathematics Subject Classification (2010). Primary 42B20, 42C40; Secondary 47B07, 47G10

1 Introduction

An operator is said to satisfy an off-diagonal estimate from $L^p(\mathbb{R}^n)$ into $L^q(\mathbb{R}^n)$ for $p, q > 0$ if there exists a function $G : [0, \infty) \to [0, \infty)$ vanishing at infinity such that

$$\|T(f\chi_E)\chi_F\|_{L^q(\mathbb{R}^n)} \lesssim G(\mathrm{dist}(E, F))\|f\|_{L^p(\mathbb{R}^n)}$$

for all Borel sets $E, F \subset \mathbb{R}^n$ and all $f \in L^p(\mathbb{R}^n)$, with implicit constant depending on the operator T and the exponents p, q. Some authors distinguish between properly off-diagonal estimates, when $E \cap F = \emptyset$, and the so-called on-diagonal estimates, when $E \cap F \neq \emptyset$. However, we will not follow such convention and instead we will always call them off-diagonal estimates.

In the specific case of singular integral operators, the study focuses on the exponents $1 \leq p = q < \infty$. Very often, off-diagonal bounds are considered in

P. Villarroya (✉)
Centre for Mathematical Sciences, University of Lund, Lund, Sweden
e-mail: paco.villarroya@maths.lth.se

M. Ruzhansky, S. Tikhonov (eds.), *Methods of Fourier Analysis and Approximation Theory*, Applied and Numerical Harmonic Analysis,
DOI 10.1007/978-3-319-27466-9_7

one of the two following dual forms:

$$\|T(\chi_I)\chi_{(\lambda I)^c}\|_{L^1(\mathbb{R}^n)} \lesssim G(\lambda\ell(I))|I|, \qquad \frac{1}{|I|}\int_I |T(\chi_{(\lambda I)^c})(x)|dx \lesssim G(\lambda\ell(I))$$

(see [1, 11]), for any cube $I \subset \mathbb{R}^n$ and $\lambda > 1$, where $|I|$ denotes the volume of the cube and λI is a concentric dilation of I.

While their use in Analysis is very classical, the interest for this type of inequalities in modern Harmonic Analysis renewed in the 90s after the publication of new proofs of the T(1) Theorem that used the wavelet decomposition approach (see [2, 8] for example). These proofs were based on the development of estimates of the form

$$|\langle T(\psi_I), \psi_J\rangle| \lesssim \Big(\frac{|J|}{|I|}\Big)^{\frac{1}{2}+\frac{\delta}{n}}\Big(1+\frac{\mathrm{dist}(I,J)}{|I|^{\frac{1}{n}}}\Big)^{-(n+\delta)} \tag{1}$$

for all cubes $I, J \subset \mathbb{R}^n$ with $|J| \leq |I|$, under the appropriate hypothesis on the operator T, the parameter $\delta > 0$ and the functions ψ_I, ψ_J involved.

The importance of off-diagonal inequalities lays mainly on two facts. On the one side, they are a satisfactory replacement for pointwise estimates of the operator kernel when these are not available or even when the operator kernel is unknown. On the other side, they completely enclose the almost orthogonality properties of the operator. For these reasons, they played a crucial role in the solution of the famous Kato's conjecture [3] about boundedness of square root of elliptic operators and are nowadays extensively used in the study of second order elliptic operators. In the field, these estimates are typically established for one-parameter collections of operators $(T_t)_{t>0}$ and the function G also depends on the parameter t in an appropriate manner (see [4–6, 9, 10, 12]). Finally, it is also worth mentioning that off-diagonal bounds provide very valuable information for the development of efficient algorithms to compress and rapidly evaluate discrete singular operators (see [7, 14]).

One of the goals of the current paper is to establish similar type of estimates for singular integral operators that can be extended compactly on $L^p(\mathbb{R}^n)$ with $1 < p < \infty$. In [13], the author proved a characterization of these operators based on a new type of off-diagonal estimates for Calderón-Zygmund operators. Now, in the current paper, we aim to improve these bounds in several ways and also obtain some new estimates. More explicitly, we show in Sect. 3 that, in a broad sense and under the right hypotheses, these operators satisfy similar inequalities to (1) but with a new factor F that encodes the extra decay obtained as a consequence of their compactness properties:

$$|\langle T(\psi_I), \psi_J\rangle| \lesssim \Big(\frac{|J|}{|I|}\Big)^{\frac{1}{2}+\frac{\delta}{n}}\Big(1+\frac{\mathrm{dist}(I,J)}{|I|^{\frac{1}{n}}}\Big)^{-(n+\delta)} F(I,J). \tag{2}$$

The focus of this work is actually placed on obtaining a sharp and as detailed as possible description of the function F in the different cases under study.

Furthermore, in Sect. 4 we establish pointwise estimates of the action of the operator over compactly supported functions. This allows to claim, in a broad sense as well, that the image of a bump function adapted and supported in a cube behaves as a bump function adapted, although not supported, to the same cube. As before, these estimates explicitly state an extra decay not present in the classical bounds that is again due to the compactness of the operator.

2 Notation and Definitions

We say that the set $I = \prod_{i=1}^{n}[a_i, b_i]$ is a cube in $\mathbb{R}^n$ if the quantity $|b_i - a_i|$ is constant when varying the index i. We denote by $\mathcal{Q}_n$ the family of all cubes in $\mathbb{R}^n$. For every cube $I \subset \mathbb{R}^n$, we denote its centre by $c(I) = ((a_i + b_i)/2)_{i=1}^{n}$, its side length by $\ell(I) = |b_i - a_i|$ and its volume by $|I| = \ell(I)^n$. For any $\lambda > 0$, we denote by λI, the cube such that $c(\lambda I) = c(I)$ and $|\lambda I| = \lambda^n |I|$.

We write $|\cdot|_p$ for the l^p-norm in $\mathbb{R}^n$ with $1 \leq p \leq \infty$ and $|\cdot|$ for the modulus of a complex number. Hopefully, the latter notation will not cause any confusion with the one used for the volume of a cube. We denote by $\mathbb{B} = [-1/2, 1/2]^n$ and $\mathbb{B}_\lambda = \lambda\mathbb{B} = [-\lambda/2, \lambda/2]^n$.

Given two cubes $I, J \subset \mathbb{R}^n$, we define $\langle I, J \rangle$ as the unique cube such that it contains $I \cup J$ with the smallest possible side length and whose center has the smallest possible first coordinate. In the last section, this notation will be applied also to points, namely $\langle x, y \rangle$, as if they were considered to be degenerate cubes.

We denote the side length of $\langle I, J \rangle$ by $\operatorname{diam}(I \cup J)$. Notice that

$$\begin{aligned} \operatorname{diam}(I \cup J) &\approx \ell(I)/2 + |c(I) - c(J)|_\infty + \ell(J)/2 \\ &\approx \ell(I) + \operatorname{dist}_\infty(I, J) + \ell(J) \end{aligned}$$

where $\operatorname{dist}_\infty(I, J)$ denotes the set distance between I and J calculated using the norm $|\cdot|_\infty$. Actually,

$$\frac{1}{2}\operatorname{diam}(I \cup J) \leq \frac{\ell(I)}{2} + |c(I) - c(J)|_\infty + \frac{\ell(J)}{2} \leq \operatorname{diam}(I \cup J).$$

We define the relative distance between I and J by

$$\operatorname{rdist}(I, J) = \frac{\operatorname{diam}(I \cup J)}{\max(\ell(I), \ell(J))},$$

which is comparable to $\max(1, n)$ where n is the smallest number of times the larger cube needs to be shifted a distance equal to its side length so that the translated cube contains the smaller one. The following equivalences hold:

$$\begin{aligned}\text{rdist}(I,J) &\approx 1+\max(\ell(I),\ell(J))^{-1}|c(I)-c(J)|_\infty\\ &\approx 1+\max(\ell(I),\ell(J))^{-1}\text{dist}_\infty(I,J).\end{aligned}$$

Finally, we define the eccentricity of I and J as

$$\text{ecc}(I,J)=\frac{\min(|I|,|J|)}{\max(|I|,|J|)}.$$

Definition 2.1 In order to characterize compactness of singular integral operators, we use two sets of auxiliary bounded functions $L,S,D:[0,\infty)\to[0,\infty)$ and $F:\mathcal{Q}_n\to[0,\infty)$ satisfying the following limits

$$\lim_{x\to\infty}L(x)=\lim_{x\to 0}S(x)=\lim_{x\to\infty}D(x)=0, \tag{3}$$

$$\lim_{\ell(I)\to\infty}F(I)=\lim_{\ell(I)\to 0}F(I)=\lim_{|c(I)|_\infty\to\infty}F(I)=0. \tag{4}$$

Remark 2.2 Since any dilation $\mathcal{D}_\lambda L(x)=L(\lambda^{-1}x)$, $\mathcal{D}_\lambda F(I)=F(\lambda^{-1}I)$ with L and F satisfying (3), (4) respectively still satisfies the same limits, we will often omit all universal constants appearing in the arguments.

Definition 2.3 Let $K:(\mathbb{R}^n\times\mathbb{R}^n)\setminus\{(t,x)\in\mathbb{R}^n\times\mathbb{R}^n:t=x\}\to\mathbb{C}$.

We say that K is a compact Calderón-Zygmund kernel if there exist constants $0<\delta\leq 1$, $C>0$ and functions L, S and D satisfying the limits in (3), such that

$$|K(t,x)-K(t',x')|\leq C\frac{(|t-t'|_\infty+|x-x'|_\infty)^\delta}{|t-x|_\infty^{n+\delta}}F_K(t,x) \tag{5}$$

whenever $2(|t-t'|_\infty+|x-x'|_\infty)<|t-x|_\infty$, with

$$F_K(t,x)=L(|t-x|_\infty)S(|t-x|_\infty)D(|t+x|_\infty).$$

We say that K is is a standard Calderón-Zygmund kernel if (5) is satisfied with $F_K\equiv 1$.

We first note that, without loss of generality, L and D can be assumed to be non-creasing while S can be assumed to be non-decreasing. This is possible because, otherwise, we can always define

$$L_1(x)=\sup_{y\in[x,\infty)}L(y)\qquad S_1(x)=\sup_{y\in[0,x]}S(y)\qquad D_1(x)=\sup_{y\in[x,\infty)}D(y)$$

which bound above L, S and D respectively, satisfy the limits in (3) and are non-creasing or non-decreasing as requested.

On the other hand, we also denote

$$F_K(t,t',x,x') = L_2(|t-x|_\infty)S_2(|t-t'|_\infty + |x-x'|_\infty)D_2\Big(1+\frac{|t+x|_\infty}{1+|t-x|_\infty}\Big)$$

and assume, in a similar way as before, that L_2 and D_2 are non-creasing while S_2 is non-decreasing. Then, as explained in [13], (5) is equivalent to the following smoothness condition

$$|K(t,x)-K(t',x')| \leq C\frac{(|t-t'|_\infty + |x-x'|_\infty)^{\delta'}}{|t-x|_\infty^{n+\delta'}}F_K(t,t',x,x') \tag{6}$$

whenever $2(|t-t'|_\infty + |x-x'|_\infty) < |x-t|_\infty$, with possibly smaller $\delta' < \delta$. This resulting parameter δ' necessarily satisfies $\delta' < 1$ since otherwise the kernel K would be a constant function.

The proof of the equivalence between both formulations appears in [13]. However, to increase readability of the current paper, we sketch the proof that (5) implies (6). For any $0 < \epsilon < \delta$, let $\delta' = \delta - \epsilon$. Then, from (5) we can write

$$|K(t,x)-K(t',x')| \leq C\frac{(|t-t'|_\infty + |x-x'|_\infty)^{\delta'}}{|t-x|_\infty^{n+\delta'}}\tilde{F}_K(t,t',x,x')$$

with

$$\tilde{F}_K(t,t',x,x') = \frac{(|t-t'|_\infty + |x-x'|_\infty)^{\epsilon}}{|t-x|_\infty^{\epsilon}}F_K(t,x).$$

Then, the functions

$$L_2(y) = \sup_{|x-t|_\infty \geq y} \tilde{F}_K(t,t',x,x')^{1/3}$$

$$S_2(y) = \sup_{|x-x'|_\infty + |t-t'|_\infty \leq y} \tilde{F}_K(t,t',x,x')^{1/3}$$

$$D_2(y) = \sup_{1+\frac{|x+t|_\infty}{1+|x-t|_\infty} \geq y} \tilde{F}_K(t,t',x,x')^{1/3}$$

satisfy all the required limits in (3) and

$$\tilde{F}_K(t,t',x,x') \leq L_2(|t-x|_\infty)S_2(|t-t'|_\infty + |x-x'|_\infty)D_2\Big(1+\frac{|t+x|_\infty}{1+|t-x|_\infty}\Big).$$

As also proved in [13], the smoothness condition (5) and the hypothesis $\lim_{|t-x|_\infty\to\infty} K(t,x) = 0$ imply the classical decay condition

$$|K(t,x)| \lesssim \frac{1}{|t-x|_\infty^n} F_K(t,x) \lesssim \frac{1}{|t-x|_\infty^n} \tag{7}$$

for all $t, x \in \mathbb{R}^n$ such that $t \neq x$. Moreover, it is easy to see that we also get the decay

$$|K(t,x)| \lesssim \frac{1}{|t-x|_\infty^n} L(|t-x|_\infty) S(|t-x|_\infty) D\Big(1 + \frac{|t+x|_\infty}{1+|t-x|_\infty}\Big), \tag{8}$$

which we will use later. Notice the change in the argument of S, which is now equal to the argument of L.

Finally, we define two more sets of auxiliary functions which we will use in the next section. First,

$$F_K(I_1, I_2, I_3) = L(\ell(I_1)) S(\ell(I_2)) D(\operatorname{rdist}(I_3, \mathbb{B})) \tag{9}$$

and $F_K(I) = F_K(I, I, I)$. Second,

$$\tilde{F}_K(I_1, I_2, I_3) = L(\ell(I_1)) S(\ell(I_2)) \tilde{D}(\operatorname{rdist}(I_3, \mathbb{B})) \tag{10}$$

and $\tilde{F}_K(I) = \tilde{F}_K(I, I, I)$, where

$$\tilde{D}(\operatorname{rdist}(I, \mathbb{B})) = \sum_{j\geq 0} 2^{-j\delta} D(\operatorname{rdist}(2^j I, \mathbb{B})). \tag{11}$$

We note that for fixed $\ell(I)$, the Lebesgue Dominated Theorem guarantees that $\tilde{D}$ satisfies $\lim_{|c(I)|_\infty\to\infty} \tilde{D}(\operatorname{rdist}(I, \mathbb{B})) = 0$.

Definition 2.4 Let $T : \mathcal{C}_0(\mathbb{R}^n) \to \mathcal{C}_0(\mathbb{R}^n)'$ be a continuous linear operator. We say that T is associated with a compact Calderón-Zygmund kernel if there exists a function K fulfilling Definition 2.3 such that the dual pairing satisfies the following integral representation

$$\langle T(f), g\rangle = \int_{\mathbb{R}^n}\int_{\mathbb{R}^n} f(t) g(x) K(t,x)\, dt\, dx$$

for all functions $f, g \in \mathcal{C}_0(\mathbb{R}^n)$ with disjoint compact supports.

Clearly, the integral converges absolutely since, by (7), we have for $d = \operatorname{dist}(\operatorname{supp} f, \operatorname{supp} g) > 0$,

$$\Big|\int_{\mathbb{R}^n}\int_{\mathbb{R}^n} f(t) g(x) K(t,x)\, dt dx\Big| \lesssim \|f\|_{L^1(\mathbb{R}^n)} \|g\|_{L^1(\mathbb{R}^n)} \frac{1}{d^n}.$$

Definition 2.5 Let $0 < p \leq \infty$. We say that a bounded function ϕ is an $L^p(\mathbb{R}^n)$-normalized bump function adapted to I with constant $C > 0$, decay $N \in \mathbb{N}$ and order 0, if for all $x \in \mathbb{R}^n$

$$|\phi(x)| \leq C|I|^{-\frac{1}{p}}\Big(1 + \frac{|r - c(I)|_\infty}{\ell(I)}\Big)^{-N}. \tag{12}$$

We say that a continuous bounded function ϕ is an $L^p(\mathbb{R}^n)$-normalized bump function adapted to I with constant $C > 0$, decay $N \in \mathbb{N}$, order 1 and parameter $0 < \alpha \leq 1$, if (12) holds and for all $t, x \in \mathbb{R}^n$

$$|\phi(t) - \phi(x)| \leq C\Big(\frac{|t - x|_\infty}{\ell(I)}\Big)^\alpha |I|^{-\frac{1}{p}} \sup_{r \in \langle t,x\rangle} \Big(1 + \frac{|r - c(I)|_\infty}{\ell(I)}\Big)^{-N}, \tag{13}$$

where $\langle t, x\rangle$ denotes the cube containing the points t and x with the smallest possible side length and whose centre has the smallest possible first coordinate.

Unless otherwise stated, we will assume the bump functions to be $L^2(\mathbb{R}^n)$-normalized.

Definition 2.6 We say that a linear operator $T : \mathcal{C}_0(\mathbb{R}^n) \to \mathcal{C}_0(\mathbb{R}^n)'$ satisfies the weak compactness condition, if there exists a bounded function F_W satisfying (4) and such that for any cube $I \subset \mathbb{R}^n$ and any bump functions ϕ_I, φ_I adapted to I with constant $C > 0$, decay N and order 0, we have

$$|\langle T(\phi_I), \varphi_I\rangle| \lesssim CF_W(I)$$

where the implicit constant only depends on the operator T.

As explained in [13], this definition admits several other reformulations, but they all essentially imply that the dual pairing $\langle T(\phi_I), \varphi_I\rangle$ tends to zero when the cube involved is large, small or far away from the origin.

Definition 2.7 We define $\mathrm{CMO}(\mathbb{R}^n)$ as the closure in $\mathrm{BMO}(\mathbb{R}^n)$ of the space of continuous functions vanishing at infinity.

The following theorem, which is the main result in [13], characterizes compactness of Calderón-Zygmund operators. This is the reason why we say that the new off-diagonal bounds appearing in the current paper apply to operators that can be extended compactly on $L^p(\mathbb{R}^n)$.

Theorem 2.8 *Let T be a linear operator associated with a standard Calderón-Zygmund kernel.*

Then, T extends to a compact operator on $L^p(\mathbb{R})$ for all $1 < p < \infty$ if and only if T is associated with a compact Calderón-Zygmund kernel and it satisfies the weak compactness condition and the cancellation conditions $T(1), T^*(1) \in \mathrm{CMO}(\mathbb{R})$.

3 Off-Diagonal Estimates for Bump Functions

In the proof of Theorem 2.8, some off-diagonal estimates were developed. Now, we improve these inequalities in several directions: by extending the result to $\mathbb{R}^n$, by weakening the smoothness requirements of the bumps, by shortening the proof and by obtaining a sharper bound for functions with compact support.

This is the purpose of the three propositions of this section, which describe the action of a compact singular integral operator over bump functions with or without zero mean properties respectively. Later, in Sect. 4, we will use these bounds to obtain several pointwise bounds and other off-estimates of a more general type.

We first set up some notation that appears in the statements of the three results. We consider K to be a compact Calderón-Zygmund kernel with parameter $0 < \delta < 1$ and T to be a linear operator with associated kernel K satisfying the weak compactness condition. We denote by $I \wedge J$ and $I \vee J$ the smallest and the largest of two given cubes I, J respectively. That is, $I \wedge J = J$, $I \vee J = I$ if $\ell(J) \leq \ell(I)$, while $I \wedge J = I$, $I \vee J = J$, otherwise. We also remind the notation of F_K, F_W and $\tilde{F}_K$ provided in the previous section.

Proposition 3.1 *If the special cancellation conditions $T(1) = T^*(1) = 0$ hold then, for all bump functions ψ_I, ψ_J adapted and supported on I, J respectively, with constant $C > 0$, order one, parameter $\alpha > \delta$ and such that $\psi_{I\wedge J}$ has mean zero,*

$$|\langle T(\psi_I), \psi_J\rangle| \lesssim C^2 \frac{\mathrm{ecc}(I,J)^{\frac{1}{2}+\frac{\delta}{n}}}{\mathrm{rdist}(I,J)^{n+\delta}} F(I,J) \tag{14}$$

where F is such that:

(i) $F(I,J) = F_K(\langle I,J\rangle, I\wedge J, \langle I,J\rangle)$ *when* $\mathrm{rdist}(I,J) > 3$,
(ii) $F(I,J) = \tilde{F}_K(I\vee J, I\wedge J, I\vee J) + F_W(I\wedge J) + F_K(I\wedge J, I\wedge J, I\vee J)$, *otherwise.*

Proposition 3.2 *For all bump functions ψ_I, ψ_J adapted and supported on I, J respectively, with constant $C > 0$, order one, parameter $\alpha > \delta$ and such that $\psi_{I\wedge J}$ has mean zero, we have*

$$|\langle T(\psi_I), \psi_J\rangle| \lesssim C^2 \frac{\mathrm{ecc}(I,J)^{\frac{1}{2}}}{\mathrm{rdist}(I,J)^{n+\delta}} F(I,J) \tag{15}$$

with $F(I,J) = \tilde{F}_K(I\wedge J) + F_W(I\wedge J) + F_K(I\wedge J, I\wedge J, I\vee J)$ *when* $\mathrm{rdist}(I,J) \leq 3$.

On the other hand, when $\mathrm{rdist}(I,J) > 3$, *inequality* (14) *still holds with the same* $F(I,J) = F_K(\langle I,J\rangle, I\wedge J, \langle I,J\rangle)$.

Proposition 3.3 *For all bump functions ψ_I, ψ_J adapted and supported on I, J respectively, with constant $C > 0$ and order zero, we have*

$$|\langle T(\psi_I), \psi_J\rangle| \lesssim C^2 \frac{\mathrm{ecc}(I,J)^{\frac{1}{2}}}{\mathrm{rdist}(I,J)^{n}} \left(1 + \left|\log \mathrm{ecc}(I,J)\right|^{\theta}\right) F(I,J) \tag{16}$$

where

(i) $F(I,J) = F_K(\langle I,J \rangle)$ *and* $\theta = 0$ *when* $\mathrm{rdist}(I,J) > 3$,
(ii) $F(I,J) = F_W(I \wedge J) + F_K(I \wedge J, I \vee J, I \vee J)$ *and* $\theta = 1$ *when* $\mathrm{rdist}(I,J) \leq 3$.

In all cases, the implicit constants depend on the operator T and the parameters δ and α but they are universal otherwise. Needless to say that the actual value appearing in the condition $\mathrm{rdist}(I,J) > 3$ plays no special role and it could be easily changed by any other value strictly larger than one.

As mentioned before, Proposition 3.1 is an improvement of the analog result in [13]. The result has been extended to non-smooth bump functions of several dimensions. At the same time, the proof has been largely simplified by using the extra hypothesis that the bump functions are compactly supported. Moreover, with this hypothesis, the last factor on the right hand side of the inequality turned out to be strictly smaller than the one appearing in [13]. In fact, when the bump functions are not longer compactly supported, as it happens in [13], the inequality (14) holds with a larger factor depending on six different cubes rather than only three cubes. Nevertheless, in both cases, the factors enjoy essentially the same properties and so, each of the two estimates suffices to prove compactness of the operator.

We also note that in Proposition 3.2, the hypotheses that $T(1), T^*(1) \in \mathrm{BMO}(\mathbb{R}^n)$ or $T(1), T^*(1) \in \mathrm{CMO}(\mathbb{R}^n)$ are not needed. Moreover, in Proposition 3.3, the assumption of T satisfying the special cancellation conditions $T(1) = T^*(1) = 0$ does not lead to any further improvement.

Notation 3.4 For the following three proofs, we provide some common notation. For every cube $I \subset \mathbb{R}^n$, we denote by $\Phi_I \in \mathcal{S}(\mathbb{R}^n)$ an L^∞-normalized function adapted to I with arbitrary large order and decay such that $0 \leq \Phi_I \leq 1$, $\Phi_I = 1$ in $2I$ and $\Phi_I = 0$ in $(4I)^c$. This implies that $\Phi_I(x) = 1$ for all $|x - c(I)|_\infty \leq \ell(I)$ while $\Phi_I(x) = 0$ for all $|x - c(I)|_\infty > 2\ell(I)$.

As customary, we define the translation and dilation operators by $\mathcal{T}_a f(x) = f(x-a)$ and $\mathcal{D}_\lambda f(x) = f(\lambda^{-1}x)$ respectively with $x, a \in \mathbb{R}^n$ and $\lambda > 0$. We also define $w_I(x) = 1 + \ell(I)^{-1}|x - c(I)|_\infty$ and for any function $\psi = \psi_1 \otimes \psi_2$ of tensor product type, we write $\Lambda(\psi) = \langle T(\psi_1), \psi_2 \rangle$.

Finally, by symmetry we can assume that $\ell(J) \leq \ell(I)$ and so, $I \wedge J = J$ while $I \vee J = I$.

Proof of Proposition 3.1 Let $\psi(t,x) = \phi_I(t)\psi_J(x)$ which, by hypothesis, is supported and adapted to $I \times J$ with constant C^2, decay N, order 1, parameter $\alpha > \delta$ and, most importantly, it has mean zero in the variable x.

(a) We first assume that $3\ell(I) < \mathrm{diam}(I \cup J)$ which implies $(5I) \cap J = \emptyset$ and so, $\mathrm{diam}(I \cup J) = \ell(I)/2 + |c(I) - c(J)|_\infty + \ell(J)/2 \leq \ell(I) + |c(I) - c(J)|_\infty$.

Then, since $|t - c(I)|_\infty \leq \ell(I)/2$, we have

$$|t - c(J)|_\infty \leq \ell(I)/2 + |c(I) - c(J)|_\infty \leq \mathrm{diam}(I \cup J) \tag{17}$$

and

$$|t-c(J)|_\infty \geq |c(I)-c(J)|_\infty - |t-c(I)|_\infty \geq \ell(I)+|c(I)-c(J)|_\infty - 3\ell(I)/2$$

$$\geq \mathrm{diam}(I\cup J) - \mathrm{diam}(I\cup J)/2 = \mathrm{diam}(I\cup J)/2.$$

On the other hand, $3\ell(I) < \mathrm{diam}(I\cup J) \leq \ell(I)+|c(I)-c(J)|_\infty$ also implies $2\ell(I) < |c(I)-c(J)|_\infty$ and since $|x-c(J)|_\infty \leq \ell(J)/2$, we get

$$|t-c(J)|_\infty \geq |c(I)-c(J)|_\infty - |t-c(I)|_\infty$$

$$\geq 2\ell(I)-\ell(I)/2 \geq 3\ell(J)/2 \geq 3|x-c(J)|_\infty.$$

The last inequality implies that the support of ψ is disjoint with the diagonal and so, we can use the Calderón-Zygmund kernel representation to write

$$\Lambda(\psi) = \int\int \psi(t,x)K(t,x)\,dtdx = \int_J\int_I \psi(t,x)(K(t,x)-K(t,c(J)))\,dtdx$$

where the second equality is due to the zero mean of ψ in the variable x. Now, we denote $Q_{I,J} = \{t\in\mathbb{R}^n : \mathrm{diam}(I\cup J)/2 < |t-c(J)|_\infty \leq \mathrm{diam}(I\cup J)\}$. Then, by the smoothness condition (6) of a compact Calderón-Zygmund kernel and the monotonicity properties of L, S and D, we bound as follows:

$$\begin{aligned}|\Lambda(\psi)| &\lesssim \int_J\int_{I\cap Q_{I,J}} |\psi(t,x)|\frac{|x-c(J)|_\infty^\delta}{|t-c(J)|_\infty^{n+\delta}} \\ &\qquad L(|t-c(J)|_\infty)S(|x-c(J)|_\infty)D\Big(1+\frac{|t+c(J)|_\infty}{1+|t-c(J)|_\infty}\Big)\,dxdt \\ &\lesssim \|\psi\|_{L^1(\mathbb{R}^{2n})}\frac{\ell(J)^\delta}{\mathrm{diam}(I\cup J)^{n+\delta}}L(\mathrm{diam}(I\cup J))S(\ell(J))D\Big(1+\frac{|c(J)|_\infty}{1+\mathrm{diam}(I\cup J)}\Big) \\ &\lesssim C^2|I|^{\frac12}|J|^{\frac12}\frac{\ell(J)^\delta}{\mathrm{diam}(I\cup J)^{n+\delta}}L(\ell(\langle I,J\rangle))S(\ell(J))D(\mathrm{rdist}(\langle I,J\rangle,\mathbb{B})) \\ &= C^2\Big(\frac{|J|}{|I|}\Big)^{\frac12+\frac{\delta}{n}}\Big(\frac{\mathrm{diam}(I\cup J)}{\ell(I)}\Big)^{-(n+\delta)}F_K(\langle I,J\rangle,J,\langle I,J\rangle)\end{aligned}$$

as stated. To completely finish this case, we explain in more detail the reasoning to obtain the bounds for D used in the second and third inequalities above. Since $|x|_\infty \leq (|x-t|_\infty + |x+t|_\infty)/2$, we have

$$1+\frac{|x|_\infty}{1+|t-x|_\infty} \leq 1+\frac12+\frac{|t+x|_\infty}{1+|t-x|_\infty} \leq \frac32\Big(1+\frac{|t+x|_\infty}{1+|t-x|_\infty}\Big).$$

Then, in the domain of integration,

$$1+\frac{|c(J)|_\infty}{1+\mathrm{diam}(I\cup J)}\leq 1+\frac{|c(J)|_\infty}{1+|t-c(J)|_\infty}\leq\frac{3}{2}\Big(1+\frac{|t+c(J)|_\infty}{1+|t-c(J)|_\infty}\Big)$$

and, since D is non-creasing, we have

$$D\Big(1+\frac{|t+c(J)|_\infty}{1+|t-c(J)|_\infty}\Big)\leq D\Big(1+\frac{|c(J)|_\infty}{1+\mathrm{diam}(I\cup J)}\Big)$$

omitting constants.

On the other hand, since $|c(I)|_\infty-|c(J)|_\infty\leq|c(I)-c(J)|_\infty\leq\mathrm{diam}(I\cup J)$, we can bound below the numerator of the argument of D in the last expression by

$$\begin{aligned}1+\mathrm{diam}(I\cup J)+|c(J)|_\infty&\geq 1+\frac{1}{2}\mathrm{diam}(I\cup J)+\frac{1}{2}(|c(I)|_\infty-|c(J)|_\infty)+|c(J)|_\infty\\&=1+\frac{1}{2}\mathrm{diam}(I\cup J)+\frac{1}{2}(|c(I)|_\infty+|c(J)|_\infty)\\&\geq\frac{1}{2}\big(1+\mathrm{diam}(I\cup J)+\frac{1}{2}|c(I)+c(J)|_\infty\big).\end{aligned}$$

Then,

$$\begin{aligned}1+\frac{|c(J)|_\infty}{1+\mathrm{diam}(I\cup J)})&\geq\frac{1}{2}\frac{1+\mathrm{diam}(I\cup J)+|c(I)+c(J)|_\infty/2}{1+\mathrm{diam}(I\cup J)}\\&\geq\frac{1}{3}\Big(\frac{3}{2}+\frac{|c(I)+c(J)|_\infty/2}{\mathrm{diam}(I\cup J)}\Big).\end{aligned}$$

Finally, since $|(c(I)+c(J))/2-c(\langle I,J\rangle)|_\infty\leq\ell(\langle I,J\rangle)/2$ and $\ell(\langle I,J\rangle)=\mathrm{diam}(I\cup J)$, we bound below previous expression by

$$\frac{1}{3}\Big(\frac{3}{2}+\frac{|c(\langle I,J\rangle)|_\infty}{\mathrm{diam}(I\cup J)}-\frac{1}{2}\Big)\geq\frac{1}{3}\Big(1+\frac{|c(\langle I,J\rangle)|_\infty}{\max(\ell(\langle I,J\rangle),1)}\Big)=\frac{1}{3}\,\mathrm{rdist}(\langle I,J\rangle,\mathbb{B}).$$

(b) We now assume that $\mathrm{diam}(I\cup J)\leq 3\ell(I)$ which implies $1\leq\mathrm{rdist}(I,J)\leq 3$. In this case, we first show that we can assume $\psi(c(J),x)=0$ for any $x\in\mathbb{R}^n$. This assumption comes from the substitution of $\psi(t,x)$ by

$$\psi(t,x)-(\mathcal{T}_{c(J)}\mathcal{D}_{\ell(I)}\Phi)(t)\psi(c(J),x)\tag{18}$$

where $\Phi=\Phi_{\mathbb{B}}$ as described in Notation 3.4. Then, we only need to prove that the subtracted term satisfies the desired bound.

We denote $\tilde{\psi}(x) = \psi(c(J), x)$. Since ψ_I and ψ_J are adapted to I and J respectively with constant $C > 0$ and decay N for any $N \in \mathbb{N}$, we have

$$|\tilde{\psi}(x)| \leq C^2|I|^{-\frac{1}{2}}|J|^{-\frac{1}{2}}w_J(x)^{-N}.$$

Then, $\|\tilde{\psi}\|_{L^1(\mathbb{R}^n)} \leq C^2|I|^{-\frac{1}{2}}|J|^{\frac{1}{2}}$. We also recall that $\tilde{\psi}$ is supported on J and has mean zero.

Now, we write $\lambda = \ell(I)/\ell(J) \geq 1$ and take $k \in \mathbb{N}$ so that $2^k \leq \lambda < 2^{k+1}$. Then,

$$\mathcal{T}_{c(J)}\mathcal{D}_{\ell(I)}\Phi = \mathcal{T}_{c(J)}\mathcal{D}_{\lambda\ell(J)}\Phi.$$

To simplify notation, we write $\Phi_0 = \mathcal{T}_{c(J)}\mathcal{D}_{\ell(I)}\Phi \in \mathcal{S}(\mathbb{R}^n)$ and $\Phi_1 = 1 - \Phi_0$. We note that Φ_1 is a smooth bounded function supported on $|t - c(J)|_\infty > \lambda\ell(J)$. By the classical theory, we know that $T(1)$ can be defined as a distribution acting on the space of compactly supported functions with mean zero in the following way

$$\begin{aligned}\langle T(1), \tilde{\psi}\rangle &= \langle T(\Phi_0), \tilde{\psi}\rangle + \int\int \Phi_1(t)\tilde{\psi}(x)K(t,x)dtdx \\ &= \langle T(\Phi_0), \tilde{\psi}\rangle + \int\int \Phi_1(t)\tilde{\psi}(x)(K(t,x) - K(t, c(J)))dtdx\end{aligned}$$

where the second equality is due to the mean zero of $\tilde{\psi}$. Notice that, since $|x - c(J)|_\infty \leq \ell(J)/2 \leq \ell(J)2^{k-1} \leq 2^{-1}|t - c(J)|_\infty$, the supports of Φ_1 and $\tilde{\psi}$ are disjoint and so the integral in the second line converges absolutely. Then, the hypothesis that $T(1) = 0$ implies

$$\langle T(\Phi_0), \tilde{\psi}\rangle = -\int\int \Phi_1(t)\tilde{\psi}(x)(K(t,x) - K(t, c(J)))dtdx.$$

Moreover, since $2|x - c(J)|_\infty < |t - c(J)|_\infty$, we can use the the smoothness condition (6) of a compact Calderón-Zygmund kernel to write

$$\begin{aligned}|\langle T(\Phi_0), \tilde{\psi}\rangle\rangle| \leq \int_J \int_{\lambda\ell(J)<|t-c(J)|_\infty} & |\Phi_1(t)||\tilde{\psi}(x)|\frac{|x - c(J)|_\infty^\delta}{|t - c(J)|_\infty^{n+\delta}} \\ & L(|t - c(J)|_\infty)S(|x - c(J)|_\infty)D\Big(1 + \frac{|t - c(J)|_\infty}{1 + |t - c(J)|_\infty}\Big)\, dtdx.\end{aligned}$$

Then, by the reasoning applied in the previous case, we have

$$\begin{aligned}|\langle T(\Phi_0), \tilde{\psi}\rangle\rangle| &\lesssim \|\tilde{\psi}\|_{L^1(\mathbb{R}^n)}\ell(J)^\delta L(\lambda\ell(J))S(\ell(J)) \\ &\int_{2^k\ell(J)<|t-c(J)|_\infty} \frac{1}{|t - c(J)|_\infty^{n+\delta}} D\Big(1 + \frac{|c(J)|_\infty}{1 + |t - c(J)|_\infty}\Big)dt.\end{aligned}$$

Now, we rewrite the last integral as

$$\begin{aligned}
&\sum_{j\geq k}\int_{2^j\ell(J)<|t-c(J)|\leq 2^{j+1}\ell(J)} \frac{1}{|t-c(J)|_\infty^{n+\delta}}D\Big(1+\frac{|c(J)|_\infty}{1+|t-c(J)|_\infty}\Big)dt \\
&\lesssim \sum_{j\geq k} D\Big(1+\frac{|c(J)|_\infty}{1+2^{j+1}\ell(J)}\Big)\frac{(2^{j+1}\ell(J))^n}{(2^j\ell(J))^{n+\delta}} \\
&\lesssim \ell(J)^{-\delta}\sum_{j\geq k} 2^{-j\delta}D\Big(1+\frac{|c(J)|_\infty}{1+2^{j+1}\ell(J)}\Big) \\
&= \ell(J)^{-\delta}2^{-k\delta}\sum_{j\geq 0} 2^{-j\delta}D\Big(1+\frac{|c(J)|_\infty}{1+2^{j+k+1}\ell(J)}\Big) \\
&\lesssim \ell(I)^{-\delta}\sum_{j\geq 0} 2^{-j\delta}D\Big(1+\frac{|c(J)|_\infty}{1+2^{j}\ell(I)}\Big) \\
&\lesssim \ell(I)^{-\delta}\sum_{j\geq 0} 2^{-j\delta}D(\,\mathrm{rdist}(2^jI,\mathbb{B})) \\
&= \ell(I)^{-\delta}\tilde{D}(\,\mathrm{rdist}(I,\mathbb{B}))
\end{aligned}$$

with $\tilde{D}$ as in (11). We now detail the step taken in the last inequality: since $|c(I)-c(J)|_\infty \leq \mathrm{diam}(I\cup J)\leq 3\ell(I)$, we have that

$$4\Big(1+\frac{|c(J)|_\infty}{1+2^j\ell(I)}\Big)\geq 4+\frac{|c(I)|_\infty}{1+2^j\ell(I)}-\frac{3}{2^j}\geq 1+\frac{|c(I)|_\infty}{1+2^j\ell(I)} = D(\,\mathrm{rdist}(2^jI,\mathbb{B})).$$

Then, we write

$$\begin{aligned}
|\langle T(\Phi_0),\tilde{\psi}\rangle\rangle| &\lesssim C^2|I|^{-\frac{1}{2}}|J|^{\frac{1}{2}}\left(\frac{\ell(J)}{\ell(I)}\right)^{\delta} L(\lambda\ell(J))S(\ell(J))\tilde{D}(\,\mathrm{rdist}(I,\mathbb{B})) \\
&= C^2\left(\frac{|J|}{|I|}\right)^{\frac{1}{2}+\frac{\delta}{n}}\tilde{F}_K(I,J,I)
\end{aligned}$$

which is the first term in the stated bound. This finishes the justification of the assumption $\psi(c(J),x)=0$ for any $x\in\mathbb{R}^n$.

Now, we decompose ψ in the following way:

$$\begin{aligned}
\psi &= \psi_{out}+\psi_{in} \\
\psi_{in}(t,x) &= \psi(t,x)\Phi_{3J}(t).
\end{aligned}$$

(b1) We first prove that ψ_{in} is adapted to $J\times J$ with order zero, decay N and constant $C^2\,(|J|/|I|)^{\frac{1}{2}+\frac{\delta}{n}}$.

By the assumption $\psi(c(J),x)=0$, the fact that ψ is supported and adapted to $I\times J$ with order one and parameter α and that ψ_{in} is supported on $3J\times J$, we have for all $t\in 3J$ and all $x\in J$,

$$\begin{aligned}
|\psi_{in}(t,x)| &= |\psi(t,x)-\psi(c(J),x)|\Phi_{3J}(t)\\
&\lesssim C\Big(\frac{|t-c(J)|_\infty}{\ell(I)}\Big)^{\alpha}|I|^{-\frac{1}{2}}\sup_{r\in\langle t,c(J)\rangle}\Big(1+\frac{|r-c(I)|_\infty}{\ell(I)}\Big)^{-N}\chi_{3J}(t)C|J|^{-\frac{1}{2}}w_J(x)^{-N}\\
&\lesssim C^2\Big(\frac{\ell(J)}{\ell(I)}\Big)^{\alpha}|I|^{-\frac{1}{2}}\chi_{3J}(t)|J|^{-\frac{1}{2}}w_J(x)^{-N}\\
&\lesssim C^2\left(\frac{|J|}{|I|}\right)^{\frac{1}{2}+\frac{\delta}{n}}|J|^{-\frac{1}{2}}w_J(t)^{-N}|J|^{-\frac{1}{2}}w_J(x)^{-N}
\end{aligned}$$

since $\delta<\alpha$ and $|J|\le|I|$. Notice that we also used $|t-c(J)|_\infty\le 3\ell(J)/2$.

Therefore ψ_{in} is adapted to $J\times J$ with order zero and the stated constant and so, by the weak compactness property of T we get

$$|\Lambda(\psi_{in})|\lesssim C^2\left(\frac{|J|}{|I|}\right)^{\frac{1}{2}+\frac{\delta}{n}}F_W(J)$$

which ends this case.

(b2) We now work with ψ_{out}. In this case, by the extra assumption again and the support of ψ, we have the following decay

$$\begin{aligned}
|\psi_{out}(t,x)| &\le |\psi(t,x)-\psi(c(J),x)|\\
&\lesssim C\Big(\frac{|t-c(J)|_\infty}{\ell(I)}\Big)^{\alpha}|I|^{-1/2}\chi_I(t)C|J|^{-1/2}w_J(x)^{-N}\\
&= C^2\Big(\frac{|t-c(J)|_\infty}{\ell(I)}\Big)^{\alpha}\varphi_{I\times J}(t,x)
\end{aligned}\tag{19}$$

by denoting $\varphi_{I\times J}(t,x)=|I|^{-1/2}\chi_I(t)|J|^{-1/2}w_J(x)^{-N}$.

Due to the support of ψ, we get $|t-c(I)|_\infty\le\ell(I)/2$ while, by the calculations in (17) and the hypothesis of this case, we also have

$$|t-c(J)|_\infty\le\operatorname{diam}(I\cup J)\le 3\ell(I).$$

Moreover, due to the support of ψ_{out}, we have $|t-c(J)|_\infty\ge 3\ell(J)$ and $|x-c(J)|_\infty\le\ell(J)/2$. The last two inequalities imply $2|x-c(J)_\infty|<|t-c(J)|_\infty$ and so, we use the integral representation and the mean zero of ψ_{out} in the variable x to write

$$\Lambda(\psi_{out}) = \int\int \psi_{out}(t,x)(K(t,x) - K(t,c(J)))\,dtdx.$$

This, together with the bound of ψ_{out} calculated in (19) and the smoothness condition (6) of a compact Calderón-Zygmund kernel, allow us to bound in the following way:

$$|\Lambda(\psi_{out})| \lesssim \frac{C^2}{\ell(I)^\alpha} \int_J \int_{3\ell(J)<|t-c(J)|_\infty \leq 3\ell(I)} |t-c(J)|_\infty^\alpha \varphi_{I\times J}(t,x)|K(t,x) - K(t,c(J))|\,dtdx$$

$$\lesssim \frac{C^2}{\ell(I)^\alpha} \|\varphi_{I\times J}\|_{L^\infty(\mathbb{R}^{2n})} \int_J \int_{3\ell(J)<|t-c(J)|_\infty \leq 3\ell(I)} |t-c(J)|_\infty^\alpha \frac{|x-c(J)|_\infty^\delta}{|t-c(J)|_\infty^{n+\delta}}$$

$$L(|t-c(J)|_\infty)S(|x-c(J)|_\infty)D\Big(1 + \frac{|t+c(J)|_\infty}{1+|t-c(J)|_\infty}\Big)\,dtdx$$

$$\lesssim \frac{C^2}{\ell(I)^\alpha} |I|^{-1/2}|J|^{-1/2} L(\ell(J))S(\ell(J))D\Big(1 + \frac{|c(J)|_\infty}{1+\ell(I)}\Big)$$

$$\int_{|x-c(J)|<\ell(J)/2} |x-c(J)|_\infty^\delta dx \int_{3\ell(J)<|t-c(J)|_\infty \leq 3\ell(I)} \frac{1}{|t-c(J)|_\infty^{n+\delta-\alpha}} dt.$$

The first integral can be bounded by

$$\int_{|x-c(J)|<\ell(J)/2} |x-c(J)|_\infty^\delta dx \lesssim |J|^{1+\frac{\delta}{n}}$$

and, since $\delta < \alpha$, the second integral is bounded by

$$\int_{3\ell(J)<|t-c(J)|_\infty \leq 3\ell(I)} \frac{1}{|t-c(J)|_\infty^{n+\delta-\alpha}} dt \lesssim (3\ell(I))^{\alpha-\delta} - (3\ell(J))^{\alpha-\delta} \lesssim \ell(I)^{\alpha-\delta}.$$

On the other hand, since $|c(I) - c(J)|_\infty \leq \mathrm{diam}(I \cup J) \leq 3\ell(I)$, we have as before

$$4\Big(1 + \frac{|c(J)|_\infty}{1+\ell(I)}\Big) \geq 4 + \frac{|c(I)|_\infty}{1+\ell(I)} - 3 \gtrsim \mathrm{rdist}(I, \mathbb{B}).$$

Finally then,

$$|\Lambda(\psi_{out})| \lesssim C^2|I|^{-1/2}|J|^{-1/2}L(\ell(J))S(\ell(J))D(\operatorname{rdist}(I,\mathbb{B}))|J|^{1+\frac{\delta}{n}}\ell(I)^{-\delta}$$

$$\leq C^2\left(\frac{|J|}{|I|}\right)^{\frac{1}{2}+\frac{\delta}{n}} F_K(J,J,I).$$

□

Proof of Proposition 3.2 As before, $\psi(t,x) = \phi_I(t)\psi_J(x)$ is supported and adapted to $I \times J$ with constant C^2, decay N, order 1, parameter $\alpha > \delta$ and it has mean zero in the variable x. We divide the proof into the same cases as before.

(a) When $3\ell(I) < \operatorname{diam}(I \cup J)$, exactly the same reasoning of case (a) in the proof of Proposition 3.2 holds since the only properties needed are the mean zero of ψ_J and the smoothness property of the compact Calderón-Zygmund kernel.
(b) We now assume that $\operatorname{diam}(I \cup J) \leq 3\ell(I)$. As before, we first show that we can assume $\psi(c(J),x) = 0$ for any $x \in \mathbb{R}^n$. This assumption comes again from the substitution of $\psi(t,x)$ by

$$\psi(t,x) - (\mathcal{T}_{c(J)}\mathcal{D}_{\ell(I)}\Phi)(t)\psi(c(J),x). \tag{20}$$

But now we need to prove that the subtracted term satisfies the desired bound without the use of the condition $T(1) = 0$. We remind the notation $\Phi_0 = \mathcal{T}_{c(J)}\mathcal{D}_{\ell(I)}\Phi$ with $\Phi = \Phi_{\mathbb{B}}$ as defined in Notation 3.4.

We denote $\tilde{\psi}(x) = \psi(c(J),x)$ which, as before, satisfies the decay

$$|\tilde{\psi}(x)| \leq C^2|I|^{-1/2}|J|^{-1/2}w_J(x)^{-N}$$

and so, it is a bump function supported and adapted to J with order zero and constant $C^2|I|^{-1/2}$.

Let $J_k = 2^kJ$ for $k \in \mathbb{N}$, $k \geq 0$ and let Φ_{J_k} be bump functions L^∞-adapted to J_k and supported on $4J_k$ as defined in Notation 3.4. We define now $\psi_0 = \Phi_{J_0}$ and $\psi_k = \Phi_{J_k} - \Phi_{J_{k-1}}$ for $k \geq 1$ which satisfy $\sum_{k\geq 0}\psi_k(x) = 1$ for all $x \in \mathbb{R}^n$. Therefore, we have

$$|\langle T(\Phi_0), \tilde{\psi}\rangle| \leq \sum_{k\geq 0} |\langle T(\Phi_0 \cdot \psi_k), \tilde{\psi}\rangle|$$

with a finite sum due to the compact support of Φ_0 which implies $2^{k-1}\ell(J) \leq |t - c(J)|_\infty \leq 2\ell(I)$.

Now, for $k = 0$, since $\Phi_0 \cdot |J|^{-1/2}\Phi_{J_0}$ is supported on $4J$ and L^2-adapted to J, we can apply weak compactness condition to obtain

$$|\langle T(\Phi_0 \cdot \Phi_{J_0}), \tilde{\psi}\rangle| \lesssim C^2|I|^{-1/2}|J|^{1/2}F_W(J).$$

When $k \geq 1$, due to the supports of ψ_k and $\tilde{\psi}$, we have that $2^{k-1}\ell(J) < |t - c(J)|_\infty < 2^{k+1}\ell(J)$ and $|x - c(J)|_\infty < \ell(J)/2$ respectively. This implies that $2|x-c(J)|_\infty \leq \ell(J) \leq |t-c(J)|_\infty$ and so, we can use the integral representation, the mean zero of $\tilde{\psi}$ and the smoothness condition (6) of a compact Calderón-Zygmund kernel to write

$$\sum_{k\geq 1} |\langle T(\Phi_0\psi_k), \tilde{\psi}\rangle| = \sum_{k\geq 1}\Big|\int\int \Phi_0(t)\psi_k(t)\tilde{\psi}(x)(K(t,x) - K(t,c(J))dtdx\Big|$$

$$\leq \sum_{k\geq 1}\int_J \int_{2^{k-1}\ell(J)<|t-c(J)|_\infty<2^{k+1}\ell(J)} |\tilde{\psi}(x)|\frac{|x-c(J)|_\infty^\delta}{|t-c(J)|_\infty^{n+\delta}}$$

$$L(|t-c(J)|_\infty)S(|x-c(J)|_\infty)D(1+\frac{|t+c(J)|_\infty}{1+|t-c(J)|_\infty})dtdx$$

$$\leq \|\tilde{\psi}\|_{L^1(\mathbb{R}^n)}\ell(J)^\delta L(\ell(J))S(\ell(J))$$

$$\sum_{k\geq 1} D\Big(1+\frac{|c(J)|_\infty}{1+2^{k+1}\ell(J)}\Big)\int_{2^{k-1}\ell(J)<|t-c(J)|_\infty}\frac{1}{|t-c(J)|_\infty^{n+\delta}}dt$$

$$\lesssim C^2\left(\frac{|J|}{|I|}\right)^{\frac{1}{2}}\ell(J)^\delta L(\ell(J))S(\ell(J))\sum_{k\geq 1}\frac{D(\,\mathrm{rdist}(2^kJ,\mathbb{B}))}{2^{k\delta}\ell(J)^\delta}$$

$$\lesssim C^2\left(\frac{|J|}{|I|}\right)^{\frac{1}{2}} L(\ell(J))S(\ell(J))\tilde{D}(\,\mathrm{rdist}(J,\mathbb{B}))$$

$$= C^2\left(\frac{|J|}{|I|}\right)^{\frac{1}{2}}\tilde{F}_K(J,J,J)$$

which is the first term of the stated bound.

This finishes the justification of the assumption $\psi(c(J),x) = 0$. From here, the proof that $\Lambda(\psi_{in})$ satisfies the required bounds follows exactly the same steps as in cases (b1) and (b2) in the proof of Proposition 3.1. □

Proof of Proposition 3.3 Now, the function $\psi(t,x) = \phi_I(t)\psi_J(x)$ is supported and adapted to $I \times J$ with constant C^2, decay N and order zero but it does not necessarily have mean zero.

(a) As before, we first assume that $3\ell(I) < \mathrm{diam}(I \cup J)$. By the calculations in the proof of Proposition 3.1, we have

$$\mathrm{diam}(I\cup J)/2 \leq |t-c(J)|_\infty \leq \mathrm{diam}(I\cup J)$$

and $|x - c(J)|_\infty \leq \ell(J)/2$, which again imply $|t-c(J)|_\infty \geq 3|x-c(J)|_\infty$. Then, the support of ψ is disjoint with the diagonal and we can use the

Calderón-Zygmund kernel representation to write

$$\Lambda(\psi) = \int\int \psi(t,x)K(t,x)\,dtdx.$$

Now, with the same notation $Q_{I,J} = \{t \in \mathbb{R}^n : \operatorname{diam}(I \cup J)/2 < |t - c(J)|_\infty \leq \operatorname{diam}(I \cup J)\}$ and the kernel decay described in (8), we bound as follows:

$$\begin{aligned}
|\Lambda(\psi)| &\leq \int_J \int_{I \cap Q_{I,J}} |\psi(t,x)||K(t,x)|\,dtdx \\
&\lesssim \int_J \int_{I \cap Q_{I,J}} |\psi(t,x)| \frac{1}{|t - c(J)|_\infty^n} \\
&\qquad L(|t - c(J)|_\infty) S(|t - c(J)|_\infty) D\Big(1 + \frac{|t + c(J)|_\infty}{1 + |t - c(J)|_\infty}\Big) dxdt \\
&\lesssim \frac{\|\psi\|_{L^1(\mathbb{R}^{2n})}}{\operatorname{diam}(I \cup J)^n} L(\operatorname{diam}(I \cup J)) S(\operatorname{diam}(I \cup J)) D\Big(1 + \frac{|c(J)|_\infty}{1 + \operatorname{diam}(I \cup J)}\Big) \\
&\lesssim \frac{C^2 |I|^{\frac{1}{2}} |J|^{\frac{1}{2}}}{\operatorname{diam}(I \cup J)^n} L(\ell(\langle I \cup J \rangle)) S(\ell(\langle I \cup J \rangle)) D(\operatorname{rdist}(\langle I \cup J \rangle, \mathbb{B})) \\
&= C^2 \Big(\frac{|J|}{|I|}\Big)^{\frac{1}{2}} \Big(\frac{\operatorname{diam}(I \cup J)}{\ell(I)}\Big)^{-n} F_K(\langle I, J \rangle, \langle I, J \rangle, \langle I, J \rangle).
\end{aligned}$$

(b) We now assume that $\operatorname{diam}(I \cup J) \leq 3\ell(I)$ and we decompose ψ in the same way as before: $\psi = \psi_{out} + \psi_{in}$ with $\psi_{in}(t,x) = \psi(t,x)\Phi_{3J}(t)$ and divide the analysis into the same cases.

(b1) We claim that ψ_{in} is adapted to $J \times J$ with order zero and constant $C^2\,(|J|/|I|)^{\frac{1}{2}}$. Since ψ is adapted to $I \times J$ and ψ_{in} is supported on $3J \times J$, we have for all $t \in 3J$ and all $x \in J$,

$$\begin{aligned}
|\psi_{in}(t,x)| = |\psi(t,x)|\Phi_{3J}(t) &\lesssim C^2 |I|^{-\frac{1}{2}} \chi_J(t) |J|^{-\frac{1}{2}} w_J(x)^{-N} \\
&\lesssim C^2 \left(\frac{|J|}{|I|}\right)^{\frac{1}{2}} |J|^{-\frac{1}{2}} w_J(t)^{-N} |J|^{-\frac{1}{2}} w_J(x)^{-N}.
\end{aligned}$$

This proves the claim and so, by the weak compactness property of T, we get

$$|\Lambda(\psi_{in})| \lesssim C^2 \left(\frac{|J|}{|I|}\right)^{\frac{1}{2}} F_W(J).$$

(b2) We now work with ψ_{out} for which we have the decay

$$|\psi_{out}(t,x)| \leq |\psi(t,x)| \lesssim |I|^{-1/2}\chi_I(t)|J|^{-1/2}w_J(x)^{-N}.$$

Moreover, the calculations in the analog case (b2) of the proof of Proposition 3.1 show that on the support of ψ we have that $|t-c(I)|_\infty \leq \ell(I)/2$ and

$$|t-c(J)|_\infty \leq \mathrm{diam}(I\cup J) \leq 3\ell(I)$$

while due to the support of ψ_{out}, we also have $|t-c(J)|_\infty \geq 3\ell(J)$ and $|x-c(J)|_\infty \leq \ell(J)/2$. Then, $2|x-c(J)_\infty| < |t-c(J)|_\infty$ and we can use the integral representation to bound in the following way:

$$\begin{aligned}|\Lambda(\psi_{out})| &= \Big|\int\int \psi_{out}(t,x)K(t,x)\,dtdx\Big|\\ &\leq \int_J \int_{3\ell(J)<|t-c(J)|_\infty\leq 3\ell(I)} |\psi_{out}(t,x)||K(t,x)-K(t,c(J))|\,dtdx\\ &\quad+\int_J\int_{3\ell(J)<|t-c(J)|_\infty\leq 3\ell(I)} |\psi_{out}(t,x)||K(t,c(J))|\,dtdx.\end{aligned}$$

The first term can be bounded by a constant times

$$C^2\left(\frac{|J|}{|I|}\right)^{\frac{1}{2}+\frac{\delta}{n}} L(\ell(J))S(\ell(J))D(\,\mathrm{rdist}(I,\mathbb{B}))$$

exactly in the same way as in case (b2) in the proof of Proposition 3.1. On the other hand, from the decay of a compact Calderón-Zygmund kernel stated in (8) and the decay for ψ_{out}, we can bound the second term by a constant times

$$\begin{aligned}&\int_J\int_{3\ell(J)<|t-c(J)|_\infty\leq 3\ell(I)} |\psi_{out}(t,x)|\frac{1}{|t-c(J)|_\infty^n}\\ &\qquad L(|t-c(J)|_\infty)S(|t-c(J)|_\infty)D\Big(1+\frac{|t+c(J)|_\infty}{1+|t-c(J)|_\infty}\Big)\,dtdx\\ &\leq \|\psi_{out}\|_{L^\infty(\mathbb{R}^{2n})}L(\ell(J))S(\ell(I))D\Big(1+\frac{|c(J)|_\infty}{1+\ell(I)}\Big)\\ &\qquad |J|\int_{3\ell(J)<|t-c(J)|_\infty\leq 3\ell(I)} \frac{1}{|t-c(J)|_\infty^n}\,dt\end{aligned}$$

$$\leq C^2|I|^{-\frac{1}{2}}|J|^{-\frac{1}{2}}L(\ell(J))S(\ell(I))D(\text{rdist}(I,\mathbb{B}))|J|(\log(3\ell(I))-\log(3\ell(J)))$$

$$= C^2\left(\frac{|J|}{|I|}\right)^{\frac{1}{2}}\log\left(\frac{\ell(I)}{\ell(J)}\right)F_K(J,I,I).$$

□

4 Pointwise and Off-Diagonal Estimates for General Functions

In this last section, we provide several pointwise estimates of the action of the operator over general functions and over bump functions, Proposition 4.5 and Corollary 4.6 respectively. Moreover, in Proposition 4.7 we prove a new off-diagonal inequality for general functions.

We start with some technical results which, despite being well-known for bounded singular operators, we reproduce here their proofs for compact singular operators in order to highlight the role played by compactness in the gain of decay and smoothness.

Lemma 4.1 *Let T be a linear operator associated with a standard Calderón-Zygmund kernel. Let $\Phi \in \mathcal{S}(\mathbb{R}^n)$ such that it is positive, it is supported on $\mathbb{B} = [-\frac{1}{2},\frac{1}{2}]^n$ and $\int \Phi(x)dx = 1$. We denote $\Phi_{x,\epsilon}(y) = \epsilon^{-n}\Phi(\epsilon^{-1}(y-x))$.*

Let f be an integrable function with compact support in a cube $I \subset \mathbb{R}^n$. Then, for all $x \notin 3I$ there exists the limit of $\langle T(f),\Phi_{x,\epsilon}\rangle$ when ϵ tends to zero.

Definition 4.2 By previous lemma, we can define

$$T(f)(x) = \lim_{\epsilon\to 0}\langle T(f),\Phi_{x,\epsilon}\rangle.$$

Proof We check that $(\langle T(f),\Phi_{x,\epsilon}\rangle)_{\epsilon>0}$ is a Cauchy sequence. Let $x \in \mathbb{R}^n\backslash(3I)$ fixed and we choose $\epsilon_1,\epsilon_2 < 2\ell(I)/5$.

Then, for all $t \in \text{supp}f$ we have $\ell(I) < |t-x|_\infty$ while for all $y \in \text{supp}\Phi_{x,\epsilon_i}$, we get $|y-x|_\infty \leq \epsilon_i/2 < \ell(I)/2$. Both inequalities imply

$$|t-y|_\infty \geq |t-x|_\infty - |x-y|_\infty \geq \ell(I)/2 > 0.$$

Hence, $f(t)$ and $\phi_{x,\epsilon_i}(y)$ have disjoint compact supports and, by the integral representation, we can write

$$\langle T(f),\Phi_{x,\epsilon_i}\rangle = \int\int f(t)\Phi_{x,\epsilon_i}(y)K(t,y)dtdy$$

and so,

$$\langle T(f), \Phi_{x,\epsilon_1}\rangle - \langle T(f), \Phi_{x,\epsilon_2}\rangle = \int\int f(t)\Phi(y)(K(t, x+\epsilon_1 y) - K(t, x+\epsilon_2 y))dtdy.$$

Now, for all $y \in \mathrm{supp}\Phi$, we have $|y|_\infty \leq 1/2$ and so,

$$\begin{aligned} 2|(\epsilon_1 - \epsilon_2)y|_\infty < \epsilon_1 + \epsilon_2 < \frac{4}{5}\ell(I) \leq \ell(I) - \frac{\epsilon_1}{2} < |t-x|_\infty - \frac{\epsilon_1}{2} \\ \leq |t-x-\epsilon_1 y|_\infty + \epsilon_1 |y|_\infty - \frac{\epsilon_1}{2} \leq |t-x-\epsilon_1 y|_\infty. \end{aligned}$$

Therefore, we can apply the smoothness condition of the kernel to bound in the following way:

$$\begin{aligned} |\langle T(f), \Phi_{x,\epsilon_1}\rangle - \langle T(f), \Phi_{x,\epsilon_2}\rangle| &\leq \int\int |f(t)||\Phi(y)| \frac{|\epsilon_1 - \epsilon_2|^\delta |y|_\infty^\delta}{|t-x-\epsilon_1 y|_\infty^{n+\delta}} dtdy \\ \leq |\epsilon_1 - \epsilon_2|^\delta \|\Phi\|_{L^\infty(\mathbb{R}^n)} \int |f(t)| &\int_{\ell(I)/2 < |t-\epsilon_1 y - x|_\infty} \frac{1}{|t-x-\epsilon_1 y|_\infty^{n+\delta}} dtdx \\ &\lesssim |\epsilon_1 - \epsilon_2|^\delta \|f\|_{L^1(\mathbb{R}^n)} \frac{1}{\ell(I)^\delta}. \end{aligned}$$

Notice we used that $|t-x-\epsilon_i y|_\infty \geq |t-x|_\infty - \epsilon_i|y|_\infty \geq \ell(I) - \epsilon_i/2 \geq \ell(I)/2$. This proves that $(\langle T(f), \Phi_{x,\epsilon}\rangle)_{\epsilon>0}$ is Cauchy. □

In all forthcoming results, we consider T to be a linear operator associated with a compact Calderón-Zygmund kernel K with parameter $0 < \delta < 1$. We do not assume on T any other hypotheses like the weak boundedness or weak compactness conditions or the membership of $T(1)$ to any space in particular.

Lemma 4.3 *Let f an integrable function with compact support in a cube I. Then, $T(f)$ admits the following representation as a function*

$$T(f)(x) = \int f(t)K(t,x)dt \tag{21}$$

for all $x \notin 3I$. Moreover, $T(f)$ is Hölder-continuous in $\mathbb{R}^n \backslash (3I)$ satisfying

$$\begin{aligned} |T(f)(x) - T(f)(x')| &\lesssim L(\ell(\langle I, x\rangle))D(\mathrm{rdist}(\langle I, x\rangle, \mathbb{B})) \\ &\frac{|x-x'|_\infty^\delta}{\ell(I)^\delta} S(|x-x'|_\infty)|I|^{-1}\|f\|_{L^1(\mathbb{R}^n)}. \end{aligned}$$

Remark 4.4 Notice that if $S(x) \leq |x|_\infty^\beta$ with $\beta > 0$, then, $T(f)$ is Hölder-continuous with parameter $\delta + \beta$ which is better than in the case when T is only a bounded singular integral operator.

Proof We first check the integral representation. We note that the integral in the right hand side of (21) converges absolutely since by hypothesis $|t-x|_\infty \geq \mathrm{dist}_\infty(x,I) \geq \ell(I)$ and so,

$$\Big|\int f(t)K(t,x)dt\Big| \leq \|f\|_1 \frac{1}{|I|}.$$

Now, let $x \in \mathbb{R}^n \backslash (3I)$, $\epsilon < 2\ell(I)/5$ be fixed and $\Phi_{x,\epsilon}$ as in Lemma 4.1. As before, for $t \in \mathrm{supp} f$, $y \in \mathrm{supp}\,\Phi_{x,\epsilon}$, we have $\ell(I) < |t-x|_\infty$, $|y-x|_\infty \leq \epsilon/2 < \ell(I)/2$ and $|t-y|_\infty \geq |t-x|_\infty - |x-y|_\infty \geq \ell(I)/2 > 0$. Hence, $f(t)$ and $\phi_{x,\epsilon}(y)$ have disjoint compact supports and we can write

$$\langle T(f), \Phi_{x,\epsilon}\rangle = \int\int f(t)\Phi_{x,\epsilon}(y)K(t,y)dtdy = \int\int f(t)\Phi(y)K(t,x+\epsilon y)dtdy.$$

Moreover, for all $y \in \mathrm{supp}\Phi$ we have $2|\epsilon y|_\infty < \epsilon < \ell(I) \leq |t-x|_\infty$ and so, by the smoothness condition of the kernel, we bound as follows:

$$\begin{aligned}
\Big|\langle T(f), \Phi_{x,\epsilon}\rangle - \int f(t)K(t,x)dt\Big| \\
&= \Big|\int\int f(t)\Phi(y)K(t,x+\epsilon y)dtdy - \int\int f(t)\Phi(y)K(t,x)dtdy\Big| \\
&\leq \int\int |f(t)||\Phi(y)||K(t,x+\epsilon y) - K(t,x)|dtdy \\
&\lesssim \int\int |f(t)| \frac{\epsilon |y|_\infty^\delta}{|t-x|_\infty^{n+\delta}} F_K(t,x)dtdy \\
&\lesssim \frac{\epsilon}{|I|^{1+\frac{\delta}{n}}} \int |f(t)|dt
\end{aligned}$$

which tends to zero as required.

Now, we check the Hölder-continuity of $T(f)$ in $\mathbb{R}^n \backslash 3I$. For all $x, x' \in \mathbb{R}^n \backslash (3I)$ with $|x-x'|_\infty < \ell(I)/2$ we have $2|x-x'|_\infty < \ell(I) \leq |x-t|_\infty$ and so,

$$\begin{aligned}
|T(f)(x) - T(f)(x')| &= \Big|\int f(t)(K(t,x) - K(t,x'))dt\Big| \\
&\lesssim \int |f(t)| \frac{|x-x'|_\infty^\delta}{|t-x|_\infty^{n+\delta}} L(|t-x|_\infty) S(|x-x'|_\infty) D\Big(1 + \frac{|t+x|_\infty}{1+|t-x|_\infty}\Big)dt \\
&\lesssim |I|^{-1} \|f\|_{L^1(\mathbb{R}^n)} \frac{|x-x'|_\infty^\delta}{\ell(I)^\delta} L(\ell(\langle I,x\rangle)) S(|x-x'|_\infty) D(\,\mathrm{rdist}(\langle I,x\rangle, \mathbb{B}))
\end{aligned}$$

Notice that, since $|t-c(I)|_\infty \leq \ell(I)/2$, $|t-x|_\infty \leq \ell(I)/2+|c(I)-x|_\infty \leq \ell(\langle I,x\rangle)$, $|t+x|_\infty \geq |x+c(I)|_\infty-\ell(I)/2$ and $|c(\langle I,x\rangle)-(x+c(I))/2| \leq \ell(\langle I,x\rangle)/2$, we have

$$\begin{aligned}1+\frac{|t+x|_\infty}{1+|t-x|_\infty} &\gtrsim \frac{1}{4}\Big(4+\frac{2|c(\langle I,x\rangle)|_\infty-\frac{3}{2}(\ell(\langle I,x\rangle)+1)}{1+\ell(\langle I,x\rangle)}\Big)\\ &\gtrsim \frac{5}{2}+2\frac{|c(\langle I,x\rangle)|_\infty}{1+\ell(\langle I,x\rangle)} \gtrsim 2\,\mathrm{rdist}(\langle I,x\rangle,\mathbb{B}).\end{aligned}$$

□

The following proposition improves the previous result and thus, it will allow us to describe, in Corollary 4.6, the behavior of $T(\phi_I)$ for any given bump function ϕ_I adapted and supported on a cube I, stating explicitly a gain in the decay and smoothness that depends on the decay and smoothness of the operator kernel.

Proposition 4.5 *Let f be an integrable function supported on a cube I. Then,*

$$|T(f)(x)| \lesssim w_I(x)^{-n}F_K(\langle I,x\rangle)|I|^{-1}\|f\|_{L^1(I)}$$

for all $x\in\mathbb{R}^n$ such that $x\notin 5I$. Moreover,

$$\begin{aligned}|T(f)(x)-T(f)(x')| \lesssim{}& \frac{|x-x'|_\infty^\delta}{\ell(I)^\delta}w_I\Big(\frac{x+x'}{2}\Big)^{-(n+\delta)}\\ &L(\ell(\langle I,x\rangle))S(|x-x'|_\infty)D(\,\mathrm{rdist}(\langle I,x\rangle,\mathbb{B}))|I|^{-1}\|f\|_{L^1(I)}\end{aligned}$$

for all $x,x'\in\mathbb{R}^n$ such that $|x-x'|_\infty<\ell(I)/2$ and $\langle x,x'\rangle\cap 5I=\emptyset$.

Proof We consider $\epsilon>0$ small enough so that it satisfies several inequalities stated along the proof. We denote by $\mathbb{B}_\epsilon=[-\epsilon/2,\epsilon/2]^n$ and, given $x\in\mathbb{R}^n$, we define $J=x+\mathbb{B}_\epsilon$. Let also $\varphi_J=C|J|^{-1}w_J(x)^{-N}$ be a positive bump function $L^1(\mathbb{R}^n)$-adapted to J with order one, decay N and constant C such that $\int\varphi_J(x)dx=1$.

First, ϵ can be taken so that $\ell(J)\leq\ell(I)$. Moreover, the hypothesis $x\notin 5I$ implies that $\mathrm{diam}(I\cup J)>3\ell(I)$. Then, from the proof of Proposition 3.3 in its case (a), we have

$$\begin{aligned}|\langle T(f),\varphi_J\rangle| &\lesssim \|f\|_{L^1(\mathbb{R}^n)}\|\varphi_J\|_{L^1(\mathbb{R}^n)}\mathrm{diam}(I\cup J)^{-n}F_K(\langle I,J\rangle)\\ &\lesssim \mathrm{rdist}(I,J)^{-n}F_K(\langle I,J\rangle)|I|^{-1}\|f\|_{L^1(\mathbb{R}^n)}.\end{aligned}$$

On the other hand, for ϵ small enough, we have

$$\ell(\langle I,x\rangle)\leq\ell(\langle I,J\rangle)\leq\ell(\langle I,x\rangle)+\epsilon/2\leq 2\ell(\langle I,x\rangle)$$

and

$$\frac{1}{2}\mathrm{rdist}(\langle I,x\rangle,\mathbb{B})\lesssim\mathrm{rdist}(\langle I,J\rangle,\mathbb{B})\leq\mathrm{rdist}(\langle I,x\rangle,\mathbb{B}).$$

These inequalities imply that

$$F_K(\langle I,J\rangle) = L(\ell(\langle I,J\rangle))S(\ell(\langle I,J\rangle))D(\operatorname{rdist}(\langle I,J\rangle,\mathbb{B})) \leq F_K(\langle I,x\rangle)$$

omitting constants.

Now, taking limit when ϵ tends to zero, we get

$$\lim_{\epsilon\to 0}\langle T(f),\varphi_J\rangle = \lim_{\epsilon\to 0}\int\int f(t)\varphi_J(y)K(t,y)dtdy = \int f(t)K(t,x)dt = T(f)(x)$$

by Lemma 4.3 . Finally then,

$$\begin{aligned}|T(f)(x)| &\lesssim \operatorname{rdist}(I,x)^{-n}F_K(\langle I,x\rangle)|I|^{-1}\|f\|_{L^1(\mathbb{R}^n)}\\ &\lesssim \Big(1+\frac{|x-c(I)|_\infty}{\ell(I)}\Big)^{-n}F_K(\langle I,x\rangle)|I|^{-1}\|f\|_{L^1(\mathbb{R}^n)}\end{aligned}$$

which proves the first inequality.

We prove now the second one. Let $x,x'\in\mathbb{R}^n$ as stated and let $c=(x+x')/2$. Again, we consider $\epsilon>0$ to be small enough for our purposes. We define $J_1=x+\mathbb{B}_\epsilon$, $J_2=x'+\mathbb{B}_\epsilon$ and the functions φ_{J_i} for $i=1,2$ as before. Then, the function $\varphi_{J_1}-\varphi_{J_2}$ is supported on $\langle J_1,J_2\rangle$ and it has mean zero.

The hypotheses on x,x' imply $|x-x'|_\infty<\ell(I)/2$ and $\operatorname{diam}(I\cup\langle x,x'\rangle)>3\ell(I)$. Then, for all $t\in I$ we have $|t-c|_\infty>\operatorname{diam}(I\cup\langle x,x'\rangle)/3>\ell(I)$. Moreover, we can take $0<\epsilon<|x-x'|_\infty/2$ small enough so that $\operatorname{diam}(I\cup\langle J_1,J_2\rangle)>3\ell(I)$, $|t-c|_\infty\geq\operatorname{diam}(I\cup\langle J_1,J_2\rangle)/3$ and $J_1\cap J_2=\emptyset$. The latter property further implies that $\ell(\langle J_1,J_2\rangle)=|x-x'|_\infty+\epsilon<\ell(I)$.

On the other hand, since $c=c(\langle J_1,J_2\rangle)$, we have $|y-c|_\infty\leq\ell(\langle J_1,J_2\rangle)/2$ for all $y\in\langle J_1,J_2\rangle$. Also notice that $\ell(J_1)=\ell(J_2)$.

Finally then, $2|x-c|_\infty=|x-x'|_\infty<\ell(I)\leq|t-c|_\infty$, which allows to use the integral representation and the smoothness property of the compact Calderón-Zygmund kernel.

After establishing all these inequalities, we can repeat the proof of case (a) in Proposition 3.1 to obtain

$$\begin{aligned}|\langle T(f),\varphi_{J_1}-\varphi_{J_2}\rangle| &= \Big|\int f(t)(\varphi_{J_1}(y)-\varphi_{J_2}(y))(K(t,y)-K(t,c))dtdy\Big|\\ &\leq\int|f(t)||\varphi_{J_1}(y)-\varphi_{J_2}(y)|\frac{|y-c|_\infty^\delta}{|t-c|_\infty^{n+\delta}}\\ &\quad L(|t-c|_\infty)S(|y-c|_\infty)D\Big(1+\frac{|t+c|_\infty}{1+|t-c|_\infty}\Big)dtdy\\ &\lesssim\|f\|_{L^1(\mathbb{R}^n)}\|\varphi_{J_1}\|_{L^1(\mathbb{R}^n)}\frac{\ell(\langle J_1,J_2\rangle)^\delta}{\operatorname{diam}(I\cup\langle J_1,J_2\rangle)^{n+\delta}}\end{aligned}$$

$$L(\ell(\langle I, J_1 \cup J_2\rangle))S(\ell(\langle J_1, J_2\rangle))D(\operatorname{rdist}(\langle I, J_1 \cup J_2\rangle), \mathbb{B}))$$
$$\lesssim \Big(\frac{\ell(\langle J_1, J_2\rangle)}{\ell(I)}\Big)^{\delta} \operatorname{rdist}(I, \langle J_1, J_2\rangle)^{-(n+\delta)}$$
$$L(\ell(\langle I, J_1 \cup J_2\rangle))S(\ell(\langle J_1, J_2\rangle))D(\operatorname{rdist}(\langle I, J_1 \cup J_2\rangle), \mathbb{B}))|I|^{-1}\|f\|_{L^1(\mathbb{R}^n)}.$$

Now, we have the following relationships:

$$\ell(\langle I, x\rangle) \le \ell(\langle I, J_1 \cup J_2\rangle) \le \ell(\langle I, x\rangle) + |x - x'|_\infty + \epsilon/2$$
$$\le \ell(\langle I, x\rangle) + \ell(I) + \ell(I)/4 \le 3\ell(\langle I, x\rangle),$$
$$\ell(\langle J_1, J_2\rangle) \le 2|x - x'|_\infty,$$
$$\frac{1}{2}\operatorname{rdist}(\langle I, \{x\}\rangle, \mathbb{B}) \lesssim \operatorname{rdist}(\langle I, J_1 \cup J_2\rangle, \mathbb{B}) \le \operatorname{rdist}(\langle I, \{x\}\rangle, \mathbb{B}),$$
$$\operatorname{rdist}(I, \langle J_1, J_2\rangle) \gtrsim \operatorname{rdist}(I, c) \gtrsim 1 + \frac{|\frac{x+x'}{2} - c(I)|_\infty}{\ell(I)}.$$

These inequalities imply

$$L(\ell(\langle I, J_1 \cup J_2\rangle))S(\ell(\langle J_1, J_2\rangle))D(\operatorname{rdist}(\langle I, J_1 \cup J_2\rangle), \mathbb{B}))$$
$$\le L(\ell(\langle I, x\rangle))S(|x - x'|_\infty)D(\operatorname{rdist}(\langle I, x\rangle), \mathbb{B}))$$

omitting constants.

Finally then, we take limit when ϵ tends to zero to get

$$|T(f)(x) - T(f)(x')| \lesssim \frac{|x - x'|_\infty^{\delta}}{\ell(I)^{\delta}}\Big(1 + \frac{|\frac{x+x'}{2} - c(I)|_\infty}{\ell(I)}\Big)^{-(n+\delta)}$$
$$L(\ell(\langle I, x\rangle))S(|x - x'|_\infty)D(\operatorname{rdist}(\langle I, x\rangle, \mathbb{B}))|I|^{-1}\|f\|_{L^1(\mathbb{R}^n)}$$

which proves the second inequality. □

Corollary 4.6 *Let ϕ_I be a bump function adapted and supported to I with constant $C > 0$, decay N and order zero. Then, in $\mathbb{R}^n \backslash 5I$, $T(\phi_I)$ satisfies the definition of a bump function adapted to I with constant $C > 0$, decay n, order one and parameter δ, plus an extra factor in decay due to compactness.*

Proof We simply rewrite the statement of Proposition 4.5:

$$|T(\phi_I)(x)| \lesssim C|I|^{-\frac{1}{2}} w_I(x)^{-n} F_K(\langle I, x\rangle)$$

and

$$|T(\phi_I)(x) - T(\phi_I)(x')| \lesssim C\frac{|x-x'|_\infty^\delta}{\ell(I)^\delta}|I|^{-\frac{1}{2}}w_I\Big(\frac{x+x'}{2}\Big)^{-(n+\delta)}$$
$$L(\ell(\langle I,x\rangle))S(|x-x'|_\infty)D(\operatorname{rdist}(\langle I,x\rangle,\mathbb{B}))$$

for all $x, x' \in \mathbb{R}^n$ such that $x \notin 5I$, $|x-x'|_\infty < \ell(I)/2$ and $\langle x, x'\rangle \cap 5I = \emptyset$. □

Notice that, even though ϕ_I has compact support and decays at infinity as fast as $|x|^{-N}$ for any large $N > 0$, $T(\phi_I)$ does not have in general compact support and its decay is only comparable to $|x|^{-n}$. Both facts are typical of bounded singular integral operators. However, if the operator is associated with a compact Calderón-Zygmund kernel, the decay of $T(\phi_I)$ improves depending on the rate of decay of the factor $L(\ell(\langle I,x\rangle))$ when x tends to infinity.

On the other hand, note the gain in smoothness with respect bounded singular integrals provided by the factor $|x-x'|_\infty^\delta S(|x-x'|_\infty)$.

We show now an off-diagonal estimate for general functions deduced directly from the previous pointwise bound.

Proposition 4.7 *Let f an integrable function supported on a cube I. Then, for all $1 < p < \infty$ and all $\lambda > 1$, we have*

$$\|T(f)\chi_{(\lambda I)^c}\|_{L^p(\mathbb{R}^n)} \lesssim \frac{1}{(1+\lambda)^{\frac{n}{p'}}}\sup_{x\in(\lambda I)^c} F_K(\langle I,x\rangle)\|f\|_{L^p(\mathbb{R}^n)}.$$

Proof From Proposition 4.5, we have

$$\begin{aligned}\|T(f)\chi_{(\lambda I)^c}\|_{L^p(\mathbb{R}^n)}^p &\lesssim \int_{(\lambda I)^c}\frac{F_K(\langle I,x\rangle)^p}{(1+\ell(I)^{-1}|x-c(I)|_\infty)^{np}}dx|I|^{-p}\|f\|_{L^1(I)}^p\\ &\lesssim \frac{\ell(I)^n}{(1+\lambda)^{n(p-1)}}\sup_{x\in(\lambda I)^c}F_K(\langle I,x\rangle)^p|I|^{-p}\|f\|_{L^1(I)}^p\\ &= \frac{1}{(1+\lambda)^{n(p-1)}}\sup_{x\in(\lambda I)^c}F_K(\langle I,x\rangle)^p|I|^{-(p-1)}\|f\|_{L^1(I)}^p\end{aligned}$$

which, by Hölder, is smaller than the right hand side of the stated inequality. □

We end the paper by adding few remarks to the previous proposition. We first note that, in the particular case of f being a bump function, we obtain

$$\|T(\phi_I)\chi_{(\lambda I)^c}\|_{L^p(\mathbb{R}^n)} \lesssim |I|^{\frac{1}{p}-\frac{1}{2}}\frac{1}{(1+\lambda)^{\frac{n}{p'}}}\sup_{x\in(\lambda I)^c}F_K(\langle I,x\rangle).$$

We also remind that, for bounded but not compact singular integral operators, the analog of Proposition 4.7 implies that for a fixed cube I and $\|f\|_{L^p(I)} \leq 1$, we have

$$\lim_{\lambda\to\infty} \|T(f)\chi_{(\lambda I)^c}\|_{L^p(\mathbb{R}^n)} = 0$$

with a rate of decay at most of order $\lambda^{\frac{n}{p'}}$. However, for compact singular integral operators, the extra factor stated in Proposition 4.7 ensures that there is always an extra gain in decay. To see this, we note that for all $x \in (\lambda I)^c$ we have $\lambda I \subset 3\langle I, x\rangle$ and so, $\lambda\ell(I) \leq 3\ell(\langle I, x\rangle)$. Hence, $F_K(\langle I, x\rangle) \lesssim L(\ell(\langle I, x\rangle)) \lesssim L(\lambda\ell(I))$ and since $\lim_{\lambda\to\infty} L(\lambda\ell(I)) = 0$, the rate of decay is now at worst as fast as $\lambda^{\frac{n}{p'}} L(\lambda\ell(I))$.

Finally, since we also have the bound

$$F_K(\langle I, x\rangle) \lesssim L(\ell(I))D\Big(1 + \frac{|c(I)|_\infty}{1 + \lambda\ell(I)}\Big)$$

we deduce that for fixed λ and $\|f\|_{L^p(I)} \leq 1$ we have that

$$\lim_{\ell(I)\to\infty} \|T(f)\chi_{(\lambda I)^c}\|_{L^p(\mathbb{R}^n)} = 0$$

while, for fixed λ, $\|f\|_{L^p(I)} \leq 1$ and fixed $\ell(I)$, we also get

$$\lim_{|c(I)|_\infty\to\infty} \|T(f)\chi_{(\lambda I)^c}\|_{L^p(\mathbb{R}^n)} = 0.$$

The last two properties do not hold in general for bounded singular integral operators.

Acknowledgements This work was completed with the support of Spanish project MTM2011-23164.

References

1. P. Auscher, Jose-Maria Martell, Weighted norm inequalities, off-diagonal estimates and elliptic operators. Part II: off-diagonal estimates on spaces of homogeneous type. J. Evol. Equ. **7**(2), 265–316 (2007)
2. P. Auscher, Ph. Tchamitchian, Bases d'Ondelettes sur les Courbes Corde-Arc, Noyau de Cauchy et Espaces de Hardy Associés. Rev. Mat. Ib. **5**(3, 4), 139–170 (1989)
3. P. Auscher, S. Hofmann, M. Lacey, A. McIntosh, Ph. Tchamitchian, The solution of the Kato square root problem for second order elliptic operators on R^n. Ann. Math. **2**, 633–654 (2002)
4. P. Auscher, C. Kriegler, S. Monniaux, P. Portal, Singular integral operators on tent spaces. J. Evol. Equ. **12**(4), 741–765 (2012)
5. A. Axelson, S. Keith, A. McIntosh, Quadratic estimates and functional calculi of perturbated dirac operators. Invent. Math. **163**(3), 455–497 (2006)

6. A. Axelson, S. Keith, A. McIntosh, The Kato square problem for mixed boundary problems. J. Lond. Math. Soc. **74**(1), 113–130 (2006)
7. G. Beylkin, R.R. Coifman, V. Rokhlin, Fast wavelet transforms and numerical algorithms I. Commun. Pure Appl. Math. **44**, 141–183 (1991)
8. R.R. Coifman, Y. Meyer, *Wavelets, Calderon-Zygmund and Multilinear Operators* (Cambridge University Press, Cambridge, 1997)
9. S. Hofmann, J.-M. Martell, L^p bounds for Riesz transforms and square roots associated to second order elliptic operators. Publ. Math. **47**, 497–515 (2003)
10. S. Hofmann, S. Mayboroda, A. McIntosh, Second order elliptic operators with complex bounded measurable coefficients in L^p, Sobolev and Hardy spaces. Ann. Sci. École Norm. Sup. **4**(44), 723–800 (2011)
11. T. Hytönen, F. Nazarov, The local T(b) theorem with rough test functions. arxiv:1201.0648 (to appear)
12. A. Rosen, Square function and maximal function estimates for operators beyond divergence form equations. J. Evol. Equ. **13**, 651–674 (2013)
13. P. Villarroya, A characterization of compactness for singular integrals. J. Math. Pures Appl. **104**(3), 485–532 (2015)
14. Q.X. Yang, Fast algorithms for Calderón-Zygmund singular integral operators. Appl. Comput. Harmon. Anal. **3**(2), 120–126 (1996)

Part II
Function Spaces of Radial Functions

Elementary Proofs of Embedding Theorems for Potential Spaces of Radial Functions

Pablo L. De Nápoli and Irene Drelichman

Abstract We present elementary proofs of weighted embedding theorems for radial potential spaces and some generalizations of Ni's and Strauss' inequalities in this setting.

Keywords Embedding theorems • Potential spaces • Power weights • Radial functions • Sobolev spaces

Mathematics Subject Classification (2000). Primary 46E35; Secondary 35A23

1 Introduction: Sobolev Spaces and Embedding Theorems

The aim of this note is twofold: to review some known results from the theory of radial functions in Sobolev (potential) spaces, and to extend some of this results to the setting of weighted radial spaces. In both cases, we provide new elementary proofs that avoid the use of interpolation theory and sophisticated tools such as atomic or wavelet decompositions.

This will have, at times, the limitation of not giving the most general possible result, in which case we will do our best to provide suitable references. However, we believe that the proofs presented here have the merit of being closer in spirit to some classical theorems (such as the Sobolev embedding theorem) and that the results obtained are good enough for common applications to the theory of partial differential equations.

Before we state the results we are interested in, let us quickly recall that for Sobolev spaces of integer order, the most classical definition is in terms of

P.L. De Nápoli • I. Drelichman (✉)
Facultad de Ciencias Exactas y Naturales, IMAS (UBA-CONICET) and Departamento de Matemática, Universidad de Buenos Aires, Ciudad Universitaria, 1428 Buenos Aires, Argentina
e-mail: pdenapo@dm.uba.ar; irene@drelichman.com

M. Ruzhansky, S. Tikhonov (eds.), *Methods of Fourier Analysis and Approximation Theory*, Applied and Numerical Harmonic Analysis,
DOI 10.1007/978-3-319-27466-9_8

derivatives: given a domain $\Omega \subset \mathbb{R}^n$, for integer $k \in \mathbb{N}$,

$$W^{k,p}(\Omega) = \{u \in L^p(\Omega) : D^\alpha u \in L^p(\Omega) \text{ for any } \alpha \in \mathbb{N}_0^n \text{ with } |\alpha| \leq k\}$$

where the derivatives $D^\alpha u$ are understood in weak (or distributional) sense. It is also usual to denote $W^{k,2}(\Omega)$ by $H^k(\Omega)$.

When it comes to Sobolev spaces of fractional order, there exist several definitions in the literature, among others:

a) Sobolev spaces of fractional order based on $L^2(\mathbb{R}^n)$ can be defined in terms of the Fourier transform: for real $s \geq 0$, let

$$H^s(\mathbb{R}^n) = \{u \in L^2(\mathbb{R}^n) : \int_{\mathbb{R}^n} |\hat{u}(\omega)|^2 (1 + |\omega|^2)^s \, d\omega < \infty\}.$$

For integer s, we have that $H^s(\mathbb{R}^n) = W^{s,2}(\mathbb{R}^n)$.

b) One way to define spaces of fractional order based on $L^p(\mathbb{R}^n)$ when $p \neq 2$ is given by the classical potential spaces $H^{s,p}(\mathbb{R}^n)$ for $s > 0$, defined by:

$$H^{s,p}(\mathbb{R}^n) = \{u : u = (I - \Delta)^{-s/2} f \text{ with } f \in L^p(\mathbb{R}^n)\}.$$

Here, $1 < p < \infty$ and the fractional power $(I - \Delta)^{-s/2}$ can be defined by means of the Fourier transform (for functions in the Schwartz class):

$$(I - \Delta)^{-s/2} f = \mathcal{F}^{-1}((1 + |\omega|^2)^{-s/2} \mathcal{F}(f)) = G_s * f$$

where

$$G_s(x) = \mathcal{F}^{-1}((1 + |x|^2)^{-s/2}) = \frac{(2\sqrt{\pi})^{-n}}{\Gamma(s/2)} \int_0^\infty e^{-t} e^{\frac{-|x|^2}{4t}} t^{\frac{s-n}{2}} \frac{dt}{t} \tag{1}$$

is called the Bessel potential (see, e.g. [43, Chap. V]). Some classical references on potential spaces are [3, 43] and [5]. Notice that this family includes all the previous spaces (when $\Omega = \mathbb{R}^n$) since we have that $H^{1,p}(\mathbb{R}^n) = W^{1,p}(\mathbb{R}^n)$ for $1 < p < \infty$ by [5, Theorem 7]. Also $H^{s,2}(\mathbb{R}^n) = H^s(\mathbb{R}^n)$ for any $s \geq 0$ by Plancherel's theorem.

c) Another (non-equivalent) definition of fractional Sobolev spaces is given by the Aronszajn-Gagliardo-Slobodeckij spaces, for $0 < s < 1$, $1 \leq p < \infty$:

$$W^{s,p}(\Omega) = \left\{ u \in L^p(\Omega) : \frac{|u(x) - u(y)|}{|x - y|^{\frac{n}{p} + s}} \in L^p(\Omega \times \Omega) \right\}.$$

See [17] for a full exposition. We have that $H^{s,2}(\mathbb{R}^n) = W^{s,2}(\mathbb{R}^n) = H^s(\mathbb{R}^n)$ for any $0 < s < 1$, but for $p \neq 2$, $H^{s,p}(\mathbb{R}^n)$ and $W^{s,p}(\mathbb{R}^n)$ are different.

d) Even more general families of functional spaces are those of Besov-Lipschitz and Triebel-Lizorkin spaces. Indeed, the Littlewood-Paley theory says that $H^{s,p}(\mathbb{R}^n)$ coincides with the Triebel-Lizorkin space $F^s_{p,2}(\mathbb{R}^n)$, whereas $W^{s,p}(\mathbb{R}^n)$ (as defined in c)) coincides with the Besov-Lipschitz space $B^s_{p,p}(\mathbb{R}^n)$.

See for instance [2] for a full exposition of the theory of Sobolev spaces, and [50] for a discussion of the relations of the different functional spaces with the scales of Besov-Lipschitz and Triebel-Lizorkin spaces.

In this note, we shall focus on potential spaces. A key result in the theory is the Sobolev embedding theorem, that reads as follows:

Theorem 1.1 (Classical Sobolev Embedding) [3, Sect. 10],[5, Theorem 6] *Assume that $sp < n$ and define the critical Sobolev exponent p^* as*

$$p^* := \frac{np}{n - sp}.$$

Then, there is a continuous embedding

$$H^{s,p}(\mathbb{R}^n) \subset L^q(\mathbb{R}^n) \quad \text{for } p \leq q \leq p^*. \tag{2}$$

Embedding theorems like this one play a central role when one wants to apply the direct methods of calculus of variations to obtain solutions of nonlinear partial differential equations. Consider, for simplicity, the following elliptic nonlinear boundary-value model problem in a smooth domain $\Omega \subset \mathbb{R}^n$:

$$\begin{cases} -\Delta u = u^{q-1} & \text{in } \Omega \\ u = 0 & \text{on } \partial\Omega \end{cases} \tag{3}$$

It is well known that solutions of this problem can be obtained as critical points of the energy functional

$$J(u) = \frac{1}{2}\int_\Omega |\nabla u|^2 - \frac{1}{q}\int_\Omega |u|^q$$

in the natural energy space, which is the Sobolev space $H^1_0(\Omega)$, i.e., the closure of $C^\infty_0(\Omega)$ in $H^1(\Omega)$ (observe that we need to restrict u to this subspace in order to reflect the Dirichlet boundary condition; this implies in turn that the functional is coercive). The Sobolev embedding gives that

$$H^1(\Omega) \subset L^q(\Omega) \quad 2 \leq q \leq 2^* = \frac{2n}{n-2}$$

which implies that J is well defined (and actually of class C^1) for q within that range.

Then, one can use minimax theorems like the mountain pass theorem of Ambrosetti and Rabinowitz (see [35] or [51]), in order to obtain critical points of J that correspond to non-trivial solutions of our elliptic problem. However, this kind of theorems usually need some compactness assumption like the Palais-Smale condition:

For any sequence u_n in the space $H_0^1(\Omega)$ such that $J(u_n) \to 0$ and $J'(u_n) \to 0$, there exist a convergent subsequence u_{n_k}

In the subcritical case $2 \leq q < 2^*$, this condition is easily verified by using the Rellich–Kondrachov theorem (see, e.g., [2, Chap. 6]):

Theorem 1.2 *If $\Omega \subset \mathbb{R}^n$ is a bounded domain, and $2 \leq q < 2^*$ then the Sobolev embedding*

$$H_0^1(\Omega) \subset L^q(\Omega)$$

is compact.

See [35] or [51] for more information on variational methods for nonlinear elliptic problems.

When Ω is unbounded, for instance $\Omega = \mathbb{R}^n$, the situation is different, since the embedding (2) is continuous but not compact, due to the invariance of the $H^{s,p}$ and L^q norms under translations. Indeed, if we choose $u \in C_0^\infty(\Omega)\backslash\{0\}$ the sequence

$$u_n(x) = u(x + nv)$$

where $v \neq 0$ is a fixed vector in $\mathbb{R}^n$, is a bounded sequence in $H^{s,p}(\mathbb{R}^n)$ without any convergent subsequence.

Assume now that we want to find solutions of the analogous subcritical problem

$$-\Delta u + u = u^{q-1} \quad 2 < q < 2^* = \frac{2n}{n-1} \tag{4}$$

in $\mathbb{R}^n$. Due to the lack of compactness of the Sobolev embedding, the previous approach does not work. However, we can still get compactness and hence non-trivial weak solutions, by restricting the energy functional

$$J(u) = \frac{1}{2}\int_{\mathbb{R}^n} |\nabla u|^2 + \frac{1}{2}\int_{\mathbb{R}^n} |u|^2 - \frac{1}{q}\int_{\mathbb{R}^n} |u|^q$$

to the subspace $H^1_{rad}(\mathbb{R}^n)$ of functions in $H^1(\mathbb{R}^n)$ with radial symmetry, i.e. such that $u(x) = u_0(|x|)$. Indeed we can make use of the following result, due to Lions [27]:

Theorem 1.3 *Let $n \geq 2$. For $2 < q < 2^*$, the embedding*

$$H^1_{rad}(\mathbb{R}^n) \subset L^q(\mathbb{R}^n)$$

is compact.

Remark 1.1 When $q = 2^*$ we don't have compactness even for radial functions, due to the rescaling invariance. The same happens for $q = 2$, since otherwise the Laplacian would have a strictly positive principal eigenvalue in $\mathbb{R}^n$.

Moreover, the orthogonal group $O(n)$ acts naturally on $H^1(\mathbb{R}^n)$ by

$$(g \cdot u)(x) = u(g \cdot x) \quad g \in O(n)$$

and the functional J is invariant by this action. Furthermore, H^1_{rad} is precisely the subspace of invariant functions under this action. Hence, the principle of symmetric criticality [34] implies that the critical points of J on $H^1_{rad}(\mathbb{R}^n)$ are also critical points of J on $H^1(\mathbb{R}^n)$, and hence weak solutions of the elliptic problem (4).

This phenomenon of having better embeddings for spaces of radially symmetric functions (or more generally, for functions invariant under some subgroup of the orthogonal group) is well known, and goes back to the pioneering work by Ni [31] and Strauss [47], who proved the following theorems:

Theorem 1.4 ([31]) *Let B be the unit ball in $\mathbb{R}^n$ $(n \geq 3)$, Then every function in $H^1_{0,rad}(B)$ is almost everywhere equal to a function in $C(\overline{B} - \{0\})$ such that*

$$|u(x)| \leq C|x|^{-(n-2)/2}\|\nabla u\|_{L^2}$$

with $C = (\omega_n(n-2))^{-1/2}$, ω_n being the surface area of ∂B.

Theorem 1.5 ([47]) *Let $n \geq 2$. Then every function in $H^1_{rad}(\mathbb{R}^n)$ is almost everywhere equal to a function in $C(\mathbb{R}^n - \{0\})$ such that*

$$|u(x)| \leq C\,|x|^{-(n-1)/2}\,\|u\|_{H^1}.$$

In both cases, by density, it is enough to establish these inequalities for smooth radial functions. Once the inequality is known for smooth functions, if u_n is a sequence of smooth functions convergent to u in H^1, the inequality implies that u_n converges uniformly on compact sets of $\mathbb{R}^n - \{0\}$ and hence, u is equal almost everywhere to a continuous function outside the origin that satisfies the same inequality. In the sequel we shall use this observation without further comments.

Even though both inequalities look quite similar at first glance, they have different features:

1. Observe that the exponent of $|x|$ in each inequality is different. As a consequence, for a function in $H^1_{rad}(\mathbb{R}^n)$ the estimate provided by Ni's inequality is sharper near the origin, whereas Strauss' inequality is better at infinity.
2. Ni's inequality is invariant under scaling. As a consequence, one has a.e.

$$|u(x)| \leq C|x|^{-(n-2)/2}\|\nabla u\|_{L^2}$$

for $u \in H^1_{rad}(\mathbb{R}^n)$. On the other hand, Strauss' inequality is not invariant by rescaling. However, replacing u by $u(\lambda x)$ and minimizing over λ, one can get the following rescaling-invariant version:

$$|u(x)| \leq C\, |x|^{-(n-1)/2}\, \|u\|_{L^2}^{1/2}\, \|\nabla u\|_{L^2}^{1/2}.$$

For elliptic problems other than (3) that involve other operators, we need different functional spaces, for instance, for the case of the p-Laplacian $\Delta_p u := \mathrm{div}(|\nabla u|^{p-2}\nabla u)$ the natural space is $W_0^{1,p}(\Omega)$ (see, e.g., [16]), and for the fractional Laplacian $(-\Delta)^{-s/2}$ the natural space is $H^s(\mathbb{R}^n)$ (see, e.g., [41]). The advantage of working with potential spaces is, as noted above, that they include all these cases in a unified framework. As in the case of H^1 and H^1_{rad}, the corresponding subspaces of radially symmetric functions of the above spaces will be denoted by the subscript *rad*.

This paper is organized as follows. In Sects 2 and 3 we recall some results on fractional integrals and derivatives, respectively. Then we give two easy proofs of Strauss' inequality for potential spaces, one for $p = 2$ (in Sect. 4) and another one for general p (in Sect. 5). Section 6 is devoted to embedding theorems with power weights for radial functions, which include a generalization of Ni's inequality in $\mathbb{R}^n$. In Sect. 7 we analyze the compactness of the embeddings, both in the unweighted case (giving an alternative proof of Lions' theorem that avoids the use of complex interpolation) and in the weighted case. Finally, in Sect. 8 we discuss a generalization of Ni's inequality and embedding theorems for potential spaces in a ball.

2 Fractional Integral Estimates and Embedding Theorems

Several ways of proving the Sobolev embedding are known. One of the most classical (which goes back to the original work by Sobolev), involves the operator

$$I^s f(x) = c(n,s) \int_{\mathbb{R}^n} \frac{f(y)}{|x-y|^{n-s}}\, dy \quad 0 < s < n$$

(where $c(n,s)$ is a normalization constant), which is known as fractional integral or Riesz potential. This operator provides an integral representation of the negative fractional powers of the Laplacian, i.e:

$$I^s(f) = (-\Delta)^{-s/2} f$$

for smooth functions $f \in \mathcal{S}(\mathbb{R}^n)$. See also [28] for an extension to a space of distributions.

A basic result on this operator is the following theorem concerning its behaviour in the classical Lebesgue spaces:

Theorem 2.1 (Hardy-Littlewood-Sobolev) *If $p > 1$ and $\frac{1}{q} = \frac{1}{p} - \frac{s}{n}$ then I^s is of strong type (p, q), i.e: there exist $C > 0$ such that*

$$\|I^s f\|_{L^q(\mathbb{R}^n)} \leq C \, \|f\|_{L^p(\mathbb{R}^n)}.$$

We recall that, using this result, the Sobolev embedding (Theorem 2) follows easily. Indeed, it is well know that the Bessel potential satisfies the following asymptotics for $s < n$ (see, e.g., [5, Theorem 4]),

$$G_s(x) \simeq C \begin{cases} |x|^{s-n} \text{ if } |x| \leq 2 \\ e^{-|x|/2} \text{ if } |x| \geq 2 \end{cases} \tag{5}$$

Since, in particular, $G_s \in L^1(\mathbb{R}^n)$, one immediately gets the embedding

$$H^{s,p}(\mathbb{R}^n) \subset L^p(\mathbb{R}^n). \tag{6}$$

On the other hand, (5) gives the pointwise bound

$$|G_s * f(x)| \leq C \, I^s(|f|)(x) \tag{7}$$

whence, using the Hardy-Littlewood-Sobolev theorem, (6) and Hölder's inequality one obtains the Sobolev embedding (2).

It is therefore natural to ask ourselves which is the corresponding extension of Theorem 2.1 in the weighted case. In this article, we are mainly interested in power weights, for which the following theorem was proved in [45] by E. Stein and G. Weiss:

Theorem 2.2 ([45, Theorem B*]) *Let $n \geq 1$, $0 < s < n$, $1 < p < \infty, \alpha < \frac{n}{p'}, \beta < \frac{n}{q}, \alpha + \beta \geq 0$, and $\frac{1}{q} = \frac{1}{p} + \frac{\alpha+\beta-s}{n}$. If $p \leq q < \infty$, then the inequality*

$$\| |x|^{-\beta} I^s f \|_{L^q(\mathbb{R}^n)} \leq C \| |x|^{\alpha} f \|_{L^p(\mathbb{R}^n)}$$

holds for any $f \in L^p(\mathbb{R}^n, |x|^{p\alpha} dx)$, where C is independent of f.

For our purposes, it will be very important to notice that when we restrict our attention to functions with radial symmetry, we can get a better result. Indeed, in [13] the authors proved:

Theorem 2.3 *Let $n \geq 1$, $0 < s < n, 1 < p < \infty, \alpha < \frac{n}{p'}, \beta < \frac{n}{q}, \alpha + \beta \geq (n-1)(\frac{1}{q} - \frac{1}{p})$, and $\frac{1}{q} = \frac{1}{p} + \frac{\alpha+\beta-s}{n}$. If $p \leq q < \infty$, then the inequality*

$$\| |x|^{-\beta} I^s f \|_{L^q(\mathbb{R}^n)} \leq C \| |x|^{\alpha} f \|_{L^p(\mathbb{R}^n)}$$

holds for all radially symmetric $f \in L^p(\mathbb{R}^n, |x|^{p\alpha}dx)$, *where* C *is independent of* f. *Moreover, the result also holds for* $p = 1$ *provided* $\alpha + \beta > (n-1)(\frac{1}{q} - \frac{1}{p})$.

Different proofs of this result were also given by Rubin in [37] (except for $p = 1$) and by J. Duoandikoetxea in [19]. A more general theorem involving inequalities with angular integrability was proved by P. D'Ancona and R. Lucà in [7].

Remark 2.1 As observed in [7], Theorem 2.3 also holds for $q = \infty$ (with essentially the same proof) provided that $\alpha + \beta > (n-1)(\frac{1}{p} - \frac{1}{q})$.

3 Riesz Fractional Derivatives

It is well known that the classical potential spaces can be characterized in terms of certain hypersingular integrals, which provide a left inverse for the Riesz potentials. We consider for that purpose, the hypersingular integral

$$D^s u(x) := \frac{1}{d_{n,l}(s)} \int_{\mathbb{R}^n} \frac{(\Delta_y^l u)(x)}{|y|^{n+s}}\, dy, \quad l > s > 0 \tag{8}$$

where

$$(\Delta_y^l u)(x) = \sum_{j=0}^{l} (-1)^j \binom{l}{j} u(x + (l-j)y)$$

denotes the finite difference of order $l \in \mathbb{N}$ of the function u, and $d_{n,l}(\alpha)$ is a certain normalization constant, which is chosen so that the construction (8) does not depend on l (see [39, Chapter 3] for details). We denote by

$$D_\varepsilon^s u(x) = \frac{1}{d_{n,l}(s)} \int_{|x|>\varepsilon} \frac{(\Delta_y^l u)(x)}{|y|^{n+s}}\, dy.$$

Then, we have the following result:

Theorem 3.1 ([40, Theorem 26.3]) *If* $u \in L^p(\mathbb{R}^n)$, *and* $1 \le p < n/s$ *then*

$$\lim_{\varepsilon \to 0} D_\varepsilon I^s u(x) = u(x)$$

in $L^p(\mathbb{R}^n)$.

Another way of inverting the Riesz potential (which holds for every s) is given by the Marchaud method (see [38]).

Then, we have the following characterization theorem due to E. M. Stein (for $0 < s < 2$, [44]) and S. Samko [40, Theorem 27.3] (see also [26, Proposition 2.1]).

Theorem 3.2 *Let $1 < p < \infty$, $0 < s < \infty$, $l > [(s+1)/2]$ if s is not an odd integer, and $l = s$ if s is an odd integer. Then $u \in H^{s,p}(\mathbb{R}^n)$ if and only if $u \in L^p(\mathbb{R}^n)$ and the limit*

$$D^s u = \lim_{\varepsilon \to 0} D^s_\varepsilon u$$

exists in the norm of $L^p(\mathbb{R}^n)$. Moreover,

$$\|u\|_{H^{s,p}} \simeq \|u\|_{L^p} + \|D^s u\|_{L^p}.$$

Remark 3.1 For $0 < s < 2$, D^s coincides with the fractional Laplacian (as defined, e.g., in [17]) for smooth functions, i.e. we have that

$$D^s u = (-\Delta)^{s/2} u.$$

Remark 3.2 For $p = 2$, we can express the L^2-norm of $D^s u$ in terms of the Fourier transform (as a consequence of Plancherel's theorem, see, e.g., [17, Proposition 3.4])

$$\|D^s u\|_{L^2} \simeq \left(\int_{\mathbb{R}^n} |\hat{u}(u)|^2 |\omega|^{2s} \, d\omega \right)^{1/2}.$$

Other related characterizations of potential spaces are given in [48] and [18].

4 A Version of Strauss' Lemma for $p = 2$ Using the Fourier Transform

In this section, we give an elementary proof of Strauss' inequality for $p = 2$, namely,

Theorem 4.1 *Let $u \in H^s_{rad}(\mathbb{R}^n)$ and $s > 1/2$. Then u is almost everywhere equal to a continuous function in $\mathbb{R}^n - \{0\}$ that satisfies that*

$$|u(x)| \leq C|x|^{-(n-1)/2} \, \|u\|_{L^2}^{1-\frac{1}{2s}} \, \|D^s u\|_{L^2}^{\frac{1}{2s}}.$$

Remark 4.1 For an application of this result to non-linear equations involving the fractional Laplacian operator see [41].

We will make use of the following well-known lemmas (see, e.g., [46, Chapter 3]):

Lemma 4.1 (Fourier Transform of Radial Functions) *Let $u \in L^1_{rad}(\mathbb{R}^n)$ be a radial function, $u(x) = u_{rad}(|x|)$. Then its Fourier transform $\hat{u}$ is also radial, and it* *is given by:*

$$\hat{u}(\omega) = (2\pi)^{n/2}|\omega|^{-\nu} \int_0^\infty u_{rad}(r)\, J_\nu(r|\omega|)\, r^{n/2}\, dr$$

where $\nu = \frac{n}{2} - 1$ *and* J_ν *denotes the Bessel function of order* ν.

Lemma 4.2 (Asymptotics of Bessel Functions) *If* $\lambda > -1/2$, *then there exists a constant* $C = C(\lambda)$ *such that*

$$|J_\lambda(r)| \leq C\, r^{-1/2} \quad \forall\, r \geq 0.$$

We observe that $u(x)$ can be reconstructed from $\hat{u}$ using the Fourier inversion formula, and that for radial functions (that are even) the Fourier transform and the inverse Fourier transform coincide up to a constant factor. Hence, from Lemma 4.1, we have that:

$$|u(x)| \leq C|\omega|^{-n/2+1} \int_0^\infty |(\hat{u})_{rad}(r)|\, |J_\nu(r|\omega|)|\, r^{n/2}\, dr.$$

Moreover, using Lemma 4.2, we have that:

$$|u(x)| \leq C\, |\omega|^{-(n-1)/2} \int_0^\infty |(\hat{u})_{rad}(r)|\, r^{(n-1)/2}\, dr.$$

In order to bound this expression, we split the integral into parts: the low frequency part from 0 to r_0, and the high frequency part from r_0 to ∞, where r_0 will be chosen later, and we bound each part using the Cauchy–Schwarz inequality.

With respect to the low frequency part, we have that:

$$\int_0^{r_0} |(\hat{u})_{rad}(r)|\, r^{(n-1)/2}\, dr \leq \left(\int_0^{r_0} |\hat{u}(x)|^2\, r^{n-1}\, dr\right)^{1/2} \left(\int_0^{r_0} dr\right)^{1/2} \leq \|u\|_{L^2}\, r_0^{1/2}.$$

With respect to the high frequency part, we have that:

$$\int_{r_0}^\infty |(\hat{u})_{rad}(r)|\, r^{(n-1)/2}\, dr = \int_{r_0}^\infty |(\hat{u})_{rad}(r)|\, r^{(n-1)/2}\, r^s\, r^{-s}\, dr$$

$$\leq \left(\int_{r_0}^\infty |(\hat{u})_{rad}(r)|^2 r^{n-1} r^{2s}\, dr\right)^{1/2} \left(\int_{r_0}^\infty r^{-2s}\, dr\right)^{1/2} \leq \|D^s u\|_{L^2} \left(\frac{r_0^{-2s-1}}{2s-1}\right)^{1/2}.$$

Summing up, we obtain the following estimate:

$$|u(x)| \leq C\, |x|^{-(n-1)/2} \left(r_0^{1/2}\, \|u\|_{L^2} + \left(\frac{r_0^{-2s-1}}{2s-1}\right)^{1/2} \|D^s u\|_{L^2} \right)$$

which implies

$$|u(x)|^2 \leq C^2 \, |x|^{-(n-1)} \left(r_0 \, \|u\|_{L^2}^2 + \left(\frac{r_0^{-2s-1}}{2s-1} \right) \|D^s u\|_{L^2}^2 \right). \tag{9}$$

Minimizing over r_0 we obtain the required inequality

$$|u(x)| \leq C|x|^{-(n-1)/2} \, \|u\|_{L^2}^{1-\frac{1}{2s}} \, \|D^s u\|_{L^2}^{\frac{1}{2s}}.$$

Remark 4.2 A similar technique was used in [6] to derive a whole family of related inequalities, including a generalization of Ni's inequality.

Remark 4.3 We observe that Lemma 4.1 actually expresses the Fourier transform of a radial function in terms of the (modified) Hankel transform. As a consequence, many results on embeddings for spaces of radial functions can also be derived by using results for the Hankel transform. For instance in [12], we have used this technique to prove Theorem 6.4 for $p = 2$, using results of De Carli [8]. Also some of the results of this paper can be alternatively derived from results obtained by Nowak and Stempak in [32].

5 Proof of Strauss' Inequality

In this section we prove a version of Strauss' inequality for potential spaces for any p. We begin by the following lemma:

Lemma 5.1 *Let $f \in L^p(\mathbb{R}^n)$ be a radial function, then*

$$|f * \chi_{B(0,R)}(x)| \leq CR^{n-1/p}|x|^{-(n-1)/p}\|f\|_{L^p(\mathbb{R}^n)}.$$

Proof Observe first that, for $y \in B(x, R)$ we have that $|x| - R \leq |y| \leq R + |x|$. Now, taking polar coordinates

$$y = ry' \quad r = |y| \quad y' \in S^{n-1}$$

$$x = \rho x' \quad \rho = |x| \quad x' \in S^{n-1}$$

we have

$$f * \chi_{B(0,R)}(x) = \int_{\rho-R}^{\rho+R} f_0(r) \left(\int_{S^{n-1}} \chi_{B(x,R)}(ry') \, dy' \right) r^{n-1} \, dr.$$

To bound the inner integral, observe that $\chi_{B(x,R)}(ry') = \chi_{[t_0,1]}(x' \cdot y')$ with $t_0 = \frac{r^2+\rho^2-R^2}{2r\rho}$, and that $t_0 \le 1$ because $\rho - R \le r \le R + \rho$.

Consider first the case $\rho > 2R$. It follows that $t_0 > -1$ and integrating over the sphere (see [13, Lemma 4.1] for details),

$$\int_{S^{n-1}} \chi_{B(x,R)}(ry')\, dy' = \int_{S^{n-1}} \chi_{[t_0,1]}(x' \cdot y')\, dy' = \omega_{n-2} \int_{-1}^{1} \chi_{[t_0,1]}(t) \left(1-t^2\right)^{\frac{n-3}{2}}\, dt.$$

where ω_{n-2} denotes the area of S^{n-2}. Therefore,

$$\begin{aligned}
|f * \chi_{B(0,R)}(x)| &= \omega_{n-2} \left| \int_{\rho-R}^{\rho+R} \int_{t_0}^{1} f_0(r) \left(1-t^2\right)^{\frac{n-3}{2}} r^{n-1}\, dt\, dr \right| \\
&\le \omega_{n-2} \left(\int_{\rho-R}^{\rho+R} |f_0(r)|^p r^{n-1}\, dr \right)^{1/p} \\
&\quad \times \left(\int_{\rho-R}^{\rho+R} \left(\int_{t_0}^{1} \left(1-t^2\right)^{\frac{n-3}{2}} dt \right)^{p'} r^{n-1}\, dr \right)^{1/p'} \\
&= \omega_{n-2}\, \|f\|_{L^p(\mathbb{R}^n)}\, A(\rho).
\end{aligned}$$

Notice that

$$\int_{t_0}^{1} \left(1-t^2\right)^{\frac{n-3}{2}} dt = \int_{t_0}^{1} (1-t)^{\frac{n-3}{2}} (1+t)^{\frac{n-3}{2}} dt \le C\,(1-t_0)^{\frac{n-1}{2}}$$

which implies

$$\begin{aligned}
A(\rho) \le& C \left(\int_{\rho-R}^{\rho+R} \left(1 - \frac{r^2+\rho^2-R^2}{2r\rho} \right)^{\frac{(n-1)p'}{2}} r^{n-1}\, dr \right)^{1/p'} \\
\le& C \left(\int_{\rho-R}^{\rho+R} \left(1 - \frac{1+(r/\rho)^2-(R/\rho)^2}{2(r/\rho)} \right)^{\frac{(n-1)p'}{2}} r^{n-1}\, dr \right)^{1/p'} \\
\le& C \left(\rho^n \int_{1-R/\rho}^{1+R/\rho} \left(1 - \frac{1+(u)^2-(R/\rho)^2}{2u} \right)^{\frac{(n-1)p'}{2}} u^{n-1}\, du \right)^{1/p'} \\
=& C \left(\rho^n \int_{1-R/\rho}^{1+R/\rho} \left(\frac{(R/\rho)^2-(1-u)^2}{2u} \right)^{\frac{(n-1)p'}{2}} u^{n-1}\, du \right)^{1/p'}
\end{aligned}$$

$$\leq C\left(\rho^n \int_{1-R/\rho}^{1+R/\rho} \left(\frac{1}{u}(R/\rho)^2\right)^{\frac{(n-1)p'}{2}} u^{n-1}\, du\right)^{1/p'}$$

$$\leq CR^{n-1}\left(\rho^{n-(n-1)p'} \int_{1-R/\rho}^{1+R/\rho} u^{(n-1)(1-\frac{p'}{2})}\, du\right)^{1/p'}$$

Since we are assuming $\rho > 2R$, we have that $1 - \frac{R}{\rho} > \frac{1}{2}$ and

$$A(\rho) \leq CR^{n-1}\left(\rho^{n-(n-1)p'}\frac{R}{\rho}\right)^{1/p'} \leq CR^{n-1/p}\rho^{-(n-1)/p}$$

whence, in this case,

$$|f * \chi_{B(0,R)}(x)| \leq CR^{n-1/p}\rho^{-(n-1)/p}\|f\|_{L^p(\mathbb{R}^n)}.$$

It remains to prove the case $\rho < 2R$, where

$$|f * \chi_{B(0,R)}(x)| \leq CR^{n/p'}\|f\|_{L^p(\mathbb{R}^n)} \leq CR^{n-1/p}\rho^{-(n-1)/p}\|f\|_{L^p(\mathbb{R}^n)}.$$

□

Theorem 5.1 (Strauss' Inequality for Potential Spaces) *Assume that* $1 < p < \infty$ *and* $1/p < s < n$. *Let* $u \in H^{s,p}_{rad}(\mathbb{R}^n)$. *Then* u *is almost everywhere equal to a continuous function in* $\mathbb{R}^n - \{0\}$ *that satisfies*

$$|u(x)| \leq C|x|^{-(n-1)/p}\|u\|_{H^{s,p}(\mathbb{R}^n)} \tag{10}$$

and, moreover,

$$|u(y)| \leq C|y|^{-(n-1)/p}\|u\|_{L^p}^{1-1/(sp)}\|D^s u\|_{L^p}^{1/(sp)}. \tag{11}$$

Proof Let $u \in H^{s,p}(\mathbb{R}^n)$ be a radial function, and $f \in L^p(\mathbb{R}^n)$ radial, such that $u = G_s * f$. Then, for $a > 0$ we have that, by Lemma 5.1

$$G_s * f(x) = \sum_{k\in\mathbb{Z}} \int_{\{2^{k-1}a\leq|y|\leq 2^k a\}} f(x-y)G_s(y)\, dy$$

$$\leq \sum_{k\in\mathbb{Z}} G_s(2^{k-1}a) \int_{\{|y|\leq 2^k a\}} f(x-y)\, dy$$

$$
\begin{aligned}
&= \sum_{k\in\mathbb{Z}} G_s(2^{k-1}a) f * \chi_{B(0,2^k a)}(x) \\
&\leq \sum_{k\in\mathbb{Z}} G_s(2^{k-1}a)\ |f * \chi_{B(0,2^k a)}(x)| \\
&\leq C\|f\|_{L^p(\mathbb{R}^n)}|x|^{-(n-1)/p} \sum_{k\in\mathbb{Z}} G_s(2^{k-1}a)(2^k a)^{n-1/p}
\end{aligned}
$$

so, letting $r_k = 2^{k-1}a$ and $\Delta r_k = r_{k+1} - r_k = 2^{k-1}a$, we can write the above sum as

$$
C \sum_{k\in\mathbb{Z}} G_s(r_k)(r_k)^{n-1/p} \Delta r_k \to \int_0^\infty G_s(r) r^{n-1/p}\, dr
$$

when $a \to 0$, so we obtain

$$
\begin{aligned}
|G_s * f(x)| &\leq C\|f\|_{L^p(\mathbb{R}^n)}|x|^{-(n-1)/p} \int_0^\infty G_s(r) r^{n-1/p}\, dr \\
&= C\|f\|_{L^p(\mathbb{R}^n)}|x|^{-(n-1)/p} \int_{\mathbb{R}^n} G_s(x)|x|^{-1/p}\, dx \\
&\leq C\|f\|_{L^p(\mathbb{R}^n)}|x|^{-(n-1)/p}
\end{aligned}
$$

where the last inequality holds since we are assuming $s > 1/p$.

Therefore,

$$
|u(x)| = |G_s * f(x)| \leq C|x|^{-(n-1)/p}\|f\|_{L^p(\mathbb{R}^n)} = C|x|^{-(n-1)/p}\|u\|_{H^{s,p}(\mathbb{R}^n)}
$$

which proves (10).

We proceed now to the proof of (11). Let $u_\lambda(x) = u(\lambda x)$. Then, using Theorem 3.2, it is easy to see that $u_\lambda \in H^{s,p}(\mathbb{R}^n)$ and, moreover, we have that

$$
D^s u_\lambda(x) = \lambda^s D^s u(\lambda x).
$$

Hence, applying (10),

$$
\begin{aligned}
|u_\lambda(x)| &\leq C|x|^{-(n-1)/p}\|u_\lambda\|_{H^{s,p}} \\
&\leq C|x|^{-(n-1)/p}\left[\ \|u_\lambda\|^p_{L^p} + \|D^s u_\lambda\|^p_{L^p}\ \right]^{1/p} \\
&\leq C|x|^{-(n-1)/p}\left[\ \lambda^{-n}\|u\|^p_{L^p} + \lambda^{sp-n}\|D^s u\|^p_{L^p}\ \right]^{1/p}.
\end{aligned}
$$

Setting $y = \lambda x$,

$$|u(y)|^p \leq C|y|^{-(n-1)}\lambda^{n-1}\left[\,\lambda^{-n}\|u\|_{L^p}^p + \lambda^{sp-n}\|D^s u\|_{L^p}^p\,\right]$$
$$\leq C|y|^{-(n-1)}\left[\,\lambda^{-1}\|u\|_{L^p}^p + \lambda^{sp-1}\|D^s u\|_{L^p}^p\,\right].$$

Now, we choose $\lambda > 0$ such that

$$\lambda^{-1}\|u\|_{L^p}^p = \lambda^{sp-1}\|D^s u\|_{L^p}^p$$

i.e.,

$$\lambda = \left(\frac{\|u\|_{L^p}}{\|D^s u\|_{L^p}}\right)^{1/s}.$$

Hence:

$$|u(y)|^p \leq 2C|y|^{-(n-1)}\|u\|_{L^p}^{p-1/s}\|D^s u\|_{L^p}^{1/s}$$
$$|u(y)| \leq 2C|y|^{-(n-1)/p}\|u\|_{L^p}^{1-1/(sp)}\|D^s u\|_{L^p}^{1/(sp)}.$$

□

Remark 5.1 Inequality (10) was proved in [42] in the more general framework of Triebel-Lizorkin spaces, using an atomic decomposition adapted to the radial situation, which is much more technically involved. A version of Strauss' inequality for a class of Orlicz-Sobolev spaces is given in [1].

6 Embedding Theorems with Power Weights for Radial Functions

We begin this section by proving a generalization of Ni's inequality.

Theorem 6.1 (Generalization of Ni's inequality) *Let* $1 < p < \infty$, $u \in H^s_{rad}(\mathbb{R}^n)$ *and* $1/p < s < n/p$. *Then* u *is almost everywhere equal to a continuous function in* $\mathbb{R}^n - \{0\}$ *that satisfies*

$$|u(x)| \leq C|x|^{-\frac{n}{p}+s}\,\|u\|_{H^{s,p}(\mathbb{R}^n)}.$$

Proof Let $u \in H^{s,p}_{rad}(\mathbb{R}^n)$, then there exists $f \in L^p_{rad}(\mathbb{R}^n)$ such that $u = G_s * f$, where G_s is as in (1). Using (7) we obtain

$$\||x|^{n/p-s}u\|_{L^\infty(\mathbb{R}^n)} = \||x|^{n/p-s}(G_s * f)\|_{L^\infty(\mathbb{R}^n)} \leq C\||x|^{n/p-s}I^s(|f|)\|_{L^\infty(\mathbb{R}^n)}.$$

The above inequality combined with Remark 2.1 gives the proof. □

For $s = 1$ and $p = 2$ this result coincides with Theorem 1.4 in $\mathbb{R}^n$, and for arbitrary p and $s = 1$ gives the result in [49, Lemma 1].

Now we can proceed to an immediate extension of the result above, that gives an embedding theorem in the critical case (cf. [49, Lemma 2] when $s = 1$).

Theorem 6.2 *Let $1 < p < \infty$, $0 < s < n/p$, $c > -n$ be such that $(1-sp)c \leq (n-1)ps$, and let $p_c^* = \frac{p(n+c)}{n-sp}$. Then*

$$\| |x|^{c/p_c^*} u \|_{L^{p_c^*}(\mathbb{R}^n)} \leq C \, \|u\|_{H^{s,p}(\mathbb{R}^n)}$$

for any $u \in H^{s,p}_{rad}(\mathbb{R}^n)$.

Proof Using the same argument as in the proof of Theorem 6.1 we obtain

$$\| |x|^{c/p_c^*} u \|_{L^{p_c^*}(\mathbb{R}^n)} \leq C \| |x|^{c/p_c^*} I^s(|f|) \|_{L^{p_c^*}(\mathbb{R}^n)}.$$

We apply Theorem 2.3 with $q = p_c^*$, $\alpha = 0$ and $\beta = \frac{-c}{p_c^*}$. Clearly, since $\alpha = 0$, it holds that $\alpha < n/p'$, and $\beta = \frac{-c}{p_c^*} < \frac{n}{p_c^*} < \frac{n}{q}$ since $c > -n$ and $q < p_c^*$. □

To prove the continuity of the embeddings in the subcritical case we will make used of the following weighted convolution theorem for radial functions, proved by the authors in [14, Theorem 5], which shows that the analogous result of Kerman [25, Theorem 3.1] for arbitrary functions can be improved in the radial case. For other results of weighted convolution inequalities see [4, 33] and references therein.

Theorem 6.3

$$\| |x|^{-\gamma} (f * g)(x) \|_{L^r(\mathbb{R}^n)} \leq C \| |x|^{\alpha} f \|_{L^p(\mathbb{R}^n)} \| |x|^{\beta} g \|_{L^q(\mathbb{R}^n)}$$

for f and g radially symmetric, provided

1. $\frac{1}{r} = \frac{1}{p} + \frac{1}{q} + \frac{\alpha+\beta+\gamma}{n} - 1, 1 < p, q, r < \infty, \frac{1}{r} \leq \frac{1}{p} + \frac{1}{q},$
2. $\alpha < \frac{n}{p'}, \beta < \frac{n}{q'}, \gamma < \frac{n}{r},$
3. $\alpha + \beta \geq (n-1)(1 - \frac{1}{p} - \frac{1}{q}), \beta + \gamma \geq (n-1)(\frac{1}{r} - \frac{1}{q}), \gamma + \alpha \geq (n-1)(\frac{1}{r} - \frac{1}{p})$
4. $\max\{\alpha, \beta, \gamma\} > 0$ *or* $\alpha = \beta = \gamma = 0.$

Theorem 6.4 *Let $1 < p < \infty$, $0 < s < \frac{n}{p}$, $p \leq r \leq p_c^* = \frac{p(n+c)}{n-sp}$. Then we have a continuous embedding*

$$H^{s,p}_{rad}(\mathbb{R}^n) \subset L^r(\mathbb{R}^n, |x|^c dx) \tag{12}$$

provided that

$$-sp < c < \frac{(n-1)(r-p)}{p}. \tag{13}$$

The case $s = 1$ was proved by Rother [36]. A different proof for the case $p = 2$ was given by the authors and Durán in [12], where this embedding theorem is used to prove existence of solutions of a weighted Hamiltonian elliptic system.

Proof Notice that the case $r = p_c^*$ corresponds to Theorem 6.1.

For the remaining cases, we can write, as before, $u = G_s * f$ with $f \in L^p_{rad}(\mathbb{R}^n)$. Using Theorem 6.3 we then have

$$\| |x|^{c/r} G_s * f \|_{L^r} \leq C \|f\|_{L^p} \| |x|^\beta G_s \|_{L^q}$$

provided $\frac{1}{r} = \frac{1}{p} + \frac{1}{q} + \frac{\beta - c/r}{n} - 1$, $\frac{1}{r} \leq \frac{1}{p} + \frac{1}{q}$, $\beta < \frac{n}{q'}$, $-\frac{c}{r} < \frac{n}{r}$, $\beta \geq (n-1)(1 - \frac{1}{p} - \frac{1}{q})$, $\beta - \frac{c}{r} \geq (n-1)(\frac{1}{r} - \frac{1}{q})$, $-\frac{c}{r} \geq (n-1)(\frac{1}{r} - \frac{1}{p})$ and either $\max\{\beta, -\frac{c}{r}\} > 0$ or $\beta = -\frac{c}{r} = 0$.

Since we are assuming $r < \frac{p(n+c)}{n-sp}$, there exists $\varepsilon > 0$ such that $r = \frac{p(n+c)}{(n-sp+\varepsilon p)}$. For this value of ε and sufficiently large $q < \infty$ to be chosen later set $\beta = \frac{n}{q'} - s + \varepsilon$.

This choice of β clearly makes the scaling condition hold, so it suffices to check the remaining conditions:

- $\frac{1}{r} \leq \frac{1}{p} + \frac{1}{q}$ follows from the fact that $p \leq r$.
- $\beta < \frac{n}{q'}$ is equivalent to $\varepsilon < s$. Let us check that this is the case: since $(\varepsilon - s)rp = pn + pc - rn$, it suffices to check that the RHS is negative, but this is equivalent to $c < \frac{n(r-p)}{p}$, which is true because $r \geq p$ and $c < \frac{(n-1)(r-p)}{p}$ by hypothesis.
- $-\frac{c}{r} < \frac{n}{r}$ is immediate from $c > -ps > -n$.
- $\beta \geq (n-1)(1 - \frac{1}{q} - \frac{1}{p})$ is equivalent to $c > \frac{r}{p} + \frac{r}{q} - r - n$. Since we already know that $c > -n$, it suffices to check that $\frac{r}{p} + \frac{r}{q} - r < 0$ which holds for sufficiently large q because $p > 1$. We pick any q that satisfies this condition.
- $\beta - \frac{c}{r} \geq (n-1)(\frac{1}{r} - \frac{1}{q})$ is equivalent to $\frac{1}{q} \leq n + \frac{1}{r}$ which trivially holds.
- $-\frac{c}{r} \geq (n-1)(\frac{1}{r} - \frac{1}{p})$ is equivalent to $c < \frac{(n-1)(r-p)}{p}$ which holds by hypothesis.
- $\max\{\beta, -\frac{c}{r}\} > 0$ obviously holds for $c < 0$, so let us assume that $c > 0$. In this case, using the scaling condition we have that $\frac{\beta}{n} > \frac{1}{r} - \frac{1}{p} - \frac{1}{q} + 1$ which is clearly positive since we chose q to satisfy $-\frac{1}{p} - \frac{1}{q} + 1 > 0$.

To complete the proof, we have to check that $\| |x|^\beta G_s \|_{L^q} < C$ which, using (5), holds provided

$$\int_0^2 r^{\beta q} r^{(s-n)q} r^{n-1} \, dr < +\infty$$

that is, $\beta > \frac{n}{q'} - s$, which is true by our choice of β. □

Remark 6.1 A more general form of Theorem 6.4 can be proved in the context of weighted Triebel-Lizorkin spaces. Indeed, for power weights $|x|^c$ with $-n < c < n(p-1)$ we have that $H^{s,p}(\mathbb{R}^n) = F^s_{p,2}(\mathbb{R}^n)$ and $L^p(\mathbb{R}^n, |x|^c\,dx) = F^0_{p,2}(\mathbb{R}^n, |x|^c\,dx)$. One way of proving the theorem in this setting is by means of a weighted Plancherel-Polya-Nikol'skij type inequality, which in turn makes use of a weighted convolution theorem. In the general setting this approach was used by Meyries and Veraar in [29]. In the radial setting, the authors proved in [14] that a better Plancherel-Polya-Nikol'skij equality holds in the case of radial functions, which in turn gives better weighted embeddings.

More technically involved proofs use atomic decompositions or wavelet decompositions but have the advantage of allowing more general weights. See, e.g., [22–24, 30]. In the case of radial functions, this is the subject of a paper by the authors and Saintier [15].

7 Compactness of the Embeddings

As announced in the introduction, we begin this section by proving the following theorem due to Lions [27], that we will need later to prove the compactness in the weighted case. Our proof is different from the original one of Lions in [27] and avoids the use of the complex method of interpolation.

Theorem 7.1 *For* $1 < p < q < p^*$, *we have a compact embedding*

$$H^{s,p}_{rad}(\mathbb{R}^n) \subset L^q(\mathbb{R}^n)$$

for any $s > 0$.

Proof Let (u_n) be a bounded sequence in $H^{s,p}_{rad}(\mathbb{R}^n)$,

$$\|u_n\|_{H^{s,p}(\mathbb{R}^n)} \leq M \tag{14}$$

i.e. $u_n = G_s * f_n$ with radial f_n such that

$$\|f_n\|_{L^p} \leq M.$$

By the Kolmogorov–Riesz theorem (see [21]) we need to check three conditions:

i) (u_n) is bounded in $L^q(\mathbb{R}^n)$. This is clear by (14) and the continuity of the Sobolev embedding (2).
ii) (equicontinuity condition) Let $\tau_h u(x) = u(x+h)$. Given $\varepsilon > 0$, there exists $\delta = \delta(\varepsilon) > 0$ such that

$$\|\tau_h u_n - u_n\|_{L^q} < \varepsilon$$

if $|h| < \delta$.

This is easy since:

$$\begin{aligned} \|\tau_h u_n - u_n\|_{L^q} &= \|\tau_h(G_s * f_n) - G_s * f_n\|_{L^q} \\ &= \|(\tau_h G_s - G_s) * f_n\|_{L^q} \leq \|\tau_h G_s - G_s\|_{L^r} \|f_n\|_{L^p} \end{aligned}$$

if

$$\frac{1}{q} + 1 = \frac{1}{r} + \frac{1}{p}$$

by Young's inequality, but, since $G_s \in L^q$,

$$\|\tau_h G_s - G_s\|_{L^r} < \frac{\varepsilon}{M}$$

if $|h| < \delta$. We conclude that

$$\|\tau_h u_n - u_n\|_{L^q} \leq \varepsilon$$

if $|h| < \delta$.

iii) (tightness condition) Given $\varepsilon > 0$, there exists $R = R(\varepsilon) > 0$ such that

$$\int_{|x|>R} |u_n|^q \, dx < \varepsilon \quad \text{for all } n \in \mathbb{N}.$$

In order to check this condition, we observe that if $s > \frac{1}{p} - \frac{1}{q}$, by Theorem 6.4, we have that

$$R^{-\gamma q} \int_{|x|>R} |u_n|^q \leq \int_{|x|>R} |x|^{-\gamma q} |u_n|^q \leq C \|u_n\|^q_{H^{s,p}(\mathbb{R}^n)}$$

where

$$\gamma = (n-1)\left(\frac{1}{q} - \frac{1}{p}\right) < 0.$$

Hence,

$$\int_{|x|>R} |u_n|^q \leq \frac{C}{R^{-\gamma q}} M^q < \varepsilon \tag{15}$$

if we choose $R = R(\varepsilon)$ large enough.

We observe that $\delta = -\gamma q$ is exactly the exponent in Lions' paper [27] (second equation on page 321), but he proves (15) using an interpolation argument.

The case $s \leq \frac{1}{p} - \frac{1}{q}$ cannot happen under the theorem hypotheses since

$$q < p^* \Rightarrow s > n\left(\frac{1}{p} - \frac{1}{q}\right) \geq \frac{1}{p} - \frac{1}{q}$$

□

Theorem 7.2 *The embedding*

$$H^{s,p}_{rad}(\mathbb{R}^n) \subset L^r(\mathbb{R}^n, |x|^c dx)$$

of Theorem 6.4 *is compact, provided that* $1 < p < \infty$, $0 < s < \frac{n}{p}$, $p < r < p^*_c = \frac{p(n+c)}{n-sp}$ *and* $-sp < c < \frac{(n-1)(r-p)}{p}$.

Proof It is enough to show that if $u_n \to 0$ weakly in $H^{s,p}_{rad}(\mathbb{R}^n)$, then $u_n \to 0$ strongly in $L^r(\mathbb{R}^n, |x|^c\, dx)$. Since

$$p < r < \frac{p(n+c)}{n-sp}$$

by hypothesis, it is possible to choose q and $\tilde{r}$ so that $p < q < r < \tilde{r} < \frac{p(n+c)}{n-sp}$. We write $r = \theta q + (1-\theta)\tilde{r}$ with $\theta \in (0,1)$ and, using Hölder's inequality, we have that

$$\int_{\mathbb{R}^n} |x|^c |u_n|^r\, dx \leq \left(\int_{\mathbb{R}^n} |u_n|^q\, dx\right)^{\theta} \left(\int_{\mathbb{R}^n} |x|^{\tilde{c}} |u_n|^{\tilde{r}}\, dx\right)^{1-\theta} \tag{16}$$

where $\tilde{c} = \frac{c}{1-\theta}$. By choosing q close enough to p (hence making θ small), we can fulfill the conditions

$$\tilde{r} < \frac{p(n+\tilde{c})}{n-ps}, \quad -ps < \tilde{c} < \frac{(n-1)(\tilde{r}-p)}{p}.$$

Therefore, by the imbedding that we have already established:

$$\left(\int_{\mathbb{R}^n} |x|^{\tilde{c}} |u_n|^{\tilde{r}}\, dx\right)^{1/\tilde{r}} \leq C\|u_n\|_{H^{s,p}} \leq C.$$

Since the imbedding $H^{s,p}_{rad}(\mathbb{R}^n) \subset L^q(\mathbb{R}^n)$ is compact by Lions' theorem, we have that $u_n \to 0$ in $L^q(\mathbb{R}^n)$. From (16) we conclude that $u_n \to 0$ strongly in $L^r(\mathbb{R}^n, |x|^c\, dx)$, which shows that the imbedding in our theorem is also compact. This concludes the proof. □

Remark 7.1 More general results related to the compactness of the embeddings in the framework of Besov and Triebel-Lizorkin spaces can be found [22–24].

8 Ni's Inequality for Potential Spaces in a Ball

We recall that the original result of Ni in [31] was for the case of the ball. Hence, it is natural to ask if our extension of Ni's inequality also holds for potential spaces of radial functions in a ball. In this section we briefly discuss this extension.

Let $B = B(0, R) = \{x \in \mathbb{R}^n : |x| < R\}$ be a ball. We denote by Δ_B the Laplacian operator with Dirichlet conditions. Let $(\lambda_k)_{k\in\mathbb{N}}$ be the Dirichlet eigenvalues, with the corresponding orthogonal basis of $L^2(B)$ of eigenfunctions $(\varphi_k)_{k\in\mathbb{N}}$. Then the negative powers of $-\Delta_B$ can be defined in $L^2(B)$ by

$$(-\Delta_B)^{-s/2}f(x) = \sum_k \lambda_k^{-s/2}\langle f, \varphi k\rangle \varphi_k(x).$$

Let

$$H_B^t(x, y) = \sum_k e^{-\lambda_k t}\varphi_k(x)\varphi_k(y)$$

be the heat kernel for B. Then, from the well-known formula,

$$(-\Delta_B)^{-s/2}f(x) = \frac{1}{\Gamma(s/2)}\int_0^\infty t^{s/2-1}\, e^{t\Delta}\, f(x)\, dt$$

we see that $(-\Delta_B)^{-s/2}$ has an integral representation

$$(-\Delta_B)^{-s/2}f(x) = \int_B K^s(x, y)\, f(y)\, dy$$

where the kernel K^s is given by

$$K^s(x, y)\frac{1}{\Gamma(s/2)}\int_0^\infty t^{s/2-1}\, H_B^t(x, y)\, f(x)\, dt$$

and this formula makes sense for $f \in L^p(B)$. Hence, we may define the potential spaces for the ball

$$H_0^{s,p}(B) = \{u : u = (-\Delta_B)^{-s/2}f \text{ with } f \in L^p(B)\}.$$

(We consider the operator $-\Delta_B$ and not $I - \Delta_B$ in the definition of these spaces since the first eigenvalue λ_1 of the Laplacian in B is strictly positive). For $p = 2$ these spaces are useful in the study of elliptic systems by variational methods (see, e.g., [10, 11]).

The parabolic maximum principle (see, e.g., [20, Sect. 7.1.4]) implies that H^t is bounded by the heat kernel of the whole space $\mathbb{R}^n$:

$$0 \leq H_B^t(x,y) \leq H_{\mathbb{R}^n}^t(x,y) = \frac{1}{(4\pi t)^{n/2}} e^{-|x-y|^2/4t}.$$

Hence, we deduce the bound

$$0 \leq K^s(x,y) \leq \frac{C(n,s)}{|x-y|^{n-s}} \quad 0 < s < n.$$

It follows that we have the pointwise estimate

$$|(-\Delta_B)^{-s/2} f(x)| \leq C(n,s)\, I^s(|\tilde{f}|)(x)$$

where we denote by $\tilde{f}$ the extension of f by zero outside B. Using Theorem 2.3 (as in the proof of Theorems 6.1 and 6.2), we immediately get

Theorem 8.1 (Generalization of Ni's Inequality for the Ball) *Let $1 < p < \infty$, $u \in H^{s,p}_{0,rad}(B)$ and $1/p < s < n/p$. Then u is almost everywhere equal to a continuous function in $B - \{0\}$ that satisfies*

$$|u(x)| \leq C|x|^{-\frac{n}{p}+s}\, \|u\|_{H_0^{s,p}(B)}.$$

Theorem 8.2 *Let $1 < p < \infty$, $0 < s < n/p$, $c > -n$ be such that $(1-sp)c \leq (n-1)ps$, and let $p_c^* = \frac{p(n+c)}{n-sp}$. Then*

$$\| |x|^{c/p_c^*} u \|_{L^{p_c^*}(\mathbb{R}^n)} \leq C\, \|u\|_{H_0^{s,p}(B)}$$

for any radial function $u \in H^{s,p}_{0,rad}(B)$.

Related inequalities for Sobolev spaces of integral order in a ball are proved in [9]. In particular, the exponent in Theorem 8.1 coincides for $p = 2$ and integral s with that of [9, Theorem 1.1(2)].

Acknowledgements Supported by ANPCyT under grant PICT 1675/2010, by CONICET under grant PIP 1420090100230 and by Universidad de Buenos Aires under grant 20020090100067. The authors are members of CONICET, Argentina.

References

1. C.O. Alves, G.M. Figueiredo, J.A. Santos, Strauss- and Lions-type results for a class of Orlicz-Sobolev spaces and applications. Topol. Method Nonlinear Anal. **44**(2), 435–456 (2014)
2. R. Adams, J.J.F. Fournier, *Sobolev Spaces*, 2nd edn. Pure and Applied Mathematics (Amsterdam), vol. 140 (Elsevier/Academic Press, Amsterdam, 2003)
3. N. Aronszajn, K.T. Smith, Theory of Bessel potentials, I. Ann. Inst. Fourier (Grenoble) **11**, 385–475 (1961)
4. H.-Q. Bui, Weighted Young's inequality and convolution theorems on weighted Besov spaces. Math. Nachr. **170**, 25–37 (1994)
5. A.P. Calderón, *Lebesgue Spaces of Differentiable Functions and Distributions*. Proceedings of Symposia in Pure Mathematics, vol. V (American Mathematical Society, Providence, RI, 1961), pp. 33–49
6. Y. Cho, T. Ozawa, Sobolev inequalities with symmetry. Commun. Contemp. Math. **11**, 355–365 (2009)
7. P. D'Ancona, R. Luca, Stein-Weiss and Caffarelli-Kohn-Nirenberg inequalities with angular integrability. J. Math. Anal. Appl. **388**, 1061–1079 (2012)
8. L. De Carli, On the $L^p - L^q$ norm of the Hankel transform and related operators. J. Math. Anal. Appl. **348**, 366–382 (2008)
9. D.G. de Figueiredo, E. Moreira dos Santos, O. Miyagaki, Sobolev spaces of symmetric functions and applications. J. Funct. Anal. **261:12**, 3735–3770 (2011)
10. D.G. de Figueiredo, P. Felmer, On superquadratic elliptic systems. Trans. Am. Math. Soc. **343**, 99–116 (1994)
11. D.G. de Figueiredo, I. Peral, J. Rossi, The critical hyperbola for a hamiltonian elliptic system with weights. Annali di Matematica Pura ed Applicata Appl. **187**, 531–545 (2008)
12. P. De Nápoli, I. Drelichman, R. G. Durán, Radial solutions for hamiltonian elliptic systems with weights. Adv. Nonlinear Stud. **9:3**, 579–594 (2009)
13. P. De Nápoli, I. Drelichman, R.G. Durán, On weighted inequalities for fractional integrals of radial functions. Ill. J. Math. **55:2**, 575–587 (2011)
14. P. De Nápoli, I. Drelichman, Weighted convolution inequalities for radial functions. Ann. Mat. Pura Appl. (4) **194**(1), 167–181 (2015). Doi: 10.1007/s10231-013-0370-6
15. P. De Nápoli, I. Drelichman, N. Saintier, Weighted embedding theorems for radial Besov and Triebel-Lizorkin spaces, preprint 2014, arXiv:1406.0542
16. G. Dinca, P. Jebelean, J. Mawhin, Variational and topological methods for Dirichlet problems with p-Laplacian. Portugaliae Mathematica **58:3**, 339–378 (2001)
17. E. Di Nezza, G. Palatucci, E. Valdinoci, Hitchhiker's guide to the fractional Sobolev spaces. Bull. Sci. Math. **136:5**, 521–573 (2012)
18. J.R. Dorronsoro, A characterization of potential spaces. Proc. Am. Math. Soc. **95**, 21–31 (1985)
19. J. Duoandikoetxea, Fractional integrals on radial functions with applications to weighted inequalities. Ann. Mat. Pura Appl. **192**(4), 553–568 (2013)
20. L.C. Evans, *Partial Differential Equations*. Graduate Studies in Mathematics, vol. 19 (American Mathematical Society, Providence, RI, 1998)
21. H. Hanche-Olsen, H. Holden, The Kolmogorov-Riesz compactness theorem. Expo. Math. **28:4**, 385–394 (2010)
22. D. Haroske, I. Piotrowska, Atomic decompositions of function spaces with Muckenhoupt weights, and some relation to fractal analysis. Math. Nachr. **281**(10), 1476–1494 (2008)
23. D. Haroske, L. Skrzypczak, Entropy and approximation numbers of embeddings of function spaces with Muckenhoupt weights. I. Rev. Mat. Complut. **21**(1), 135–177 (2008)
24. D. Haroske, L. Skrzypczak, Entropy and approximation numbers of embeddings of function spaces with Muckenhoupt weights, II. General weights. Ann. Acad. Sci. Fenn. Math. **36**(1), (2011), 111–138.
25. R.A. Kerman, Convolution theorems with weights. Trans. Am. Math. Soc. **280**(1), 207–219 (1983)

26. T. Kurokawa, On the relations between Bessel potential spaces and Riesz potential spaces. Potential Anal. **12**(3), 299–323 (2000)
27. P.L. Lions, Symétrie e compacité dans les espaces de Sobolev. J. Funct. Anal. **49**, 315–334 (1982)
28. C. Martínez, M. Sanz, F. Periago, Distributional fractional powers of the Laplacean. Riesz potentials. Stud. Math. **135**(3), 253–271 (1999)
29. M. Meyries, M. Veraar, Sharp embedding results for spaces of smooth functions with power weights. Stud. Math. **208**(3), 257–293 (2012)
30. M. Meyries, M. Veraar, Characterization of a class of embeddings for function spaces with Muckenhoupt weights. Arch. Math. **103**, 435–449 (2014)
31. W.M. Ni, A nonlinear Dirichlet problem on the unit ball and its applications. Indiana Univ. Math. J. **31**(6), 801–807 (1982)
32. A. Nowak, K. Stempak, Potential operators associated with Hankel and Hankel-Dunkl transforms. J. Anal. Math. (to appear), arXiv: 1402.3399
33. E. Nursultanov, S. Tikhonov, Weighted norm inequalities for convolution and riesz potential. Potential Anal. **42**, 435–456 (2015). Doi: 10.1007/s11118-014-9440-7.
34. R.S. Palais, The principle of symmetric criticality. Commun. Math. Phys. **69**, 19–30 (1979)
35. P. Rabinowitz, *Minimax Methods in Criticalpoint Theory with Applications to Differential Equations*. CBMS Regional Conference Series in Mathematics, vol. 65 (American Mathematical Society, RI, 1986)
36. W. Rother, Some existence theorems for the equation $-\Delta u + K(x)u^p = 0$. Commun. Partial Differ. Eq. **15**, 1461–1473 (1990)
37. B.S. Rubin, One-dimensional representation, inversion and certain properties of Riesz potentials of radial functions (Russian). Math. Zametki **34**(4), 521–533 (1983) [English translation: Math. Notes **34**(3–4) 751–757 (1983)]
38. B.S. Rubin, *Fractional Integrals and Potentials*. Monographs and Surveys in Pure and Applied Mathematics (Chapman and Hall/CRC, Boca Raton, 1996)
39. G. Samko, *Hypersingular Integrals and Their Applications*, Series Analytical Methods and Special Functions (Taylor & Francis, London, 2002)
40. S.G. Samko, A.A. Kilbas, O.I. Marichev, *Fractional Integrals and Derivatives* (Gordon and Breach Science, New York, 1983)
41. S. Secchi, On fractional Schrödinger equations in $\mathbb{R}^n$ without the Ambrosetti-Rabinowitz condition, preprint 2014, arXiv:1210.0755
42. W. Sickel, L. Skrzypczak, Radial subspaces of Besov and Lizorkin-Triebel classes: extended Strauss lemma and compactness of embeddings. J. Fourier Anal. Appl. **6**, 639–662 (2000)
43. E.M. Stein, *Singular Integrals and Differentiability Properties of Functions* (Princeton University Press, Princeton, 1970)
44. E.M. Stein, The characterization of functions arising as potentials. Bull. Am. Math. Soc. **67**(1), 1–163 (1961)
45. E.M. Stein, G. Weiss, Fractional integrals on n-dimensional Euclidean space. J. Math. Mech. **7**, 503–514 (1958)
46. E.M. Stein, G. Weiss, *Introduction to Fourier Analysis on Euclidean Spaces* (Princeton University Press, Princeton, 1971)
47. W.A. Strauss, Existence of solitary waves in higher dimensions. Comm. Math. Phys. **55**, 149–162 (1977)
48. R.S. Strichartz, Multipliers on fractional Sobolev spaces. J. Math. Mech. **16**, 1031–1060 (1967)
49. J. Su, Z. Wang, M. Willem, Weighted Sobolev imbedding with unbounded and decaying radial potentials. J. Differ. Eq. **238**, 201–219 (2007)
50. H. Triebel, *Theory of Function Spaces II*. Monographs in Mathematics, vol. 84 (Birkhäuser Verlag, Basel, 1992)
51. M. Willem, *Minimax Theorems* (Birkhäuser, Boston, 1996)

On Leray's Formula

E. Liflyand and S. Samko

Abstract We transform Leray's formula on the Fourier transform of a radial function in such a way that it preserves the form of a one-dimensional Fourier transform and the transformed function is close to the initial function as much as possible.

Keywords Bessel function • Fourier transform • Fractional integral/derivative • Radial function

Mathematics Subject Classification (2000). Primary 42B10, Secondary 42B35, 26A33

1 Introduction

In the study of the multidimensional Fourier transform that of a radial function is of special interest. This is partially because of importance of radial functions in many other problems. One of the peculiarities of the Fourier transform of a radial function is that it is also radial. More precisely, if f is an integrable radial function, that is, $f(x) = f_0(|x|)$, where $|x|$ denotes the Euclidean norm of $x = (x_1, \dots, x_n) \in \mathbb{R}^n$, the n-dimensional Fourier integral

$$\hat{f}(x) = \int_{\mathbb{R}^n} f(u)\, e^{-ix\cdot u} du,$$

E. Liflyand (✉)
Department of Mathematics, Bar-Ilan University, 52900 Ramat-Gan, Israel
e-mail: liflyand@math.biu.ac.il

S. Samko
Faculdade de Ciências e Tecnologia, Departamento de Matemática, Universidade do Algarve, Campus de Gambelas, Faro 8005-139, Portugal
e-mail: ssamko@ualg.pt

M. Ruzhansky, S. Tikhonov (eds.), *Methods of Fourier Analysis and Approximation Theory*, Applied and Numerical Harmonic Analysis,
DOI 10.1007/978-3-319-27466-9_9

where $x \cdot u = x_1u_1 + \cdots + x_nu_n$ is the scalar product of $u, x \in \mathbb{R}^n$, reduces to the one-dimensional integral (see, e.g., [9, Chap. IV]):

$$\hat{f}(x) = \hat{f}_0(|x|) = (2\pi)^{\frac{n}{2}} \int_0^\infty f_0(t)(|x|t)^{-\frac{n}{2}-1} J_{\frac{n}{2}-1}(|x|t) t^{n-1}\, dt,$$

where J_μ is the Bessel function of first type and order μ. This makes sense also for radial L^p-functions with $1 < p < \frac{2n}{n+1}$: in that case $\hat{f}(x)$ is everywhere continuous away from the origin. However, working with Bessel functions is quite tedious business sometimes, and certain attempts are known to simplify this formula towards obtaining a "genuine" one-dimensional Fourier transform of some function related to f_0; let us mention [7] and [6].

We discuss one such formula going back to Leray. It is known (see [3, Chap. II, Sect. 2] or Lemma 25.1' in [8]) that when

$$\int_0^\infty t^{n-1}(1+t)^{\frac{1-n}{2}} |f_0(t)|\, dt < \infty, \tag{1}$$

the following relation holds

$$\hat{f}(x) = 2\pi^{\frac{n-1}{2}} \int_0^\infty I_{\frac{n-1}{2}} f_0\,(t) \cos|x|t\, dt, \tag{2}$$

where the fractional integral I_α is defined by

$$I_\alpha f_0\,(t) = \frac{2}{\Gamma(\alpha)} \int_t^\infty s f_0(s)(s^2 - t^2)^{\alpha-1} ds.$$

This formula seems to be very attractive, however much is hidden in the fractional integral of f_0. We change (2) in such a way that it reduces to the one-dimensional Fourier transform of a function "closer" to f_0. More precisely, in Theorem 2.1 we transform a linear combination of fractional integrals (cf. Lemma 2.2) contrary to (2). The highest one is of order $[\frac{n-1}{4}]$, where $[a]$ is the integer part of a, while the least one is the function itself (times an exponential function) in odd dimension and the integral of order $\frac{1}{2}$ (times an exponential function as well) in even dimension. We also gain in getting the factor $|x|^{-\frac{n-1}{2}}$ before the Fourier transform. The price for this are somewhat more restrictive assumptions than (1). Observe that for $n = 2$ the initial formula (2) remains untouched.

This is the matter of the next section. In the last section we discuss possible applications.

2 Main Result

Let us define $m = \frac{n-1}{2}$; as above $[m]$ means the integer part of m.

Theorem 2.1 *Let for some $\varepsilon > 0$ there hold*

$$\int_0^\infty t^{n-1-2[\frac{[m]}{2}]-\varepsilon}(1+t)^{2[\frac{[m]}{2}]-1+\varepsilon}|f_0(t)|\,dt < \infty. \tag{3}$$

Then for $r = |x|$

$$\hat{f}(x) = (-1)^{[m]} 2\pi^m r^{-[m]} \int_0^\infty I_m^{([m])} f_0(t) \cos(\frac{\pi[m]}{2} - rt)\,dt, \tag{4}$$

where $I_m^{([m])} f_0(t)$ is integrable over $(0,\infty)$. In particular,

$$\hat{f}(x) = 2\sqrt{\pi} \int_0^\infty I_{\frac{1}{2}} f_0(t)\ \cos(rt)\,dt, \quad n = 2, \tag{5}$$

$$\hat{f}(x) = \frac{4\pi}{r} \int_0^\infty t f_0(t)\ \sin(rt)\,dt, \quad n = 3. \tag{6}$$

Proof We write $I_m := I_m f_0$ for brevity. Integrating in (2) by parts $[m]$ times, we get

$$\begin{aligned} 2\pi^{\frac{1-n}{2}} \hat{f}(x) &= \int_0^\infty I_m(t) \cos rt\,dt \\ &= \sum_{p=1}^{[m]} (-1)^{p-1} r^{-p} I_m^{(p-1)}(t) \cos(\frac{\pi p}{2} - rt)\Big|_0^\infty \\ &\quad + (-1)^{[m]} r^{-[m]} \int_0^\infty I_m^{([m])}(t) \cos(\frac{\pi[m]}{2} - rt)\,dt. \end{aligned} \tag{7}$$

It remains to figure out which conditions ensure vanishing of integrated terms. Since we started with (2), they must fit (1).

We need a formula for $I_m^{(p)}$, $p = 1, 2, \ldots, [m]$.

Lemma 2.2 *The following formula is valid:*

$$I_m^{(p)}(t) = (p-1)! \sum_{j=0}^{[\frac{p}{2}]} \frac{(-1)^{p-k}}{j!(p-2j)!} (2t)^{p-j} I_{m-p+j}(t), \tag{8}$$

for all $p = 0, 1, 2, \ldots, [m]$, where m may be an arbitrary positive number.

Proof of Lemma 2.2 Let

$$I_\alpha f(t) = \frac{2}{\Gamma(\alpha)} \int_t^\infty s f_0(s)(s^2 - t^2)^{\alpha-1} ds.$$

We may use the relation

$$I_\alpha f = Q I_-^\alpha Q^{-1} f \tag{9}$$

where

$$Qf(t) = f(t^2), \quad Q^{-1} f(t) = f(\sqrt{t})$$

and

$$I_-^\alpha f(t) = \frac{1}{\Gamma(\alpha)} \int_t^\infty \frac{f(s) ds}{(s-t)^{1-\alpha}}$$

is the usual Liouville fractional integral.

The formula for the p-th derivative of a composite function $F(t^2)$ is known and is given in [2, 0.432]. In a slightly modified form it runs as follows:

$$D^p Qf(t) = (p-1)! \sum_{j=0}^{[\frac{p}{2}]} \frac{(2t)^{p-j}}{j!(p-2j)!} (2t)^{p-2j} Q D^{p-j} f \tag{10}$$

where $D = \frac{d}{dt}$. Therefore, by (9) and by the formula

$$D^{p-j} I_-^\alpha = (-1)^{p-j} I_-^{\alpha-(p-j)}, \qquad p - j \le \alpha,$$

we get

$$I_m^{(p)}(t) = (p-1)! \sum_{j=0}^{[\frac{p}{2}]} \frac{(-1)^{p-j}}{j!(p-2j)!} (2t)^{p-j} Q I_-^{m-(p-j)} Q^{-1} f.$$

Making use of the relation (9) again, we arrive at (8). □

The following obvious corollary ensures the existence of (2) under assumptions of Theorem 2.1.

Corollary 2.3 *If for some* $\gamma, \delta \ge 0$ *we have*

$$\int_0^\infty t^{n-1-\gamma} (1+t)^{\frac{1-n}{2}+\delta+\gamma} |f_0(t)| \, dt < \infty, \tag{11}$$

then (1) is fulfilled.

Observe now that if we have

$$\int_0^\infty t^a(1+t)^b |f_0(t)|\, dt < \infty,$$

then for $\delta, \gamma \geq 0$ also

$$\int_0^\infty t^{a+\gamma}(1+t)^{b-\delta-\gamma} |f_0(t)|\, dt < \infty.$$

Indeed, it suffices to notice that $\frac{t^\gamma}{(1+t)^{\delta+\gamma}} \leq 1$.

With this and Lemma 2.2 in hand, we can prove that indeed the integrated terms in (7) vanish under (3). Let us first check when

$$I_m(t) \sin rt \Big|_0^\infty = 0.$$

First,

$$\begin{aligned}\lim_{t\to\infty} |I_m(t)| &\leq \lim_{t\to\infty} \int_t^\infty s|f_0(s)|(s^2-t^2)^{m-1} ds \\ &\leq \lim_{t\to\infty} \int_t^\infty s^{n-2}|f_0(s)|\, ds = 0\end{aligned}$$

provided

$$\int_1^\infty s^{n-2}|f_0(s)|\, ds < \infty. \tag{12}$$

Further,

$$\begin{aligned}\lim_{t\to 0} |I_m(t)| \leq C \lim_{t\to 0}\Bigg(& t^\varepsilon \int_t^1 s^{2-\varepsilon}|f_0(s)|(s^2-t^2)^{m-1} ds \\ & + t\int_1^\infty s|f_0(s)|(s^2-t^2)^{m-1} ds\Bigg),\end{aligned}$$

where C denotes some absolute constant. The integral over $(1, \infty)$ times t vanishes provided (12) holds true. For the integral over $(t, 1)$, we have

$$\lim_{t\to 0} t^\varepsilon \int_t^1 s^{2-\varepsilon}|f_0(s)|(s^2-t^2)^{m-1} ds \leq \lim_{t\to 0} t^\varepsilon \int_0^1 s^{n-1-\varepsilon}|f_0(s)|\, ds = 0$$

provided

$$\int_0^1 s^{n-1-\varepsilon}|f_0(s)|\,ds < \infty. \tag{13}$$

We can combine (12) and (13) in the unified form

$$\int_0^\infty s^{n-1-\varepsilon}(1+s)^{-1+\varepsilon}|f_0(s)|\,ds < \infty. \tag{14}$$

By Corollary 2.3, this condition is more restrictive than (1), that is, a function satisfying (14) moreover satisfies (1) for $n \geq 3$.

Let us then check when

$$tI_{m-1}(t)\Big|_0^\infty = 0.$$

First,

$$t\int_t^\infty s|f_0(s)|(s^2-t^2)^{m-2}ds \leq \int_t^\infty s^2|f_0(s)|(s^2-t^2)^{m-2}ds$$
$$\leq \int_t^\infty s^{n-3}|f_0(s)|\,ds$$

vanishes as $t \to \infty$ provided

$$\int_1^\infty s^{n-3}|f_0(s)|\,ds < \infty. \tag{15}$$

The latter is less restrictive than (12). It is also clear that vanishing of the following integrated terms at infinity will demand less restrictive conditions than (12). We will thus restrict ourselves to finding conditions of vanishing of this and the following integrated terms at zero. Like above, it suffices to study the integrals over $(t, 1)$, since those over $(1, \infty)$ are estimated easily. We have

$$\lim_{t\to 0} t\int_t^1 s|f_0(s)|(s^2-t^2)^{m-2}ds \leq \lim_{t\to 0} t^\varepsilon \int_t^1 s^{2-\varepsilon}|f_0(s)|(s^2-t^2)^{m-2}ds$$
$$\leq \lim_{t\to 0} t^\varepsilon \int_0^1 s^{n-3-\varepsilon}|f_0(s)|\,ds.$$

The right-hand side vanishes provided

$$\int_0^1 s^{n-3-\varepsilon}|f_0(s)|\,ds < \infty. \tag{16}$$

We can combine (12) and (16) in the unified form

$$\int_0^\infty s^{n-3-\varepsilon}(1+s)^{1+\varepsilon}|f_0(s)|\,ds < \infty. \tag{17}$$

Continuing with the next term, we have to estimate

$$\lim_{t\to 0} t \int_t^1 s|f_0(s)|(s^2-t^2)^{m-2}ds$$

and

$$\lim_{t\to 0} t^3 \int_t^1 s|f_0(s)|(s^2-t^2)^{m-3}ds,$$

where additional t has appeared from the bound $\sin rt \leq rt$. The first value was estimated above. The second one vanishes provided (18) holds, and we again have (17).

Let us take a look at the next term. We have to estimate

$$\lim_{t\to 0} t \int_t^1 s|f_0(s)|(s^2-t^2)^{m-3}ds$$

and

$$\lim_{t\to 0} t^3 \int_t^1 s|f_0(s)|(s^2-t^2)^{m-4}ds.$$

Estimates similar to those above yield

$$\int_0^1 s^{n-5-\varepsilon}|f_0(s)|\,ds < \infty. \tag{18}$$

We can combine (12) and (18) in the unified form

$$\int_0^\infty s^{n-5-\varepsilon}(1+s)^{3+\varepsilon}|f_0(s)|\,ds < \infty. \tag{19}$$

We can continue this process up to $p = [m]$.

Similarly, (3) ensures integrability of $I_m^{([m])}(t)$. By this, (7) reduces to (4), exactly as claimed. □

3 Concluding Remarks

The formula (2) has already been used; see, e.g., [4, 5]. It proved to be very convenient in the sense that already obtained one-dimensional results for the Fourier transform were applied more or less immediately. On the other hand, assumptions have mostly been posed on the fractional integral rather than on the function itself. The reason is that aside from certain special situations, it is quite difficult to figure out how the fractional integral inherits certain properties assumed for the integrated function. Another problem is that the multidimensional results are mainly not sharp if compared with those obtained directly, of course after much harder work (cf. [4, 5] and [1] with regard to this). Keeping in mind the Jacobian of the polar coordinates, one may expect that the results will be sharper if one has the factor r^{1-n} before the one-dimensional Fourier transform. For this, one must deal with higher derivatives of I_m. Technically, this will involve not only the function itself but also its derivatives. This will apply to different tools and is supposed to be a matter of future work.

References

1. D. Gorbachev, E. Liflyand, S. Tikhonov, Weighted fourier inequalities: Boas conjecture in $\mathbb{R}^n$. J. d'Analyse Math. **CXIV**, 99–120 (2011)
2. I.S. Gradstein, I.M. Ryzhik, *Table of Integrals, Series, and Products* (Academic Press, New York, 2000)
3. J. Leray, *Hyperbolic Differential Equations* (Princeton, New Jersey, 1953)
4. E. Liflyand, S. Tikhonov, Extended solution of Boas' conjecture on Fourier transforms. C. R. Math. Acad. Sci. Paris **346**, 1137–1142 (2008)
5. E. Liflyand, S. Tikhonov, Two-sided weighted Fourier inequalities. Ann. Sc. Norm. Super. Pisa CI. Sci. (5) **XI**, 341–362 (2012)
6. E. Liflyand, W. Trebels, On asymptotics for a class of radial Fourier transforms. Z. Anal. Anwendungen **17**, 103–114 (1998)
7. A.N. Podkorytov, *Linear Means of Spherical Fourier Sums*. Operator Theory and Function Theory, ed. by M.Z. Solomyak, vol. 1 (Leningrad University, Leningrad, 1983), pp. 171–177 (Russian)
8. S.G. Samko, A.A. Kilbas, O.I. Marichev, *Fractional Integrals and Derivatives: Theory and Applications* (Gordon and Breach, New York, 1993)
9. E. Stein, G. Weiss, *Introduction to Fourier Analysis on Euclidean Spaces* (Princeton University Press, Princeton, 1971)

Part III
Approximation Theory

Order of Approximation of Besov Classes in the Metric of Anisotropic Lorentz Spaces

K.A. Bekmaganbetov

Abstract In this work the sharp estimate of the order of approximation of the Besov classes $B^{\alpha\theta}_{\mathbf{pr}(\mathbb{T}^{\mathbf{d}})}$ in the metric of anisotropic Lorentz spaces $L_{q\theta(\mathbb{T}^{\mathbf{d}})}$ is obtained.

Keywords Anisotropic Lorentz spaces • Besov classes • Order of approximation

Mathematics Subject Classification (2010). Primary 41A46, 41A63, Secondary 42C40

1 Introduction

We introduce some notation needed for the succinct formulation and proof of our results. Given a multi-index $\mathbf{p} = (p_1, \ldots, p_n)$, we write $\mathbf{1/p} = (1/p_1, \ldots, 1/p_n)$. The expression $\mathbf{p} + \mathbf{p_1}$ stands for the coordinatewise sum of $\mathbf{p}$ and $\mathbf{p_1}$, and the relations $\mathbf{p} = \mathbf{p_1}, \mathbf{p} \geq \mathbf{p_1}, \mathbf{p} > \mathbf{p_1}$ mean the validity of the relations $p_i = p_i^1, p_i \geq p_i^1$, $p_i > p_i^1$, respectively, for $i = 1, \ldots, n$.

Let $\mathbf{d} = (d_1, \ldots, d_n) \in \mathbb{N}^n$, $\mathbb{T}^{\mathbf{d}} = \{\mathbf{x} = (\mathbf{x}_1, \ldots, \mathbf{x}_n) : \mathbf{x}_i = (x_1^i, \ldots, x_{d_i}^i) \in [0, 2\pi)^{d_i}, i = 1, \ldots, n\}$ and $f(\mathbf{x}) = f(\mathbf{x}_1, \ldots, \mathbf{x}_n)$ be a measurable function defined on $\mathbb{T}^{\mathbf{d}}$. Let us denote by $f^*(\mathbf{t}) = f^{*_1,\ldots,*_n}(t_1, \ldots, t_n)$ the function obtained by applying non-increasing rearrangement first with respect to first multi-variable with fixed other multi-variables, then sequentially with respect to second multi-variable with fixed others and etc.

Let multi-indexes $\mathbf{p} = (p_1, \ldots, p_n)$, $\mathbf{q} = (q_1, \ldots, q_n)$ satisfy the condition: if $0 < p_j < \infty$, then $0 < q_j \leq \infty$, if $p_j = \infty$, then $q_j = \infty$ for $j = 1, \ldots, n$. Anisotropic Lorentz space $L_{\mathbf{pq}}(\mathbb{T}^{\mathbf{d}})$ (see [1, 2]) is a set of functions such that the

K.A. Bekmaganbetov (✉)
Lomonosov Moscow State University, Kazakhstan branch, Kazhymukan Street 11, 010010 Astana, Kazakhstan
e-mail: bekmaganbetov-ka@yandex.ru

M. Ruzhansky, S. Tikhonov (eds.), *Methods of Fourier Analysis and Approximation Theory*, Applied and Numerical Harmonic Analysis,
DOI 10.1007/978-3-319-27466-9_10

following quasinorm is finite

$$\|f\|_{L_{\mathbf{pq}}(\mathbb{T}^{\mathbf{d}})} =$$

$$\left(\int_0^{(2\pi)^{d_n}} \dots \left(\int_0^{(2\pi)^{d_1}} \left(t_1^{\frac{1}{p_1}} \dots t_n^{\frac{1}{p_n}} f^{*_1,\dots,*_n}(t_1,\dots,t_n)\right)^{q_1} \frac{dt_1}{t_1}\right)^{\frac{q_2}{q_1}} \dots \frac{dt_n}{t_n}\right)^{\frac{1}{q_n}}.$$

For $q = \infty$ by expression $\left(\int_0^T (G(s))^q \frac{ds}{s}\right)^{\frac{1}{q}}$ we mean $\sup_{s>0} G(s)$.

For function $f \in L_{\mathbf{pr}}(\mathbb{T}^{\mathbf{d}})$ let us denote

$$\Delta_{\mathbf{s}}(f,\mathbf{x}) = \sum_{\mathbf{k}\in\mathbf{j}(\mathbf{s})} a_{\mathbf{k}}(f) e^{i(\mathbf{k},\mathbf{x})},$$

where $\{a_{\mathbf{k}}(f)\}_{\mathbf{k}\in\mathbb{Z}^{\mathbf{d}}}$ be the Fourier coefficients of function f with respect to the multiple trigonometric system, $(\mathbf{k},\mathbf{x}) = \sum_{i=1}^{n}\sum_{j=1}^{d_i} k_j^i x_j^i$, $\rho(\mathbf{s}) = \{\mathbf{k} = (\mathbf{k}_1,\dots,\mathbf{k}_n) \in \mathbb{Z}^{\mathbf{d}} : \left[2^{s_i-1}\right] \le \max_{j=1,\dots,d_i} |k_j^i| < 2^{s_i}, i = 1,\dots,n\}$.

Let $-\infty < \boldsymbol{\alpha} = (\alpha_1,\dots,\alpha_n) < \infty$, $\mathbf{0} < \boldsymbol{\theta} = (\theta_1,\dots,\theta_n) \le \infty$. The set of functions $f \in L_{\mathbf{pr}}(\mathbb{T}^{\mathbf{d}})$ is called anisotropic Besov space $B_{\mathbf{pr}}^{\boldsymbol{\alpha\theta}}(\mathbb{T}^{\mathbf{d}})$ [3, 4] if

$$\|f\|_{B_{\mathbf{pr}}^{\boldsymbol{\alpha\theta}}(\mathbb{T}^{\mathbf{d}})} = \left\| \left\{ \mathbf{2}^{(\boldsymbol{\alpha},\mathbf{s})} \|\Delta_{\mathbf{s}}(f)\|_{L_{\mathbf{pr}}(\mathbb{T}^{\mathbf{d}})} \right\}_{\mathbf{s}\in\mathbb{Z}_+^n} \right\|_{l_{\boldsymbol{\theta}}} < \infty,$$

where $\|\cdot\|_{l_{\boldsymbol{\theta}}}$ is a norm of the discrete Lebesgue spaces with mixed metric $l_{\boldsymbol{\theta}}$.

Remark 1 For $n = 1$ the spaces $B_{\mathbf{pr}}^{\boldsymbol{\alpha\theta}}(\mathbb{T}^{\mathbf{d}})$ coincide with periodic Nikol'skii-Besov type spaces $B_{pr}^{\alpha\theta}(\mathbb{T}^{d_1})$ (see [5]); for $d = (1,\dots,1)$ the spaces $B_{\mathbf{pr}}^{\boldsymbol{\alpha\theta}}(\mathbb{T}^{\mathbf{d}})$ are the anisotropic Nikol'skii–Besov spaces with dominating mixed derivate $B_{\mathbf{pr}}^{\boldsymbol{\alpha\theta}}(\mathbb{T}^n)$ (see [6]).

Let $\boldsymbol{\gamma} = (\gamma_1,\dots,\gamma_n)$, $\mathbf{s} = (s_1,\dots,s_n)$, where $\gamma_j > 0$, $s_j \in \mathbb{Z}_+$ for all $j = 1,\dots,n$ and

$$Q^n(\boldsymbol{\gamma d},N) = \bigcup_{(\boldsymbol{\gamma d},\mathbf{s})<N} \rho(\mathbf{s}), \quad T_{Q^n(\boldsymbol{\gamma d},N)} = \left\{ t(\mathbf{x}) = \sum_{\mathbf{k}\in Q^n(\boldsymbol{\gamma d},N)} b_{\mathbf{k}} e^{2\pi i(\mathbf{k},\mathbf{x})} \right\},$$

where $\boldsymbol{\gamma d} = (\gamma_1 d_1,\dots,\gamma_n d_n)$.

Let $E_{\gamma\mathbf{d},N}(f)_{L_{\mathbf{pr}}}$ be the best approximation of $f \in L_{\mathbf{pr}}(\mathbb{T}^{\mathbf{d}})$ by polynomials from $T_{Q^n(\boldsymbol{\gamma d},N)}$, $S_{\gamma\mathbf{d},N}(f,\mathbf{x}) = \sum_{\mathbf{k}\in Q^n(\boldsymbol{\gamma d},N)} a_{\mathbf{k}}(f) e^{2\pi i(\mathbf{k},\mathbf{x})}$ be the partial sum of the Fourier series of function f.

The first author suggesting approximation of functions of several variables by polynomials with harmonics in hyperbolic crosses was Babenko [7]. After that, approximation of functions of various classes of smooth functions by this method was considered by Telyakovskii [8], Mityagin [9], Bugrov [10], Nikol'skaya [11], Galeev [12], Dinh Dung [13], Temlyakov [14], Romanuk [15], Nursultanov and Tleukhanova [16], Akishev [4] and the author [17].

Let us note, that problems of the approximation theory of function in the metric of isotropic Lebesgue space was studied in works [7–15], and for the metric of anisotropic Lorentz space based on the space with mixed derivative this problem was considered in works [4, 17].

Theorem 1 *Let* $\mathbf{d} = (d_1, \dots, d_n) \in \mathbb{N}^n$, $\mathbf{0} < \boldsymbol{\alpha} = (\alpha_1, \dots, \alpha_n) < \infty$, $\mathbf{1} < \mathbf{p} = (p_1, \dots, p_n) < \mathbf{q} = (q_1, \dots, q_n) < \infty$, $\mathbf{1} \le \boldsymbol{\theta} = (\theta_1, \dots, \theta_n), \boldsymbol{\tau} = (\tau_1, \dots, \tau_n), \mathbf{r} = (r_1, \dots, r_n) \le \infty$, $0 < \alpha_{j_0}/d_{j_0} + 1/q_{j_0} - 1/p_{j_0} = \min\{\alpha_j/d_j + 1/q_j - 1/p_j : j = 1, \dots, n\}$ *and* $\gamma_j = \dfrac{\alpha_j/d_j + 1/q_j - 1/p_j}{\alpha_{j_0}/d_{j_0} + 1/q_{j_0} - 1/p_{j_0}}$, $1 \le \gamma_j' \le \gamma_j$, $j = 1, \dots, n$. *Then*

$$E_{\gamma' \mathbf{d}, N} \left(B^{\boldsymbol{\alpha\tau}}_{\mathbf{pr}} \right)_{L_{\mathbf{q}\boldsymbol{\theta}}} \asymp 2^{-(\alpha_{j_0}/d_{j_0} + 1/q_{j_0} - 1/p_{j_0})N} N^{\sum_{j \in A \setminus \{j_1\}} (1/\theta_j - 1/\tau_j)_+}, \tag{1}$$

where $E_{\gamma' \mathbf{d}, N} \left(B^{\boldsymbol{\alpha\tau}}_{\mathbf{p\tau}} \right)_{L_{\mathbf{q}\boldsymbol{\theta}}} = \sup\limits_{\|f\|_{B^{\boldsymbol{\alpha\tau}}_{\mathbf{p\tau}}(\mathbb{T}^{\mathbf{d}})} \le 1} E_{\gamma' \mathbf{d}, N}(f)_{L_{\mathbf{q},,}}$, $A = \{j : \gamma_j' = \gamma_j, j = 1, \dots, n\}$, $j_1 = \min\{j : j \in A\}$, $(a)_+ = \max(a, 0)$.

Remark 2 In the case when $\mathbf{d} = (1, \dots, 1)$ the result of theorem was received by the author in [17], and in particular cases of Akishev in [4].

2 Auxiliary Results

Lemma 1 ([18]) *Let* $\mathbf{1} < \mathbf{q} = (q_1, \dots, q_n) < \mathbf{p} = (p_1, \dots, p_n) < \infty$, $\mathbf{0} < \boldsymbol{\theta} = (\theta_1, \dots, \theta_n), \mathbf{r} = (r_1, \dots, r_n) \le \infty$ *and* $\sigma = \mathbf{d}\,(1/\mathbf{p} - 1/\mathbf{q})$. *Then*

$$L_{\mathbf{q}\boldsymbol{\theta}}(\mathbb{T}^{\mathbf{d}}) \hookrightarrow B^{\sigma\boldsymbol{\theta}}_{\mathbf{pr}}(\mathbb{T}^{\mathbf{d}}).$$

Further we will need in the following sets

$$Y^n(\boldsymbol{\gamma}, N) = \left\{ \mathbf{s} = (s_1, \dots, s_n) \in \mathbb{Z}^n_+ : \sum_{j=1}^n \gamma_j s_j \ge N \right\},$$

$$\aleph^n(\boldsymbol{\gamma}, N) = \left\{ \mathbf{s} = (s_1, \dots, s_n) \in \mathbb{Z}^n_+ : \sum_{j=1}^n \gamma_j s_j = N \right\}.$$

Let $\mathbf{b}$ be a multi-index $(b_1, \ldots, b_{n-1}, b_n)$. By notion $\bar{\mathbf{b}}$ we mean multiindex $(b_1, \ldots, b_{n-1})$.

Lemma 2 *Let* $n \in \mathbb{N}$, $n \geq 2$, $\mathbf{0} < \boldsymbol{\gamma}' = (\gamma_1', \ldots, \gamma_n') \leq \boldsymbol{\gamma} = (\gamma_1, \ldots, \gamma_n) < \infty$, $\beta > 0$ *and* $\mathbf{0} < \boldsymbol{\varepsilon} = (\varepsilon_1, \ldots, \varepsilon_n) \leq \infty$. *Then*

$$\left\| \{2^{-\beta(\boldsymbol{\gamma},\mathbf{s})}\}_{\mathbf{s}\in Y^n(\boldsymbol{\gamma}',N)} \right\|_{l_{\boldsymbol{\varepsilon}}(\mathbb{Z}_+^n)} \leq C 2^{-\beta\delta N} N^{\sum_{j\in A\setminus\{j_1\}} 1/\varepsilon_j},$$

where $\delta = \min\{\frac{\gamma_j}{\gamma_j'} : j = 1, \ldots, n\}$, $A = \{j : \frac{\gamma_j}{\gamma_j'} = \delta, j = 1, \ldots, n\}$, $j_1 = \min\{j : j \in A\}$.

Proof We prove by using induction method on the dimension n.

Let $n = 2$. By definition of the set $Y^2(\boldsymbol{\gamma}', N)$ and using Minkowski inequality we get

$$\left\| \{2^{-\beta(\boldsymbol{\gamma},\mathbf{s})}\}_{\mathbf{s}\in Y^2(\boldsymbol{\gamma}',N)} \right\|_{l_{\boldsymbol{\varepsilon}}(\mathbb{Z}_+^2)} \leq$$

$$C_1 \left\{ \left(\sum_{s_2 < N/\gamma_2'} \left(\sum_{s_1 \geq (N-\gamma_2' s_2)/\gamma_1'} 2^{-\beta(\gamma_1 s_1 + \gamma_2 s_2)\varepsilon_1} \right)^{\frac{\varepsilon_2}{\varepsilon_1}} \right)^{\frac{1}{\varepsilon_2}} + \right.$$

$$\left. \left(\sum_{s_2 \geq N/\gamma_2'} \left(\sum_{s_1 \geq 0} 2^{-\beta(\gamma_1 s_1 + \gamma_2 s_2)\varepsilon_1} \right)^{\frac{\varepsilon_2}{\varepsilon_1}} \right)^{\frac{1}{\varepsilon_2}} \right\} \leq$$

$$C_2 \left\{ \left(\sum_{s_2 < N/\gamma_2'} 2^{-\beta(\frac{\gamma_1}{\gamma_1'}(N-\gamma_2' s_2) + \gamma_2 s_2)\varepsilon_2} \right)^{\frac{1}{\varepsilon_2}} + \left(\sum_{s_2 \geq N/\gamma_2'} 2^{-\beta\gamma_2 s_2 \varepsilon_2} \right)^{\frac{1}{\varepsilon_2}} \right\} =$$

$$C_2 \{J_1 + J_2\}. \tag{2}$$

Let us estimate each term separately. To estimate J_1 we use

$$\sum_{k<N} 2^{\alpha k}(N-k)^\tau \asymp \begin{cases} N^\tau & \text{when } \alpha < 0 \\ N^{\tau+1} & \text{when } \alpha = 0 \\ 2^{\alpha N} & \text{when } \alpha > 0 \end{cases}, \quad (\tau \geq 0). \tag{3}$$

$$J_1 = \left(\sum_{s_2 < N/\gamma_2'} 2^{-\beta(\frac{\gamma_1}{\gamma_1'}(N-\gamma_2' s_2) + \gamma_2 s_2)\varepsilon_2} \right)^{\frac{1}{\varepsilon_2}} =$$

$$2^{-\beta\frac{\gamma_1}{\gamma_1'}N}\left(\sum_{s_2<N/\gamma_2'}2^{-\beta(\frac{\gamma_2}{\gamma_2'}-\frac{\gamma_1}{\gamma_1'})\gamma_2's_2\varepsilon_2}\right)^{\frac{1}{\varepsilon_2}}\asymp 2^{-\beta\delta_2 N}N^{\sum_{j\in A_2\setminus\{j_1\}}1/\varepsilon_j},\qquad(4)$$

$$J_2=\left(\sum_{s_2\geq N/\gamma_2'}2^{-\beta\gamma_2 s_2\varepsilon_2}\right)^{\frac{1}{\varepsilon_2}}\asymp 2^{-\beta\frac{\gamma_2}{\gamma_2'}N}\leq C_3 2^{-\beta\delta_2 N},\qquad(5)$$

here $\delta_2=\min\{\frac{\gamma_j}{\gamma_j'},j=1,2\}, A_2=\{j:\frac{\gamma_j}{\gamma_j'}=\delta_2,j=1,2\}, j_1=\min\{j:j\in A_2\}$.

Substituting (4) and (5) to (2) we verify the validity of the Lemma for $n=2$.

We assume that the Lemma holds for $n-1\geq 2$, i.e.

$$\left\|\{2^{-\beta(\boldsymbol{\gamma},\mathbf{s})}\}_{\mathbf{s}\in Y^{n-1}(\boldsymbol{\gamma},N)}\right\|_{l_{\boldsymbol{\varepsilon}}(\mathbb{Z}_+^{n-1})}\leq C_4 2^{-\beta\delta_{n-1}N}N^{\sum_{j\in A_{n-1}\setminus\{j_1\}}1/\varepsilon_j},\qquad(6)$$

where $\delta_{n-1}=\min\{\frac{\gamma_j}{\gamma_j'},j=1,\ldots,n-1\}, A_{n-1}=\{j:\frac{\gamma_j}{\gamma_j'}=\delta_{n-1},j=1,\ldots,n-1\}$, $j_1=\min\{j:j\in A_{n-1}\}$, and we prove the statement of Lemma for n. By definition of the set $Y^n(\boldsymbol{\gamma}',N)$ and according to Minkowski inequality we obtain

$$\left\|\{2^{-\beta(\boldsymbol{\gamma},\mathbf{s})}\}_{\mathbf{s}\in Y^n(\boldsymbol{\gamma}',N)}\right\|_{l_{\boldsymbol{\varepsilon}}(\mathbb{Z}_+^n)}\leq$$

$$C_1\left\{\left(\sum_{s_n<N/\gamma_n'}\left(2^{-\beta\gamma_n s_n}\left\|\{2^{-\beta(\bar{\gamma},\bar{\mathbf{s}})}\}_{\bar{\mathbf{s}}\in Y^{n-1}(\bar{\boldsymbol{\gamma}}',N-\gamma_n's_n)}\right\|_{l_{\bar{\boldsymbol{\varepsilon}}}(\mathbb{Z}_+^{n-1})}\right)^{\varepsilon_n}\right)^{\frac{1}{\varepsilon_n}}+\right.$$

$$\left.\left(\sum_{s_n\geq N/\gamma_n'}\left(2^{-\beta\gamma_n s_n}\left\|\{2^{-\beta(\bar{\gamma},\bar{\mathbf{s}})}\}_{\bar{\mathbf{s}}\in\mathbb{Z}_+^{n-1}}\right\|_{l_{\bar{\boldsymbol{\varepsilon}}}(\mathbb{Z}_+^{n-1})}\right)^{\varepsilon_n}\right)^{\frac{1}{\varepsilon_n}}\right\}=$$

$$C_1\left\{J_3+J_4\right\}.\qquad(7)$$

We estimate each term separately. To estimate J_3 let us use inequality (6) and (3), then

$$J_3\leq$$

$$\left(\sum_{s_n<N/\gamma_n'}\left(2^{-\beta\gamma_n s_n}C_4 2^{-\beta\delta_{n-1}(N-\gamma_n's_n)}(N-\gamma_n's_n)^{\sum_{j\in A_{n-1}\setminus\{j_1\}}\frac{1}{\varepsilon_j}}\right)^{\varepsilon_n}\right)^{\frac{1}{\varepsilon_n}}=$$

$$C_4 2^{-\beta\delta_{n-1}N}\left(\sum_{s_n<N/\gamma_n'}\left(2^{-\beta(\frac{\gamma_n}{\gamma_n'}-\delta_{n-1})\gamma_n's_n}(N-\gamma_n's_n)^{\sum_{j\in A_{n-1}\setminus\{j_1\}}\frac{1}{\varepsilon_j}}\right)^{\varepsilon_n}\right)^{\frac{1}{\varepsilon_n}}\asymp$$

$$2^{-\beta\delta_n N} N^{\sum_{j\in A_n\setminus\{j_1\}} 1/\varepsilon_j}, \tag{8}$$

$$J_4 \le C_5 \left(\sum_{s_n \ge N/\gamma_n'} 2^{-\beta\gamma_n\varepsilon_n s_n} \right)^{\frac{1}{\varepsilon_n}} \asymp 2^{-\beta\frac{\gamma_n}{\gamma_n'}N} \le C_6 2^{-\beta\delta_n N}, \tag{9}$$

here $\delta_n = \min\{\frac{\gamma_j}{\gamma_j'}, j = 1, \dots, n\}$, $A_n = \{j : \frac{\gamma_j}{\gamma_j'} = \delta_n, j = 1, \dots, n\}$, $j_1 = \min\{j : j \in A_n\}$.

Substituting (8) and (9) to (7), we get

$$\left\| \{2^{-\beta(\boldsymbol{\gamma},\mathbf{s})}\}_{\mathbf{s}\in Y^n(\boldsymbol{\gamma},N)} \right\|_{l_{\boldsymbol{\varepsilon}}(\mathbb{Z}_+^n)} \le C_7 2^{-\beta\delta_n N} N^{\sum_{j\in A_n\setminus\{j_1\}} 1/\varepsilon_j},$$

where $\delta_n = \min\{\frac{\gamma_j}{\gamma_j'}, j = 1, \dots, n\}$, $A_n = \{j : \frac{\gamma_j}{\gamma_j'} = \delta_n, j = 1, \dots, n\}$, $j_1 = \min\{j : j \in A_n\}$. □

Lemma 3 *Let $n \in \mathbb{N}$, $n \ge 2$, $\mathbf{0} < \boldsymbol{\gamma} = (\gamma_1, \dots, \gamma_n) < \infty$, $\beta \in \mathbb{R}$ and $\mathbf{0} < \boldsymbol{\varepsilon} = (\varepsilon_1, \dots, \varepsilon_n) \le \infty$. Then*

$$\left\| \left\{2^{-\beta(\boldsymbol{\gamma},\mathbf{s})}\right\}_{\mathbf{s}\in\aleph^n(\boldsymbol{\gamma},N)} \right\|_{l_{\boldsymbol{\varepsilon}}(\mathbb{Z}_+^n)} \asymp 2^{-\beta N} N^{\sum_{j=2}^n 1/\varepsilon_j}.$$

Proof Let us prove by using induction method on dimension n.

Let $n = 2$. By definition of set $\aleph^2(\boldsymbol{\gamma}, N)$ we have

$$\left\| \{2^{-\beta(\boldsymbol{\gamma},\mathbf{s})}\}_{\mathbf{s}\in\aleph^2(\boldsymbol{\gamma},N)} \right\|_{l_{\boldsymbol{\varepsilon}}(\mathbb{Z}_+^2)} = \left(\sum_{s_2 \le N/\gamma_2} \sum_{s_1 = (N-\gamma_2 s_2)/\gamma_1} 2^{-\beta\varepsilon_2(\gamma_1 s_1 + \gamma_2 s_2)} \right)^{\frac{1}{\varepsilon_2}} =$$

$$2^{-\beta N} \left(\sum_{s_2 \le N/\gamma_2} 1 \right)^{\frac{1}{\varepsilon_2}} \asymp 2^{-\beta N} N^{1/\varepsilon_2}.$$

Suppose that the statement of Lemma holds for $n - 1 \ge 2$, i.e.

$$\left\| \left\{2^{-\beta(\boldsymbol{\gamma},\mathbf{s})}\right\}_{\mathbf{s}\in\aleph^{n-1}(\boldsymbol{\gamma},N)} \right\|_{l_{\boldsymbol{\varepsilon}}(\mathbb{Z}_+^{n-1})} \asymp 2^{-\beta N} N^{\sum_{j=2}^{n-1} 1/\varepsilon_j}, \tag{10}$$

and let us prove Lemma for n. By definition of set $\aleph^n(\boldsymbol{\gamma}, N)$ and according to (10) we receive

$$\left\| \left\{ 2^{-\beta(\boldsymbol{\gamma},\mathbf{s})} \right\}_{\mathbf{s}\in\aleph^n(\boldsymbol{\gamma},N)} \right\|_{l_{\boldsymbol{\varepsilon}}(\mathbb{Z}^n_+)} =$$

$$\left(\sum_{s_n \le N/\gamma_n} \left(2^{-\beta\gamma_n s_n} \left\| \left\{ 2^{-\beta \sum_{j=1}^{n-1} \gamma_j s_j} \right\}_{\bar{\mathbf{s}}\in\aleph^{n-1}(\bar{\boldsymbol{\gamma}},N-\gamma_n s_n)} \right\|_{l_{\bar{\boldsymbol{\varepsilon}}}(\mathbb{Z}^{n-1}_+)} \right)^{\varepsilon_n} \right)^{\frac{1}{\varepsilon_n}} \asymp$$

$$\left(\sum_{s_n \le N/\gamma_n} \left(2^{-\beta\gamma_n s_n} 2^{-\beta(N-\gamma_n s_n)} (N-\gamma_n s_n)^{\sum_{j=2}^{n-1} 1/\varepsilon_j} \right)^{\varepsilon_n} \right)^{\frac{1}{\varepsilon_n}} \asymp$$

$$2^{-\beta N} \left(\sum_{s_n \le N/\gamma_n} (N-\gamma_n s_n)^{\varepsilon_n \sum_{j=2}^{n-1} 1/\varepsilon_j} \right)^{\frac{1}{\varepsilon_n}} \asymp 2^{-\beta N} N^{\sum_{j=2}^{n} 1/\varepsilon_j}.$$

□

Remark 3 In the case when $\gamma_j' = \gamma_j = 1, j = 1, \ldots, \nu$ and $1 < \gamma_j' < \gamma_j, j = \nu + 1, \ldots, n$ of Lemmas 2 and 3, which are earlier proved in [4].

3 Proof of Theorem 1

Proof Let $f \in B^{\boldsymbol{\alpha}\tau}_{\mathbf{pr}}(\mathbb{T}^{\mathbf{d}})$. Since

$$\Delta_{\mathbf{s}}\left(f - S_{\gamma'\mathbf{d},N}(f)\right) = \begin{cases} 0, & \mathbf{s} \notin Y^n(\gamma'\mathbf{d}, N) \\ \Delta_{\mathbf{s}}(f), & \mathbf{s} \in Y^n(\gamma'\mathbf{d}, N) \end{cases},$$

then by Lemma 1 and by Hölder inequality we get

$$\left\| f - S_{\gamma'\mathbf{d},N}(f) \right\|_{L_{\mathbf{q},}(\mathbb{T}^{\mathbf{d}})} = \left\| \sum_{\mathbf{s}\in Y^n(\gamma'\mathbf{d},N)} \Delta_{\mathbf{s}}(f) \right\|_{L_{\mathbf{q},}(\mathbb{T}^{\mathbf{d}})} \le$$

$$C_8 \left\| \sum_{\mathbf{s}\in Y^n(\gamma'\mathbf{d},N)} \Delta_{\mathbf{s}}(f) \right\|_{B^{\mathbf{d}(1/\mathbf{p}-1/\mathbf{q})\,\theta}_{\mathbf{p}\quad\ \mathbf{r}}(\mathbb{T}^{\mathbf{d}})} =$$

$$C_8 \left\| \left\{ 2^{(\mathbf{d}(1/\mathbf{p}-1/\mathbf{q}),\mathbf{s})} \left\| \Delta_{\mathbf{s}}(f) \right\|_{L_{\mathbf{pr}}(\mathbb{T}^{\mathbf{d}})} \right\}_{\mathbf{s}\in Y^n(\gamma'\mathbf{d},N)} \right\|_{l_\theta} =$$

$$C_8 \left\| \left\{ 2^{-(\alpha+\mathbf{d}(1/\mathbf{q}-1/\mathbf{p}),\mathbf{s})} 2^{(\alpha,\mathbf{s})} \left\| \Delta_{\mathbf{s}}(f) \right\|_{L_{\mathbf{pr}}(\mathbb{T}^{\mathbf{d}})} \right\}_{\mathbf{s}\in Y^n(\gamma'\mathbf{d},N)} \right\|_{l_\theta} \leq$$

$$C_8 \left\| \left\{ 2^{(\alpha,\mathbf{s})} \left\| \Delta_{\mathbf{s}}(f) \right\|_{L_{\mathbf{pr}}(\mathbb{T}^{\mathbf{d}})} \right\}_{\mathbf{s}\in Y^n(\gamma'\mathbf{d},N)} \right\|_{l_\tau} \times$$

$$\left\| \left\{ 2^{-(\alpha+\mathbf{d}(1/\mathbf{q}-1/\mathbf{p}),\mathbf{s})} \right\}_{\mathbf{s}\in Y^n(\gamma'\mathbf{d},N)} \right\|_{l_\varepsilon} \leq$$

$$C_8 \left\| \left\{ 2^{-(\alpha_{j_0}/d_{j_0}+1/q_{j_0}-1/p_{j_0})(\gamma\mathbf{d},\mathbf{s})} \right\}_{\mathbf{s}\in Y^n(\gamma'\mathbf{d},N)} \right\|_{l_\varepsilon} \|f\|_{B^{\alpha\tau}_{\mathbf{pr}}(\mathbb{T}^{\mathbf{d}})},$$

here ε such that $1/\varepsilon = (1/\theta - 1/\tau)_+$.

Applying Lemma 2, we obtain

$$\left\| f - S_{\gamma'\mathbf{d},N}(f) \right\|_{L_{\mathbf{q},}} \leq$$

$$\leq C_9 \|f\|_{B^{\alpha\tau}_{\mathbf{pr}}} \cdot 2^{-\left(\alpha_{j_0}/d_{j_0}+1/q_{j_0}-1/p_{j_0}\right)N} N^{\sum_{j\in A\setminus\{j_1\}}\left(1/\theta_j-1/\tau_j\right)_+},$$

where $A = \{j : \gamma'_j = \gamma_j, j = 1, \ldots, n\}$, $j_1 = \min\{j : j \in A\}$, $(a)_+ = \max(a, 0)$.

The upper estimation is proved.

Let us prove the lower estimation. Let $A = \{j : \gamma'_j = \gamma_j, j = 1, \ldots, n\}$, $j_1 = \min\{j : j \in A\}$, $B = \{j : \theta_j < \tau_j, j = 1, \ldots, n\}$ and $B' = A \cap B \cup \{j_1\}$. Suppose $\mathbf{s_0} = (s^0_1, \ldots, s^0_n)$, where $s^0_j = s_j$ for $j \in B'$ and $s^0_j = 0$ for $j \notin B'$, $\tilde{\mathbf{s}} = (s_{j_1}, \ldots, s_{j_{|B'|}})$, $\tilde{\mathbf{d}} = (d_{j_1}, \ldots, d_{j_{|B'|}})$, $\tilde{\boldsymbol{\gamma}}' = (\gamma'_{j_1}, \ldots, \gamma'_{j_{|B'|}})$, where $j_i \in B'$, $i = 1, \ldots, |B'|$ and $j_1 < \ldots < j_{|B'|}$. Let us consider function

$$f(\mathbf{x}) = N^{-\sum_{j\in B'\setminus\{j_1\}} 1/\tau_j} \sum_{(\gamma'\mathbf{d},\mathbf{s_0})=N} \prod_{j=1}^{n} 2^{-\left(\alpha_j+d_j(1-1/p_j)\right)s^0_j} \sum_{\mathbf{k}\in\rho(\mathbf{s_0})} e^{i(\mathbf{k},\mathbf{x})}.$$

Function $f \in L_{\mathbf{pr}}$, since it is polynomial with bounded spectrum. According to Hardy-Littlewood theorem

$$\left\| \sum_{\mathbf{k}\in\rho(\mathbf{s})} e^{i(\mathbf{k},\mathbf{x})} \right\|_{L_{\mathbf{pr}}(\mathbb{T}^{\mathbf{d}})} \asymp \prod_{j=1}^{n} 2^{d_j\left(1-1/p_j\right)s_j}. \tag{11}$$

In a case when $|B'| = 1$ by definition of function $f(\mathbf{x})$ and (11) it is easy to receive the lower estimation. Let $|B'| > 1$, then according to (11) and Lemma 3 for $\beta = 0$, we obtain

$$\|f\|_{B^{\alpha\tau}_{\mathbf{pr}}(\mathbb{T}^{\mathbf{d}})} = \left\| \left\{ \prod_{j=1}^{n} 2^{\alpha_j s_j} \|\Delta_{\mathbf{s}}(f)\|_{L_{\mathbf{pr}}(\mathbb{T}^{\mathbf{d}})} \right\}_{\mathbf{s}\in\mathbb{Z}^n_+} \right\|_{l_\tau} =$$

$$N^{-\sum_{j\in B'\setminus\{j_1\}} 1/\tau_j} \times$$

$$\left\| \left\{ \prod_{j=1}^{n} 2^{\alpha_j s_j^0} 2^{-(\alpha_j + d_j(1-1/p_j))s_j^0} \left\| \sum_{\mathbf{k}\in\rho(\mathbf{s_0})} e^{2\pi i(\mathbf{k},\mathbf{x})} \right\|_{L_{\mathbf{pr}}(\mathbb{T}^{\mathbf{d}})} \right\}_{\mathbf{s_0}\in\aleph^n(\gamma' d,N)} \right\|_{l_\tau} \asymp$$

$$N^{-\sum_{j\in B'\setminus\{j_1\}} 1/\tau_j} \left\| \left\{ \chi_{\aleph^n(\gamma d,N)}(\mathbf{s_0}) \right\}_{\mathbf{s_0}\in\aleph^n(\gamma d,N)} \right\|_{l_\tau} =$$

$$N^{-\sum_{j\in B'\setminus\{j_1\}} 1/\tau_j} \left\| \left\{ \chi_{\aleph^{|B'|}(\tilde{\gamma}\tilde{\mathbf{d}},N)}(\tilde{\mathbf{s}}) \right\}_{\tilde{\mathbf{s}}\in\aleph^{|B'|}(\tilde{\gamma}\tilde{\mathbf{d}},N)} \right\|_{l_\tau} \asymp$$

$$N^{-\sum_{j\in B'\setminus\{j_1\}} 1/\tau_j} \cdot N^{\sum_{j\in B'\setminus\{j_1\}} 1/\tau_j} = C_{10}. \tag{12}$$

On the other hand. Let $\mathbf{1} < \mathbf{q} = (q_1, \ldots, q_n) < \lambda = (\lambda_1, \ldots, \lambda_n) < \infty$ and $\mathbf{0} < \kappa = (\kappa_1, \ldots, \kappa_n) \le \infty$, according to definition of function $f(\mathbf{x})$, and by Lemma 1, Lemma 3 and (11), we have

$$\|f - S_{\gamma\mathbf{d},N}(f)\|_{L_{\mathbf{q}\theta}(\mathbb{T}^{\mathbf{d}})} \ge C_{11} \|f - S_{\gamma\mathbf{d},N}(f)\|_{B^{\mathbf{d}(1/\lambda - 1/\mathbf{q})\,\theta}_{\lambda\ \ \kappa}(\mathbb{T}^{\mathbf{d}})} =$$

$$C_{11} \left\| \left\{ \prod_{j=1}^{n} 2^{d_j(1/\lambda_j - 1/q_j)s_j} \|\Delta_{\mathbf{s}}(f - S_{\gamma\mathbf{d},N}(f))\|_{L_{\lambda\kappa}} \right\}_{\mathbf{s}\in\mathbb{Z}^n_+} \right\|_{l_\theta} =$$

$$C_{11} N^{-\sum_{j\in B'\setminus\{j_1\}} 1/\tau_j} \times$$

$$\left\| \left\{ \prod_{j=1}^{n} 2^{d_j(1/\lambda_j - 1/q_j)s_j^0} 2^{-(\alpha_j + d_j(1-1/p_j))s_j^0} \left\| \sum_{\mathbf{k}\in\rho(\mathbf{s_0})} e^{i(\mathbf{k},\mathbf{x})} \right\|_{L_{\lambda\kappa}(\mathbb{T}^{\mathbf{d}})} \right\}_{\mathbf{s}\in\mathbb{Z}^n_+} \right\|_{l_\theta} \asymp$$

$$N^{-\sum_{j\in B'\setminus\{j_1\}} 1/\tau_j} \times$$

$$\left\| \left\{ \prod_{j=1}^{n} 2^{d_j(1/\lambda_j - 1/q_j)s_j^0} 2^{-(\alpha_j + d_j(1-1/p_j))s_j^0} 2^{d_j(1-1/\lambda_j)s_j^0} \right\}_{\mathbf{s_0}\in\aleph^n(\gamma' d,N)} \right\|_{l_\theta} =$$

$$N^{-\sum_{j\in B'\setminus\{j_1\}}1/\tau_j}\left\|\left\{\prod_{j=1}^{n}2^{-\left(\alpha_j+d_j(1/q_j-1/p_j)\right)s_j^0}\right\}_{\mathbf{s_0}\in\aleph^n(\boldsymbol{\gamma}'\boldsymbol{d},N)}\right\|_{l_\theta}=$$

$$N^{-\sum_{j\in B'\setminus\{j_1\}}1/\tau_j}\left\|\left\{2^{-\left(\alpha_{j_0}/d_{j_0}+1/q_{j_0}-1/p_{j_0}\right)(\tilde{\gamma}\tilde{\mathbf{d}},\tilde{\mathbf{s_0}})}\right\}_{\tilde{\mathbf{s_0}}\in\aleph^n(\tilde{\boldsymbol{\gamma}}'\tilde{\mathbf{d}},N)}\right\|_{l_{\tilde{\theta}}}=$$

$$N^{-\sum_{j\in B'\setminus\{j_1\}}1/\tau_j}\left\|\left\{2^{-\left(\alpha_{j_0}/d_{j_0}+1/q_{j_0}-1/p_{j_0}\right)(\tilde{\gamma}'\tilde{\mathbf{d}},\tilde{\mathbf{s}})}\right\}_{\tilde{\mathbf{s}}\in\aleph^{|B'|}(\tilde{\boldsymbol{\gamma}}'\tilde{\mathbf{d}},N)}\right\|_{l_{\tilde{\theta}}}\asymp$$

$$N^{-\sum_{j\in B'\setminus\{j_1\}}1/\tau_j}\cdot 2^{-N\left(\alpha_{j_0}/d_{j_0}+1/q_{j_0}-1/p_{j_0}\right)}N^{\sum_{j\in B'\setminus\{j_1\}}1/\theta_j}=$$

$$C_{12}2^{-N\left(\alpha_{j_0}/d_{j_0}+1/q_{j_0}-1/p_{j_0}\right)}N^{\sum_{j\in A\setminus\{j_1\}}\left(1/\theta_j-1/\tau_j\right)_+}. \tag{13}$$

By (12) and (13) we have the lower estimation of (1). □

Acknowledgements The research was supported by the Committee of Science of the Ministry of Education and Science of Republic of Kazakhstan (Grant 0816/GF4).

References

1. E.D. Nursultanov, Interpolation properties of some anisotropic spaces and Hardy-Littlewood type inequalities. East J. Approx. **4**(2), 243–275 (1998)
2. E.D. Nursultanov, S.M. Nikol'skii's inequality for different metrics, and properties of the sequence of norms of Fourier sums of a function in the Lorentz space. Proc. Steklov Inst. Math. **255**(4), 185–202 (2006)
3. E.D. Nursultanov, Interpolation theorems for anisotropic function spaces and their applications. Dokl. Akad. Nauk **394**(1), 22–25 (2004, in Russian)
4. G.A. Akishev, Approximation of function classes in spaces with mixed norm. Sbornik Math. **197**(7–8), 1121–1144 (2006)
5. J. Bergh, J. L ofstr oom, *Interpolation Spaces. An Introduction.* (Springer, New York, 1976)
6. K.A. Bekmaganbetov, E.D. Nursultanov, Embedding theorems of anisotropic Besov spaces $B_{\mathbf{pr}}{}^{\alpha}\mathbf{q}([0,2\pi)^n)$. Izvestiya Math. **73**(4), 655–668 (2009)
7. K.I. Babenko, Approximation by trigonometric polynomials in a certain class of periodic functions of several variables. Sov. Math. Dokl. **1**, 672–675 (1960)
8. S.A. Teljakovskii, Some bounds for trigonometric series with quasi-convex coefficients. Math. Sb. (N.S.) **63**, 426–444 (1964, in Russian)
9. B.S. Mitjagin, Approximation of functions in L_p and C spaces on the torus. Math. Sb. (N.S.) **58**, 397–414 (1962, in Russian)
10. Ja.S. Bugrov, Approximation of a class of functions with dominant mixed derivative. Mat. Sb. (N.S.) **64**, 410–418 (1964, in Russian)
11. N.S. Nikol'skaya, Approximation of differentiable functions of several variables by Fourier sums in the L_p-metric. Sib. Math. J. **15**, 395–412 (1974)
12. E.M. Galeev, Kolmogorov widths in the space $\widetilde{L}_q$ of the classes $\widetilde{W}_p^{\bar{\alpha}}$ and $\widetilde{H}_p^{\bar{\alpha}}$ of periodic functions of several variables. Math. USSR-Izvestiya **27**(2), 219–237 (1986)
13. D. Dung, Approximation by trigonometric polynomials of functions of several variables on the torus. Math. USSR-Sbornik **59**(1), 247–267 (1988)

14. V.N. Temlyakov, Approximations of functions with bounded mixed derivative. Proc. Steklov Inst. Math. **178**, 1–121 (1989)
15. A.S. Romanyuk, Approximation of the Besov classes of periodic functions of several variables in a space L_q. Ukr. Math. J. **43**(10), 1297–1306 (1991)
16. E.D. Nursultanov, N.T. Tleukhanova, On the approximate computation of integrals for functions in the spaces $W_p^\alpha([0,1]^n)$. Russ. Math. Surv. **55**(6), 1165–1167 (2000)
17. K.A. Bekmaganbetov, About order of approximation of Besov classes in metric of anisotropic Lorentz spaces. Ufimsk. Mat. Zh. bf 1(2), 9–16 (2009, in Russian)
18. K. Bekmaganbetov, E. Orazgaliev, *Embedding Theorems for Nikol'Skii-Besov Type Spaces.* Inverse Problems: Modeling and Simulation - VI (Izmir University, 2012)

Analogues of Ulyanov Inequalities for Mixed Moduli of Smoothness

M.K. Potapov and B.V. Simonov

Abstract In this paper analogues of well-known Ulyanov inequality about connection between moduli of smoothness in different metrics for the mixed moduli of smoothness are proved.

Keywords Mixed moduli of smoothness • Ulyanov inequality

Mathematics Subject Classification (2000). Primary 26B05, Secondary 46E35

1 Notations and Auxiliary Results

We define by

- L_p, $1 \leq p \leq \infty$, the set of measurable functions $f(x, y)$, 2π—periodic in each variable, such that $||f||_p < \infty$, where

$$||f||_p = \left(\int_0^{2\pi} \int_0^{2\pi} |f(x, y)|^p dxdy \right)^{\frac{1}{p}} \text{ for } 1 \leq p < \infty,$$

and

$$||f||_p = \sup_{\substack{0 \leq x \leq 2\pi \\ 0 \leq y \leq 2\pi}} \text{vrai} |f(x, y)| \text{ for } p = \infty.$$

M.K. Potapov (✉)
Department of Mechanics and Mathematics, Moscow State University, Vorobyevy Gory, Moscow 119992, Russia
e-mail: mkpotapov@mail.ru

B.V. Simonov
Volgograd State Technical University, Volgograd 400131, Russia
e-mail: simonov-b2002@yandex.ru

M. Ruzhansky, S. Tikhonov (eds.), *Methods of Fourier Analysis and Approximation Theory*, Applied and Numerical Harmonic Analysis,
DOI 10.1007/978-3-319-27466-9_11

- L_p^0, the set of functions $f \in L_p$, such that $\int\limits_0^{2\pi} f(x,y)dy = 0$ for a.e. x and $\int\limits_0^{2\pi} f(x,y)dx = 0$ for a.e. y,
- $V_{m_1,\infty}(f), V_{\infty,m_2}(f), V_{m_1,m_2}(f), m_i = 0,1,2,\dots, i = 1,2$, de la Vallee Poussin sums of the Fourier series of $f(x,y)$, i.e.

$$V_{m_1,\infty}(f) = \frac{1}{\pi}\int\limits_0^{2\pi} f(x+t_1,y)V_{m_1}^{2m_1}(t_1)dt_1,$$

$$V_{\infty,m_2}(f) = \frac{1}{\pi}\int\limits_0^{2\pi} f(x,y+t_2)V_{m_2}^{2m_2}(t_2)dt_2,$$

$$V_{m_1,m_2}(f) = \frac{1}{\pi^2}\int\limits_0^{2\pi}\int\limits_0^{2\pi} f(x+t_1,y+t_2)V_{m_1}^{2m_1}(t_1)V_{m_2}^{2m_2}(t_2)dt_1dt_2,$$

where $V_0^0(t) = D_0(t), V_n^{2n}(t) = \frac{1}{n}\big(D_n(t)+\ldots+D_{2n-1}(t)\big), n = 1,2,\dots, D_k(t) = \frac{\sin((k+\frac{1}{2})t)}{2\sin\frac{t}{2}}, k = 0,1,2,\dots,$

- $f^{(\rho_1,\rho_2)}(x,y)$, derivative in the sense of Weyl of the function $f(x,y) \in L_p^0$ of order $\rho_1 \geq 0$ with respect to x and of order $\rho_2 \geq 0$ with respect to y (see [2, p. 238]),
- $T_{n_1,n_2}(x,y)$, a trigonometric polynomial of degree at most n_1, in x and of degree at most n_2, in y,
- $T_{n_1,\infty}(x,y)$, a trigonometric polynomial of degree at most n_1, in x,
- $T_{\infty,n_2}(x,y)$, a trigonometric polynomial of degree at most n_2, in y,
- $Y_{n_1,n_2}(f)_p$, the best (two-dimensional) angular approximation of the function $f \in L_p$, i.e.

$$Y_{n_1,n_2}(f)_p = \inf_{T_{n_1,\infty},T_{\infty,n_2}} ||f - T_{\infty,n_2} - T_{n_1,\infty}||_p,$$

- $[a]$, the integer part of number a.

For a function $f \in L_p$, the difference of order $\alpha_1 > 0$ with respect to the variable x and the difference of order $\alpha_2 > 0$ with respect to the variable y are defined as follows:

$$\Delta_{h_1}^{\alpha_1}(f) = \sum_{\nu_1=0}^{\infty}(-1)^{\nu_1}\binom{\alpha_1}{\nu_1}f(x+(\alpha_1-\nu_1)h_1,y),$$

$$\Delta_{h_2}^{\alpha_2}(f) = \sum_{\nu_2=0}^{\infty}(-1)^{\nu_2}\binom{\alpha_2}{\nu_2}f(x,y+(\alpha_2-\nu_2)h_2),$$

where $\binom{\alpha}{\nu} = 1$ for $\nu = 0$, $\binom{\alpha}{\nu} = 1$ for $\nu = 1$, $\binom{\alpha}{\nu} = \frac{\alpha(\alpha-1)\cdots(\alpha-\nu+1)}{\nu!}$ for $\nu \geq 2$.

Denote by $\omega_{\alpha_1,\alpha_2}(f,\delta_1,\delta_2)_p$ the mixed modulus of smoothness of a function $f \in L_p$ of order $\alpha_1 > 0$ and $\alpha_2 > 0$ with respect to the variables x and y, respectively, i.e.,

$$\omega_{\alpha_1,\alpha_2}(f,\delta_1,\delta_2)_p = \sup_{|h_i|\le\delta_i, i=1,2} ||\Delta_{h_1}^{\alpha_1}(\Delta_{h_2}^{\alpha_2}(f))||_p.$$

If $F(f,\delta_1,\delta_2) > 0$ and $G(f,\delta_1,\delta_2) > 0$ for all $\delta_1,\delta_2 > 0$, then writing $F(f,\delta_1,\delta_2) \ll G(f,\delta_1,\delta_2)$ means that there exists a constant C, independent of f,δ_1,δ_2 such that $F(f,\delta_1,\delta_2) \le CG(f,\delta_1,\delta_2)$. If $F(f,\delta_1,\delta_2) \ll G(f,\delta_1,\delta_2)$ and $G(f,\delta_1,\delta_2) \ll F(f,\delta_1,\delta_2)$ simultaneously, then we will write $F(f,\delta_1,\delta_2) \asymp G(f,\delta_1,\delta_2)$.

Let us consider further that $q^* = q$ if $q < \infty$ and $q^* = 1$ if $q = \infty$.

Lemma 1.1 ([8, 9]) *Let $f \in L_q^0$, $1 \le q \le \infty$, $n_i = 0,1,2,\dots, i = 1,2$. Then*

$$I \equiv \omega_{\alpha_1,\alpha_2}\Big(f,\frac{1}{2^{n_1}},\frac{1}{2^{n_2}}\Big)_q \asymp \big\|f - V_{2^{n_1},\infty}(f) - V_{\infty,2^{n_2}}(f) + V_{2^{n_1},2^{n_2}}(f)\big\|_q +$$

$$+\frac{1}{2^{n_1\alpha_1}}\big\|V_{2^{n_1},\infty}^{(\alpha_1,0)}(f - V_{\infty,2^{n_2}}(f))\big\|_q + \frac{1}{2^{n_2\alpha_2}}\big\|V_{\infty,2^{n_2}}^{(0,\alpha_2)}(f - V_{2^{n_1},\infty}(f))\big\|_q +$$

$$+\frac{1}{2^{(n_1\alpha_1+n_2\alpha_2)}}\big\|V_{2^{n_1},2^{n_2}}^{(\alpha_1,\alpha_2)}(f)\big\|_q \equiv I_1 + 2^{-n_1\alpha_1}I_2 + 2^{-n_2\alpha_2}I_3 + 2^{-n_1\alpha_1-n_2\alpha_2}I_4.$$

Lemma 1.2 ([3]) *Let $f \in L_p$, $1 \le p \le \infty$, $m_i = 0,1,2,\dots, i = 1,2$. Then*

a) $||f - V_{m_1,\infty}(f) - V_{\infty,m_2}(f) + V_{m_1,m_2}(f)||_p \ll Y_{m_1,m_2}(f)_p$,
b) $||V_{m_1,\infty}(f)||_p \ll ||f||_p$,
c) $||V_{\infty,m_2}(f)||_p \ll ||f||_p$.

Lemma 1.3 ([3, 5]) *Let $f \in L_p^0$, $1 \le p < q \le \infty$, $n_i = 0,1,2,\dots, i = 1,2$. Then*

$$Y_{2^{n_1}-1,2^{n_2}-1}(f)_q \ll \left(\sum_{\nu_1=n_1}^{\infty}\sum_{\nu_2=n_2}^{\infty} 2^{(\nu_1+\nu_2)\left(\frac{1}{p}-\frac{1}{q}\right)q^*} Y_{2^{\nu_1}-1,2^{\nu_2}-1}^{q^*}(f)_p\right)^{\frac{1}{q^*}}.$$

Lemma 1.4 ([8, 9]) *Let $f \in L_p^0$, $g \in L_p^0$, $1 \le p \le \infty$, $\alpha_i > 0$, $m_i = 0,1,2,\dots, i = 1,2$. Then*

a) $\omega_{\alpha_1,\alpha_2}(f,\delta_1,0)_p = \omega_{\alpha_1,\alpha_2}(f,0,\delta_2)_p = \omega_{\alpha_1,\alpha_2}(f,0,0)_p = 0$,
b) $\omega_{\alpha_1,\alpha_2}(f+g,\delta_1,\delta_2)_p \ll \omega_{\alpha_1,\alpha_2}(f,\delta_1,\delta_2)_p + \omega_{\alpha_1,\alpha_2}(g,\delta_1,\delta_2)_p$,
c) $\omega_{\alpha_1,\alpha_2}(f,\delta_1,\delta_2)_p \ll \omega_{\alpha_1,\alpha_2}(f,t_1,t_2)_p$, *if* $0 \le \delta_i \le t_i \le 1, i = 1,2$,
d) $\frac{\omega_{\alpha_1,\alpha_2}(f,\delta_1,\delta_2)_p}{\delta_1^{\alpha_1}\delta_2^{\alpha_2}} \ll \frac{\omega_{\alpha_1,\alpha_2}(f,t_1,t_2)_p}{t_1^{\alpha_1}t_2^{\alpha_2}}$, *if* $0 < t_i \le \delta_i \le 1, i = 1,2$,
e) $\omega_{\alpha_1,\alpha_2}(f,\lambda_1\delta_1,\lambda_2\delta_2)_p \ll (\lambda_1+1)^{\alpha_1}(\lambda_2+1)^{\alpha_2}\omega_{\alpha_1,\alpha_2}(f,\delta_1,\delta_2)_p$, *if* $\lambda_i > 0, i = 1,2$,
f) $Y_{m_1,m_2} \ll \omega_{\alpha_1,\alpha_2}\Big(f,\frac{1}{m_1+1},\frac{1}{m_2+1}\Big)_p$.

Lemma 1.5 ([8, 9]) *Let* $1 \le p \le \infty, \alpha_i > 0, n_i \in \mathbb{N}, i = 1, 2$. *Then*

a) if $T_{n_1,\infty} \in L_p^0$, *then* $||T_{n_1,\infty}^{(\alpha_1,0)}||_p \ll n_1^{\alpha_1}||T_{n_1,\infty}||_p$,
b) if $T_{\infty,n_2} \in L_p^0$, *then* $||T_{\infty,n_2}^{(0,\alpha_2)}||_p \ll n_2^{\alpha_2}||T_{\infty,n_2}||_p$,
c) if $T_{n_1,n_2} \in L_p^0$, *then* $||T_{n_1,n_2}^{(\alpha_1,\alpha_2)}||_p \ll n_1^{\alpha_1}n_2^{\alpha_2}||T_{n_1,\infty}||_p$.

Lemma 1.6 ([4]) *Let* $a_n \ge 0, b_n \ge 0, \sum_{k=1}^{n} a_k = a_n\gamma_n, 1 \le p < \infty$. *Then*

$$\sum_{k=1}^{\infty} a_k\Big(\sum_{n=k}^{\infty} b_n\Big)^p \ll \sum_{k=1}^{\infty} a_k(b_k\gamma_k)^p.$$

Lemma 1.7 ([1, p. 99]) *There in a constant C, independent from n and x, such that*

$$\Big|\sum_{k=1}^{n} \frac{\sin kx}{k}\Big| \le C.$$

Lemma 1.8 *Let* $T_{m_1,m_2} \in L_1^0$, *where* $m_i \in \mathbb{N}, i = 1, 2$. *Then inequalities are true*

a) $\big|T_{m_1,m_2}(x, y)\big| \ll m_2||T_{m_1,m_2}^{(1,0)}||_1$,
b) $\big|T_{m_1,m_2}(x, y)\big| \ll m_1||T_{m_1,m_2}^{(0,1)}||_1$,
c) $\big|T_{m_1,m_2}(x, y)\big| \ll ||T_{m_1,m_2}^{(1,1)}||_1$.

Proof Denote by $k_n(t) = \sum_{k=1}^{n} \frac{\sin kx}{k}, D_n(t) = \frac{1}{2} + \sum_{\nu=1}^{n} \cos \nu t$. Then

$$|D_n(t)| \le (n + 1).$$

a) One can show that

$$T_{m_1,m_2}(x, y) = -\frac{1}{\pi^2}\int_0^{2\pi}\int_0^{2\pi} T_{m_1,m_2}^{(1,0)}(x + t_1, y + t_2)k_{m_1}(t_1)D_{m_2}(t_2)dt_1dt_2.$$

Whence, using Lemma 1.7, we have

$$\big|T_{m_1,m_2}(x, y)\big| \le \frac{1}{\pi^2}\int_0^{2\pi}\int_0^{2\pi} \big|T_{m_1,m_2}^{(1,0)}(x + t_1, y + t_2)\big|\big|k_{m_1}(t_1)\big|\big|D_{m_2}(t_2)\big|dt_1dt_2 \ll$$

$$\ll (m_2 + 1)\int_0^{2\pi}\int_0^{2\pi} \big|T_{m_1,m_2}^{(1,0)}(x + t_1, y + t_2)\big|dt_1dt_2 \ll m_2||T_{m_1,m_2}^{(1,0)}||_1.$$

The item a) is proved.

b) One can show that

$$T_{m_1,m_2}(x,y) = -\frac{1}{\pi^2}\int_0^{2\pi}\int_0^{2\pi} T^{(0,1)}_{m_1,m_2}(x+t_1,y+t_2)D_{m_1}(t_1)k_{m_2}(t_2)dt_1dt_2.$$

Whence, using Lemma 1.7, we have

$$\left|T_{m_1,m_2}(x,y)\right| \le \frac{1}{\pi^2}\int_0^{2\pi}\int_0^{2\pi} \left|T^{(0,1)}_{m_1,m_2}(x+t_1,y+t_2)\right|\left|D_{m_1}(t_1)\right|\left|k_{m_2}(t_2)\right|dt_1dt_2 \ll$$

$$\ll (m_1+1)\int_0^{2\pi}\int_0^{2\pi} \left|T^{(0,1)}_{m_1,m_2}(x+t_1,y+t_2)\right|dt_1dt_2 \ll m_1||T^{(0,1)}_{m_1m_2}||_1.$$

The item b) is proved.

c) One can show that

$$T_{m_1,m_2}(x,y) = \frac{1}{\pi^2}\int_0^{2\pi}\int_0^{2\pi} T^{(1,1)}_{m_1,m_2}(x+t_1,y+t_2)k_{m_1}(t_1)k_{m_2}(t_2)dt_1dt_2.$$

Whence, using Lemma 1.7, we have

$$\left|T_{m_1,m_2}(x,y)\right| \le \frac{1}{\pi^2}\int_0^{2\pi}\int_0^{2\pi} \left|T^{(1,1)}_{m_1,m_2}(x+t_1,y+t_2)\right|\left|k_{m_1}(t_1)\right|\left|k_{m_2}(t_2)\right|dt_1dt_2 \ll$$

$$\ll \int_0^{2\pi}\int_0^{2\pi} \left|T^{(1,1)}_{m_1,m_2}(x+t_1,y+t_2)\right|dt_1dt_2 = ||T^{(1,1)}_{m_1,m_2}||_1.$$

The item c) is proved.

2 Interrelation Between the Mixed Moduli of Smoothness in Various Metrics

Theorem 2.1 *Let* $f \in L_p^0$, $1 \le p < q \le \infty, n_i = 0, 1, 2, \ldots, \alpha_i > 0, i = 1, 2$. *Then*

$$I \equiv \omega_{\alpha_1,\alpha_2}\Big(f, \frac{1}{2^{n_1}}, \frac{1}{2^{n_2}}\Big)_q \ll \Big(\sum_{\nu_1=n_1}^{\infty}\sum_{\nu_2=n_2}^{\infty}\Big(2^{(\nu_1+\nu_2)\left(\frac{1}{p}-\frac{1}{q}\right)}Y_{[2^{\nu_1}-1],[2^{\nu_2}-1]}(f)_p\Big)^{q^*}\Big)^{\frac{1}{q^*}} +$$

$$+\frac{1}{2^{n_1\alpha_1}}\Big(\sum_{\nu_1=0}^{n_1}\sum_{\nu_2=n_2}^{\infty}\Big(2^{\nu_1\left(\alpha_1+\frac{1}{p}-\frac{1}{q}\right)}2^{\nu_2\left(\frac{1}{p}-\frac{1}{q}\right)}Y_{[2^{\nu_1}-1],[2^{\nu_2}-1]}(f)_p\Big)^{q^*}\Big)^{\frac{1}{q^*}} +$$

$$+\frac{1}{2^{n_2\alpha_2}}\Big(\sum_{\nu_1=n_1}^{\infty}\sum_{\nu_2=0}^{n_2}\Big(2^{\nu_1\left(\frac{1}{p}-\frac{1}{q}\right)}2^{\nu_2\left(\alpha_2+\frac{1}{p}-\frac{1}{q}\right)}Y_{[2^{\nu_1}-1],[2^{\nu_2}-1]}(f)_p\Big)^{q^*}\Big)^{\frac{1}{q^*}} +$$

$$+\frac{1}{2^{n_1\alpha_1}}\frac{1}{2^{n_2\alpha_2}}\Big(\sum_{\nu_1=0}^{n_1}\sum_{\nu_2=0}^{n_2}\Big(2^{\nu_1\left(\alpha_1+\frac{1}{p}-\frac{1}{q}\right)}2^{\nu_2\left(\alpha_2+\frac{1}{p}-\frac{1}{q}\right)}Y_{[2^{\nu_1}-1],[2^{\nu_2}-1]}(f)_p\Big)^{q^*}\Big)^{\frac{1}{q^*}} \equiv$$

$$\equiv A_1 + 2^{-n_1\alpha_1}A_2 + 2^{-n_2\alpha_2}A_3 + 2^{-n_1\alpha_1-n_2\alpha_2}A_4.$$

Proof Using Lemma 1.1, we have

$$I \asymp I_1 + 2^{-n_1\alpha_1}I_2 + 2^{-n_2\alpha_2}I_3 + 2^{-n_1\alpha_1-n_2\alpha_2}I_4.$$

Using Lemma 1.2, we get $I_1 \ll Y_{2^{n_1},2^{n_2}}(f)_q$. Using Lemma 1.3, we have

$$I_1 \ll \Big(\sum_{\nu_1=n_1}^{\infty}\sum_{\nu_2=n_2}^{\infty}2^{(\nu_1+\nu_2)\left(\frac{1}{p}-\frac{1}{q}\right)q^*}Y^{q^*}_{2^{\nu_1}-1,2^{\nu_2}-1}(f)_p\Big)^{\frac{1}{q^*}} \ll$$

$$\ll \Big(\sum_{\nu_1=n_1}^{\infty}\sum_{\nu_2=n_2}^{\infty}2^{(\nu_1+\nu_2)\left(\frac{1}{p}-\frac{1}{q}\right)q^*}Y^{q^*}_{[2^{\nu_1}-1],[2^{\nu_2}-1]}(f)_p\Big)^{\frac{1}{q^*}} \equiv A_1.$$

Let as estimate I_2. We denote $\varphi_1(x,y) = V^{(\alpha_1,0)}_{2^{n_1},\infty}(f), \psi_1(x,y) = V^{(\alpha_1,0)}_{2^{n_1},\infty}(f - V_{\infty,2^{n_2}}(f)) == \varphi_1 - V_{\infty,2^{n_2}}(\varphi_1)$. As $f \in L_p^0$, then $V_{0,\infty}(\varphi_1) = V_{0,2^{n_2}}(\varphi_1) = 0$. But then

$$I_2 = ||\psi_1(x,y)||_q = ||\varphi_1 - V_{\infty,2^{n_2}}(\varphi_1) - V_{0,\infty}(\varphi_1) + V_{0,2^{n_2}}(\varphi_1)||_q \ll Y_{0,2^{n_2}}(\varphi_1)_q.$$

Using Lemma 1.3, we have

$$I_2 \ll \Big(\sum_{\nu_1=0}^{\infty}\sum_{\nu_2=n_2}^{\infty}2^{(\nu_1+\nu_2)\left(\frac{1}{p}-\frac{1}{q}\right)q^*}Y^{q^*}_{2^{\nu_1}-1,2^{\nu_2}-1}(\varphi_1)_p\Big)^{\frac{1}{q^*}}.$$

If $\nu_1 \geq n_1 + 1$, taking into account that $\varphi_1 = V^{(\alpha_1,0)}_{2^{n_1},\infty}(f)$, we get that for anyone $\nu_2 \geq 0$ $Y_{2^{\nu_1}-1,2^{\nu_2}-1}(\varphi_1)_p = 0$. Therefore

$$I_2 \ll \Big(\sum_{\nu_1=0}^{n_1}\sum_{\nu_2=n_2}^{\infty} 2^{(\nu_1+\nu_2)\left(\frac{1}{p}-\frac{1}{q}\right)q^*} Y^{q^*}_{2^{\nu_1}-1,2^{\nu_2}-1}(\varphi_1)_p\Big)^{\frac{1}{q^*}}.$$

Using definition of "approximation by an angle", for $\nu_1 \leq n_1$ and $\nu_2 \geq n_2$ we have

$$I_5 = Y_{2^{\nu_1}-1,2^{\nu_2}-1}(\varphi_1)_p \leq ||\varphi_1 - V_{[2^{\nu_1}-1],\infty}(\varphi_1) - V_{\infty,[2^{\nu_2}-1]}(\varphi_1)$$
$$+V_{[2^{\nu_1}-1],[2^{\nu_2}-1]}(\varphi_1)||_p =$$
$$= ||V^{(\alpha_1,0)}_{2^{n_1},\infty}(f) - V^{(\alpha_1,0)}_{[2^{\nu_1}-1],\infty}(f) - V^{(\alpha_1,0)}_{2^{n_1},[2^{\nu_2}-1]}(f) + V^{(\alpha_1,0)}_{[2^{\nu_1}-1],[2^{\nu_2}-1]}(f)||_p =$$
$$= ||V^{(\alpha_1,0)}_{2^{n_1},\infty}(f - V_{\infty[2^{\nu_2}-1]}(f)) - V^{(\alpha_1,0)}_{[2^{\nu_1}-1],\infty}(f - V_{\infty,[2^{\nu_2}-1]}(f))||_p =$$
$$= \Big\| \sum_{\mu_1=\nu_1}^{n_1} (V^{(\alpha_1,0)}_{2^{\mu_1},\infty}(f - V_{\infty[2^{\nu_2}-1]}(f)) - V^{(\alpha_1,0)}_{[2^{\mu_1}-1],\infty}(f - V_{\infty,[2^{\nu_2}-1]}(f)))\Big\|_p \ll$$
$$\ll \sum_{\mu_1=\nu_1}^{n_1} \Big\|\Big[V_{2^{\mu_1},\infty}(f - V_{\infty,[2^{\nu_2}-1]}(f)) - V_{[2^{\mu_1}-1],\infty}(f - V_{\infty,[2^{\nu_2}-1]}(f))\Big]^{(\alpha_1,0)}\Big\|_p.$$

Using lemmas 1.5 a) and 1.2 a), we get

$$I_5 \ll \sum_{\mu_1=\nu_1}^{n_1} 2^{\mu_1\alpha_1}\Big\|V_{2^{\mu_1},\infty}(f - V_{\infty,[2^{\nu_2}-1]}(f)) - V_{[2^{\mu_1}-1],\infty}(f - V_{\infty,[2^{\nu_2}-1]}(f))\Big\|_p \ll$$
$$\ll \sum_{\mu_1=\nu_1}^{n_1} 2^{\mu_1\alpha_1}\Big\|f - V_{\infty,[2^{\nu_2}-1]}(f) - V_{2^{\mu_1},\infty}(f) + V_{2^{\mu_1},[2^{\nu_2}-1]}(f))\Big\|_p +$$
$$+ \sum_{\mu_1=\nu_1}^{n_1} 2^{\mu_1\alpha_1}\Big\|f - V_{\infty,[2^{\nu_2}-1]}(f) - V_{[2^{\mu_1}-1],\infty}(f) + V_{[2^{\mu_1}-1],[2^{\nu_2}-1]}(f))\Big\|_p \ll$$
$$\ll \sum_{\mu_1=\nu_1}^{n_1} 2^{\mu_1\alpha_1} Y_{[2^{\mu_1}-1],[2^{\nu_2}-1]}(f)_p.$$

Using this estimate, we have

$$I_2^{q^*} \ll \sum_{\nu_1=0}^{n_1}\sum_{\nu_2=n_2}^{\infty} 2^{(\nu_1+\nu_2)\left(\frac{1}{p}-\frac{1}{q}\right)q^*}\Big(\sum_{\mu_1=\nu_1}^{n_1} 2^{\mu_1\alpha_1} Y_{[2^{\mu_1}-1],[2^{\nu_2}-1]}(f)_p\Big)^{q^*}.$$

Using Lemma 1.6, we get

$$I_2^{q^*} \ll \sum_{\nu_1=0}^{n_1} \sum_{\nu_2=n_2}^{\infty} 2^{(\nu_1+\nu_2)\left(\frac{1}{p}-\frac{1}{q}\right)q^*} 2^{\nu_1\alpha_1 q^*} Y^{q^*}_{[2^{\nu_1}-1][2^{\nu_2}-1]}(f)_p \ll A_2^{q^*}.$$

Similarly, we obtain $I_3^{q^*} \ll A_3^{q^*}$.

Now we shall estimate I_4. Denote $\varphi(x,y) = V^{(\alpha_1,\alpha_2)}_{2^{n_1},2^{n_2}}(f)$. As $f \in L_p^0$, then $V_{0,\infty}(\varphi) = V_{\infty,0}(\varphi) = V_{0,0}(\varphi) = 0$. Therefore

$$I_4 = ||V^{(\alpha_1,\alpha_2)}_{2^{n_1},2^{n_2}}(f)||_q = ||\varphi - V_{0,\infty}(\varphi) - V_{\infty,0}(\varphi) + V_{0,0}(\varphi)||_q \ll Y_{0,0}(\varphi)_q.$$

Using Lemma 1.3, we have

$$I_4 \ll \left(\sum_{\nu_1=0}^{\infty} \sum_{\nu_2=0}^{\infty} 2^{(\nu_1+\nu_2)\left(\frac{1}{p}-\frac{1}{q}\right)q^*} Y^{q^*}_{2^{\nu_1}-1,2^{\nu_2}-1}(\varphi)_p \right)^{\frac{1}{q^*}}.$$

If $\nu_1 \geq n_1+1$ and $\nu_2 \geq n_2+1$ that, taking into account, that $\varphi = V^{(\alpha_1,\alpha_2)}_{2^{n_1},2^{n_2}}(f)$, we get, that $Y_{2^{\nu_1}-1,2^{\nu_2}-1}(\varphi)_p = 0$. Therefore

$$I_4^{q^*} \ll \sum_{\nu_1=0}^{n_1} \sum_{\nu_2=0}^{n_2} 2^{(\nu_1+\nu_2)\left(\frac{1}{p}-\frac{1}{q}\right)q^*} Y^{q^*}_{2^{\nu_1}-1,2^{\nu_2}-1}(\varphi)_p.$$

Using definition of "approximation by an angle", for $\nu_1 \leq n_1$ and $\nu_2 \leq n_2$, we have

$$\begin{aligned} I_6 \equiv Y_{2^{\nu_1}-1,2^{\nu_2}-1}(\varphi)_p &\leq ||\varphi - V_{[2^{\nu_1}-1],\infty}(\varphi) - V_{\infty,[2^{\nu_2}-1]}(\varphi) + V_{[2^{\nu_1}-1],[2^{\nu_2}-1]}(\varphi)||_p \\ &= ||V^{(\alpha_1,\alpha_2)}_{2^{n_1},2^{n_2}}(f) - V^{(\alpha_1,\alpha_2)}_{[2^{\nu_1}-1],2^{n_2}}(f) - V^{(\alpha_1,\alpha_2)}_{2^{n_1},[2^{\nu_2}-1]}(f) + V^{(\alpha_1,\alpha_2)}_{[2^{\nu_1}-1],[2^{\nu_2}-1]}(f)||_p \end{aligned}$$

$$= \left\| \sum_{\mu_1=\nu_1}^{n_1} \sum_{\mu_2=\nu_2}^{n_2} \left(V^{(\alpha_1,\alpha_2)}_{2^{\mu_1},2^{\mu_2}}(f) - V^{(\alpha_1,\alpha_2)}_{[2^{\mu_1}-1],2^{\mu_2}}(f) - V^{(\alpha_1,\alpha_2)}_{2^{\mu_1},[2^{\mu_2}-1]}(f) + V^{(\alpha_1,\alpha_2)}_{[2^{\mu_1}-1],[2^{\mu_2}-1]}(f) \right) \right\|_p$$

$$\begin{aligned} \ll \sum_{\mu_1=\nu_1}^{n_1} \sum_{\mu_2=\nu_2}^{n_2} \Big\| \Big[& V_{2^{\mu_1},2^{\mu_2}}(f) - V_{[2^{\mu_1}-1]2^{\mu_2}}(f) - V_{2^{\mu_1},[2^{\mu_2}-1]}(f) \\ & + V_{[2^{\mu_1}-1][2^{\mu_2}-1]}(f)) \Big]^{(\alpha,\alpha_2)} \Big\|_p. \end{aligned}$$

Using lemmas 1.5 c) and 1.2a), we get

$$\begin{aligned} I_6 \ll \sum_{\mu_1=\nu_1}^{n_1} \sum_{\mu_2=\nu_2}^{n_2} 2^{\mu_1\alpha_1+\mu_2\alpha_2} \Big\| & V_{2^{\mu_1},2^{\mu_2}}(f) - V_{[2^{\mu_1}-1],2^{\mu_2}}(f) - V_{2^{\mu_1},[2^{\mu_2}-1]}(f) \\ & + V_{[2^{\mu_1}-1],[2^{\mu_2}-1]}(f)) \Big\|_p \end{aligned}$$

$$\ll \sum_{\mu_1=\nu_1}^{n_1} \sum_{\mu_2=\nu_2}^{n_2} 2^{\mu_1\alpha_1+\mu_2\alpha_2}\Big\{\Big\| f - V_{2^{\mu_1},\infty}(f) - V_{\infty,2^{\mu_2}}(f) + V_{2^{\mu_1},2^{\mu_2}}(f)\Big\|_p$$

$$+\Big\| f - V_{[2^{\mu_1-1}],\infty}(f) - V_{\infty,2^{\mu_2}}(f) + V_{[2^{\mu_1-1}],2^{\mu_2}}(f)\Big\|_p$$

$$+\Big\| f - V_{2^{\mu_1},\infty}(f) - V_{\infty,[2^{\mu_2-1}]}(f) + V_{2^{\mu_1},[2^{\mu_2-1}]}(f)\Big\|_p$$

$$+\Big\| f - V_{[2^{\mu_1-1}],\infty}(f) - V_{\infty,[2^{\mu_2-1}]}(f) + V_{[2^{\mu_1-1}],[2^{\mu_2-1}]}(f)\Big\|_p\Big\}$$

$$\ll \sum_{\mu_1=\nu_1}^{n_1} \sum_{\mu_2=\nu_2}^{n_2} 2^{\mu_1\alpha_1+\mu_2\alpha_2} Y_{[2^{\mu_1-1}],[2^{\mu_2-1}]}(f)_p.$$

Using this estimate, we have

$$I_4^{q^*} \ll \sum_{\nu_1=0}^{n_1} \sum_{\nu_2=0}^{n_2} 2^{(\nu_1+\nu_2)\left(\frac{1}{p}-\frac{1}{q}\right)q^*}\Big(\sum_{\mu_1=\nu_1}^{n_1} \sum_{\mu_2=\nu_2}^{n_2} 2^{(\mu_1\alpha_1+\mu_2\alpha_2)} Y_{[2^{\mu_1-1}],[2^{\mu_2-1}]}(f)_p\Big)^{q^*}.$$

Using Lemma 1.6, we have

$$I_4^{q^*} \ll \sum_{\nu_1=0}^{n_1} \sum_{\nu_2=n_2}^{\infty} 2^{(\nu_1+\nu_2)\left(\frac{1}{p}-\frac{1}{q}\right)q^*} 2^{(\nu_1\alpha_1+\nu_2\alpha_2)q^*} Y^{q^*}_{[2^{\nu_1-1}],[2^{\nu_2-1}]}(f)_p \ll A_4^{q^*}.$$

Finally, combining estimates for I_1, I_2, I_3 and I_4, we conclude the proof of Theorem 2.1. □

Theorem 2.2 *Let* $f \in L_p^0, 1 \le p < q \le \infty, \alpha_i > 0, \delta_i \in (0,1), i = 1,2, \rho_j > 0, j = \overline{1,8}$. *Then*

$$\omega_{\alpha_1,\alpha_2}\big(f,\delta_1,\delta_2\big)_q \ll \Big(\int_0^{\delta_1}\int_0^{\delta_2} \Big[(t_1t_2)^{-\frac{1}{p}+\frac{1}{q}}\omega_{\rho_1,\rho_2}(f,t_1,t_2)_p\Big]^{q^*}\frac{dt_1}{t_1}\frac{dt_2}{t_2}\Big)^{\frac{1}{q^*}}$$

$$+\delta_1^{\alpha_1}\Big(\int_{\delta_1}^{1}\int_0^{\delta_2} \Big[t_1^{-\left(\alpha_1+\frac{1}{p}-\frac{1}{q}\right)} t_2^{-\frac{1}{p}+\frac{1}{q}}\omega_{\rho_3,\rho_4}(f,t_1,t_2)_p\Big]^{q^*}\frac{dt_1}{t_1}\frac{dt_2}{t_2}\Big)^{\frac{1}{q^*}}$$

$$+\delta_2^{\alpha_2}\Big(\int_0^{\delta_1}\int_{\delta_2}^{1} \Big[t_1^{-\frac{1}{p}+\frac{1}{q}} t_2^{-\left(\alpha_2+\frac{1}{p}-\frac{1}{q}\right)}\omega_{\rho_5,\rho_6}(f,t_1,t_2)_p\Big]^{q^*}\frac{dt_1}{t_1}\frac{dt_2}{t_2}\Big)^{\frac{1}{q^*}}$$

$$+\delta_1^{\alpha_1}\delta_2^{\alpha_2}\Big(\int_{\delta_1}^{1}\int_{\delta_2}^{1} \Big[t_1^{-\left(\alpha_1+\frac{1}{p}-\frac{1}{q}\right)} t_2^{-\left(\alpha_2+\frac{1}{p}-\frac{1}{q}\right)}\omega_{\rho_7,\rho_8}(f,t_1,t_2)_p\Big]^{q^*}\frac{dt_1}{t_1}\frac{dt_2}{t_2}\Big)^{\frac{1}{q^*}}$$

$$\equiv B_1 + B_2 + B_3 + B_4.$$

Proof For everyone $\delta_i \in (0,1)$ there exists an integer non-negative number n_i such that $\frac{1}{2^{n_i+1}} \le \delta_i < \frac{1}{2^{n_i}}, i = 1, 2$. Using Lemma 1.4, we have

$$I = \omega_{\alpha_1,\alpha_2}\big(f,\delta_1,\delta_2\big)_q \ll \omega_{\alpha_1,\alpha_2}\Big(f,\frac{1}{2^{n_1}},\frac{1}{2^{n_2}}\Big)_q.$$

Using Theorem 2.1, and then further Lemma 1.4 (property f), we get

$$I \equiv \omega_{\alpha_1,\alpha_2}\Big(f,\frac{1}{2^{n_1}},\frac{1}{2^{n_2}}\Big)_q \ll \Big(\sum_{\nu_1=n_1}^{\infty}\sum_{\nu_2=n_2}^{\infty}\Big[2^{(\nu_1+\nu_2)\left(\frac{1}{p}-\frac{1}{q}\right)}\omega_{\rho_1,\rho_2}\Big(f,\frac{1}{[2^{\nu_1-1}]+1},\frac{1}{[2^{\nu_2-1}]+1}\Big)_p\Big]^{q^*}\Big)^{\frac{1}{q^*}}+$$

$$+\frac{1}{2^{n_1\alpha_1}}\Big(\sum_{\nu_1=0}^{n_1}\sum_{\nu_2=n_2}^{\infty}\Big[2^{\nu_1\left(\alpha_1+\frac{1}{p}-\frac{1}{q}\right)}2^{\nu_2\left(\frac{1}{p}-\frac{1}{q}\right)}\omega_{\rho_3,\rho_4}\Big(f,\frac{1}{[2^{\nu_1-1}]+1},\frac{1}{[2^{\nu_2-1}]+1}\Big)_p\Big]^{q^*}\Big)^{\frac{1}{q^*}}+$$

$$+\frac{1}{2^{n_2\alpha_2}}\Big(\sum_{\nu_1=n_1}^{\infty}\sum_{\nu_2=0}^{n_2}\Big[2^{\nu_1\left(\frac{1}{p}-\frac{1}{q}\right)}2^{\nu_2\left(\alpha_2+\frac{1}{p}-\frac{1}{q}\right)}\omega_{\rho_5,\rho_6}\Big(f,\frac{1}{[2^{\nu_1-1}]+1},\frac{1}{[2^{\nu_2-1}]+1}\Big)_p\Big]^{q^*}\Big)^{\frac{1}{q^*}}+$$

$$+\frac{1}{2^{n_1\alpha_1}}\frac{1}{2^{n_2\alpha_2}}\Big(\sum_{\nu_1=0}^{n_1}\sum_{\nu_2=0}^{n_2}\Big[2^{\nu_1\left(\alpha_1+\frac{1}{p}-\frac{1}{q}\right)}2^{\nu_2\left(\alpha_2+\frac{1}{p}-\frac{1}{q}\right)}\omega_{\rho_7,\rho_8}\Big(f,\frac{1}{[2^{\nu_1-1}]+1},\frac{1}{[2^{\nu_2-1}]+1}\Big)_p\Big]^{q^*}\Big)^{\frac{1}{q^*}}.$$

Using properties of the mixed module of smoothness (Lemma 1.4) and a fact that $2^{-\nu\eta} \asymp \int_{\frac{1}{2^{\nu+1}}}^{\frac{1}{2^{\nu}}} t^{\eta}\frac{dt}{t}$, we get $I \ll B_1 + B_2 + B_3 + B_4$. Thus, the proof of Theorem 2.2 is complete. □

Let us note a special case of Theorem 2.2.

Theorem 2.3 *Let* $f \in L_p^0$, $1 \le p < q \le \infty$, $\beta_i > \alpha_i > 0$, $\delta_i \in (0,1)$, $i = 1, 2$. *Then*

$$I = \omega_{\alpha_1,\alpha_2}\big(f,\delta_1,\delta_2\big)_q \ll \Big(\int_0^{\delta_1}\int_0^{\delta_2}\Big[(t_1t_2)^{-\frac{1}{p}+\frac{1}{q}}\omega_{\alpha_1+\frac{1}{p}-\frac{1}{q},\alpha_2+\frac{1}{p}-\frac{1}{q}}(f,t_1,t_2)_p\Big]^{q^*}\frac{dt_1}{t_1}\frac{dt_2}{t_2}\Big)^{\frac{1}{q^*}}+$$

$$+\delta_1^{\alpha_1}\Big(\int_{\delta_1}^{1}\int_0^{\delta_2}\Big[t_1^{-\left(\alpha_1+\frac{1}{p}-\frac{1}{q}\right)}t_2^{-\frac{1}{p}+\frac{1}{q}}\omega_{\beta_1+\frac{1}{p}-\frac{1}{q},\alpha_2+\frac{1}{p}-\frac{1}{q}}(f,t_1,t_2)_p\Big]^{q^*}\frac{dt_1}{t_1}\frac{dt_2}{t_2}\Big)^{\frac{1}{q^*}}$$

$$+\delta_2^{\alpha_2}\Big(\int_0^{\delta_1}\int_{\delta_2}^{1}\Big[t_1^{-\frac{1}{p}+\frac{1}{q}}t_2^{-\left(\alpha_2+\frac{1}{p}-\frac{1}{q}\right)}\omega_{\alpha_1+\frac{1}{p}-\frac{1}{q},\beta_2+\frac{1}{p}-\frac{1}{q}}(f,t_1,t_2)_p\Big]^{q^*}\frac{dt_1}{t_1}\frac{dt_2}{t_2}\Big)^{\frac{1}{q^*}}$$

$$+\delta_1^{\alpha_1}\delta_2^{\alpha_2}\Big(\int_{\delta_1}^{1}\int_{\delta_2}^{1}\Big[t_1^{-\left(\alpha_1+\frac{1}{p}-\frac{1}{q}\right)}t_2^{-\left(\alpha_2+\frac{1}{p}-\frac{1}{q}\right)}\omega_{\beta_1+\frac{1}{p}-\frac{1}{q},\beta_2+\frac{1}{p}-\frac{1}{q}}(f,t_1,t_2)_p\Big]^{q^*}\frac{dt_1}{t_1}\frac{dt_2}{t_2}\Big)^{\frac{1}{q^*}} \tag{1}$$

$$\equiv A_1(f,\delta_1,\delta_2) + A_2(f,\delta_1,\delta_2) + A_3(f,\delta_1,\delta_2) + A_4(f,\delta_1,\delta_2).$$

The Theorem 2.3 is sharp in the sense that there is a function $f_0(x,y)$ such that the symbol $\ll$ in (1) can be replaced by $\asymp$ for $f = f_0$.

Let us prove this fact. We shall consider the function $f_0(x,y) = \sin x \sin y$. For everyone $\delta_i \in (0,1)$ there exists integer non-negative number n_i such, that $\frac{1}{2^{n_i+1}} \le \delta_i < \frac{1}{2^{n_i}}, i = 1,2$. Using lemmas 1.1 and 1.4, for any $r_i > 0, i = 1,2$, and any $p \in [1,\infty]$ we have

$$\omega_{r_1,r_2}\big(f_0,\delta_1,\delta_2\big)_p \asymp \omega_{r_1,r_2}\Big(f_0,\frac{1}{2^{n_1}},\frac{1}{2^{n_2}}\Big)_p \asymp \frac{1}{2^{n_1 r_1 + n_2 r_2}} \asymp \delta_1^{r_1}\delta_2^{r_2}.$$

But then

$$A_0(f_0,\delta_1,\delta_2) \equiv \omega_{\alpha_1,\alpha_2}\big(f_0,\delta_1,\delta_2\big)_q \asymp \delta_1^{\alpha_1}\delta_2^{\alpha_2}, A_1(f_0,\delta_1,\delta_2) \asymp \delta_1^{\alpha_1}\delta_2^{\alpha_2},$$
$$A_2(f_0,\delta_1,\delta_2) \asymp \delta_1^{\alpha_1}\delta_2^{\alpha_2}, A_3(f_0,\delta_1,\delta_2) \asymp \delta_1^{\alpha_1}\delta_2^{\alpha_2}, A_4(f_0,\delta_1,\delta_2) \asymp \delta_1^{\alpha_1}\delta_2^{\alpha_2}.$$

From this estimates it follows that

$$A_0(f_0,\delta_1,\delta_2) \asymp A_1(f_0,\delta_1,\delta_2) + A_2(f_0,\delta_1,\delta_2) + A_3(f_0,\delta_1,\delta_2) + A_4(f_0,\delta_1,\delta_2).$$

This means that for the function $f_0(x,y)$ symbol $\ll$ can be replaced by $\asymp$ in (1).

Note that for some p and q the second, third and fourth terms in the right-hand side part of inequality (1) in the Theorem 2.3 can be omitted. It follows from the theorems stated below.

3 Analogues of Ul'ynov Inequalities for Mixed Moduli of Smoothness

Theorem 3.1 *Let $f \in L_p^0$, where $\alpha_i > 0, \delta_i \in (0,1), i = 1,2$, and $1 = p < q = \infty$. Then*

$$\omega_{\alpha_1,\alpha_2}\big(f,\delta_1,\delta_2\big)_q \ll \Big(\int_0^{\delta_1}\int_0^{\delta_2}\Big[(t_1 t_2)^{-\frac{1}{p}+\frac{1}{q}}\omega_{\alpha_1+\frac{1}{p}-\frac{1}{q},\alpha_2+\frac{1}{p}-\frac{1}{q}}(f,t_1,t_2)_p\Big]^{q^*}\frac{dt_1}{t_1}\frac{dt_2}{t_2}\Big)^{\frac{1}{q^*}}. \tag{2}$$

The Theorem 3.1 is sharp in the sense that there is a function $f_0(x,y)$ such that symbol $\ll$ in (2) can be replace by $\asymp$ for this function.

Sharp Ulyanov inequality (2) in the case $1 < p < q < \infty$ was proved in [9] and [7]. Here we study the case when $1 = p < q = \infty$. Similar theorem is proved in [11] for functions of one variable in the case of $1 = p < q = \infty$.

Proof For everyone $\delta_i \in (0,1)$ there exists an integer non-negative number n_i such, that $\frac{1}{2^{n_i+1}} \le \delta_i < \frac{1}{2^{n_i}}$, $i = 1, 2$. Therefore

$$I = \omega_{\alpha_1,\alpha_2}\big(f, \delta_1, \delta_2\big)_\infty \ll \omega_{\alpha_1,\alpha_2}\Big(f_0, \frac{1}{2^{n_1}}, \frac{1}{2^{n_2}}\Big)_\infty.$$

Using Lemma 1.1, we have

$$I \ll I_1 + 2^{-n_1\alpha_1} I_2 + 2^{-n_2\alpha_2} I_3 + 2^{-n_1\alpha_1 - n_2\alpha_2} I_4.$$

Using lemmas 1.2 a), 1.3 a), and 1.4 f), we get

$$I_1 \ll Y_{2^{n_1},2^{n_2}}(f)_\infty \ll \sum_{\nu_1=n_1}^{\infty} \sum_{\nu_2=n_2}^{\infty} 2^{\nu_1+\nu_2} Y_{2^{\nu_1},2^{\nu_2}}(f)_1$$

$$\ll \sum_{\nu_1=n_1}^{\infty} \sum_{\nu_2=n_2}^{\infty} 2^{\nu_1+\nu_2} \omega_{\alpha_1+1,\alpha_2+1}\Big(f, \frac{1}{2^{\nu_1}}, \frac{1}{2^{\nu_2}}\Big)_1.$$

Now we shall estimate $I_2 = \big\| V^{(\alpha_1,0)}_{2^{n_1},\infty}\big(f - V_{\infty,2^{n_2}}(f)\big)\big\|_\infty$. Let us denote

$$A \equiv V^{(\alpha_1,0)}_{2^{n_1},\infty}\big(f - V_{\infty,2^{n_2}}(f)\big) = V^{(\alpha_1,0)}_{2^{n_1},\infty}(f) - V^{(\alpha_1,0)}_{2^{n_1},2^{n_2}}(f).$$

For $f \in L_1^0$ and for almost all x and y and any $N_2 > n_2$ we have

$$\sum_{\nu_2=n_2}^{N_2} \big(V^{(\alpha_1,0)}_{2^{n_1},2^{\nu_2+1}}(f) - V^{(\alpha_1,0)}_{2^{n_1},2^{\nu_2}}(f)\big) = V^{(\alpha_1,0)}_{2^{n_1},2^{N_2+1}}(f) - V^{(\alpha_1,0)}_{2^{n_1},2^{n_2}}(f) =$$

$$= \big(V^{(\alpha_1,0)}_{2^{n_1},\infty}(f) - V^{(\alpha_1,0)}_{2^{n_1},2^{n_2}}(f)\big) - \big(V^{(\alpha_1,0)}_{2^{n_1},\infty}(f) - V^{(\alpha_1,0)}_{2^{n_1},2^{N_2+1}}(f)\big) \equiv A - B.$$

This implies that

$$||A||_\infty \le \sum_{\nu_2=n_2}^{N_2} \big\| V^{(\alpha_1,0)}_{2^{n_1},2^{\nu_2+1}}(f) - V^{(\alpha_1,0)}_{2^{n_1},2^{\nu_2}}(f)\big\|_\infty + \|B\|_\infty. \tag{3}$$

Let us estimate $||B||_\infty$. Using Lemma 1.5, we have

$$||B||_\infty \equiv \|V^{(\alpha_1,0)}_{2^{n_1},\infty}(f - V_{\infty,2^{N_2+1}}(f))\|_\infty \ll 2^{n_1\alpha_1} \|V_{2^{n_1},\infty}(f - V_{\infty,2^{N_2+1}}(f))\|_\infty.$$

Using Lemma 1.2 b), we get

$$||B||_\infty \ll 2^{n_1\alpha_1} \|f - V_{\infty,2^{N_2+1}}(f)\|_\infty.$$

Since $f \in L_1^0$, then $V_{0\infty}(f) = V_{02^{N_2+1}}(f) = 0$. Therefore, using Lemma 1.2 a), we have

$$||B||_\infty \ll 2^{n_1\alpha_1}\|f - V_{0,\infty}(f) - V_{\infty,2^{N_2+1}}(f) + V_{0,2^{N_2+1}}(f)\|_\infty \ll 2^{n_1\alpha_1}Y_{0,2^{N_2+1}}(f)_\infty.$$

Using Lemma 1.3, we have

$$||B||_\infty \ll 2^{n_1\alpha_1}\sum_{\nu_1=0}^{\infty}\sum_{\nu_2=N_2+1}^{\infty} 2^{\nu_1+\nu_2}Y_{2^{\nu_1}-1,2^{\nu_2}-1}(f)_1.$$

Using Lemma 1.4 f), we get

$$||B||_\infty \ll 2^{n_1\alpha_1}\sum_{\nu_1=0}^{\infty}\sum_{\nu_2=N_2+1}^{\infty} 2^{\nu_1+\nu_2}\omega_{\alpha_1+1,\alpha_2+1}\left(f, \frac{1}{2^{\nu_1}}, \frac{1}{2^{\nu_2}}\right)_1 \ll$$

$$\ll 2^{n_1\alpha_1}\int_0^1\int_0^{\frac{1}{2^{N_2+1}}}(t_1t_2)^{-1}\omega_{\alpha_1+1,\alpha_2+1}(f, t_1, t_2)_1\frac{dt_1}{t_1}\frac{dt_2}{t_2}.$$

This implies that $||B||_\infty \to 0$ while $N_2 \to \infty$. Since $||B||_\infty \to 0$ while $N_2 \to \infty$ then implies the following inequality

$$||A||_\infty \le \sum_{\nu_2=n_2}^{\infty}\left\|V_{2^{n_1},2^{\nu_2+1}}^{(\alpha_1,0)}(f) - V_{2^{n_1},2^{\nu_2}}^{(\alpha_1,0)}(f)\right\|_\infty.$$

Using Lemma 1.8 a), we get

$$||A||_\infty \ll \sum_{\nu_2=n_2}^{\infty} 2^{\nu_2}\left\|V_{2^{n_1},2^{\nu_2+1}}^{(\alpha_1+1,0)}(f) - V_{2^{n_1},2^{\nu_2}}^{(\alpha_1+1,0)}(f)\right\|_1 \le$$

$$\le \sum_{\nu_2=n_2}^{\infty} 2^{\nu_2}\left\|V_{2^{n_1},\infty}^{(\alpha_1+1,0)}(f) - V_{2^{n_1},\infty}^{(\alpha_1+1,0)}(V_{\infty,2^{\nu_2}}(f))\right\|_1 +$$

$$+ \sum_{\nu_2=n_2}^{\infty} 2^{\nu_2}\left\|V_{2^{n_1},\infty}^{(\alpha_1+1,0)}(f) - V_{2^{n_1},\infty}^{(\alpha_1+1,0)}(V_{\infty,2^{\nu_2+1}}(f))\right\|_1 \ll$$

$$\ll \sum_{\nu_2=n_2}^{\infty} 2^{\nu_2}\left\|V_{2^{n_1},\infty}^{(\alpha_1+1,0)}(f - V_{\infty,2^{\nu_2}}(f))\right\|_1 + \sum_{\nu_2=n_2}^{\infty} 2^{\nu_2}\left\|V_{2^{n_1},\infty}^{(\alpha_1+1,0)}(f - V_{\infty,2^{\nu_2+1}}(f))\right\|_1.$$

Using Lemma 1.1, we have

$$||A||_\infty \ll \sum_{\nu_2=n_2}^{\infty} 2^{\nu_2} 2^{n_1(\alpha_1+1)} \omega_{\alpha_1+1,\alpha_2+1}\left(f, \frac{1}{2^{n_1}}, \frac{1}{2^{\nu_2}}\right)_1,$$

i.e.

$$I_2 \ll \sum_{\nu_2=n_2}^{\infty} 2^{\nu_2} 2^{n_1} \omega_{\alpha_1+1,\alpha_2+1}\left(f, \frac{1}{2^{n_1}}, \frac{1}{2^{\nu_2}}\right)_1 2^{n_1\alpha_1}.$$

Similarly, we get an estimate

$$I_3 \ll \sum_{\nu_1=n_1}^{\infty} 2^{\nu_1} 2^{n_2} \omega_{\alpha_1+1,\alpha_2+1}\left(f, \frac{1}{2^{\nu_1}}, \frac{1}{2^{n_2}}\right)_1 2^{n_2\alpha_2}.$$

Let us estimate I_4. Using Lemma 1.8 c), we have

$$I_4 \ll \left\| V_{2^{n_1},2^{n_2}}^{(\alpha_1+1,\alpha_2+1)}(f) \right\|_1.$$

Using Lemma 1.1, we get

$$I_4 \ll 2^{n_1(\alpha_1+1)+n_2(\alpha_2+1)} \omega_{\alpha_1+1,\alpha_2+1}\left(f, \frac{1}{2^{n_1}}, \frac{1}{2^{n_2}}\right)_1.$$

Finally, combining estimates for I_1, I_2, I_3 and I_4, we have

$$I \ll \sum_{\nu_1=n_1}^{\infty} \sum_{\nu_2=n_2}^{\infty} 2^{\nu_1+\nu_2} \omega_{\alpha_1+1,\alpha_2+1}\left(f, \frac{1}{2^{\nu_1}}, \frac{1}{2^{\nu_2}}\right)_1$$

$$\ll \int_0^{\delta_1}\int_0^{\delta_2} (t_1t_2)^{-1} \omega_{\alpha_1+1,\alpha_2+1}(f, t_1, t_2)_1 \frac{dt_1}{t_1}\frac{dt_2}{t_2},$$

i.e., the proof of Theorem 3.1 is complete for $1 = p < q = \infty$. □

Theorem 3.2 *Let $f \in L_p^0$, where $\alpha_i > \gamma_i > 0, \delta_i \in (0,1), i = 1,2$, and $1 = p < q < \infty$ or $1 < p < q = \infty$. Then*

$$\omega_{\alpha_1,\alpha_2}(f, \delta_1, \delta_2)_q \ll \left(\int_0^{\delta_1}\int_0^{\delta_2} \left[(t_1t_2)^{-\frac{1}{p}+\frac{1}{q}} \omega_{\gamma_1+\frac{1}{p}-\frac{1}{q},\gamma_2+\frac{1}{p}-\frac{1}{q}}(f, t_1, t_2)_p\right]^{q^*} \frac{dt_1}{t_1}\frac{dt_2}{t_2}\right)^{\frac{1}{q^*}}. \tag{4}$$

The Theorem 3.2 cannot be improved in the sense that if we replace even one of γ_i by α_i in the right-hand side part of inequality (4), then the received inequality will be false.

Remark that similar results for functions of one variable where proved in [13] for $1 \le p < q < \infty$ and in [6] for $1 < p < q = \infty$.

Proof Let us take $\rho_1 = \rho_3 = \rho_5 = \rho_7 = \gamma_1 + \frac{1}{p} - \frac{1}{q}$, $\rho_2 = \rho_4 = \rho_6 = \rho_8 = \gamma_2 + \frac{1}{p} - \frac{1}{q}$ in Theorem 2.2. Then we have

$$I = \omega_{\alpha_1,\alpha_2}(f,\delta_1,\delta_2)_q \ll \left(\int_0^{\delta_1}\int_0^{\delta_2}\left[(t_1t_2)^{-\frac{1}{p}+\frac{1}{q}}\omega_{\gamma_1+\frac{1}{p}-\frac{1}{q},\gamma_2+\frac{1}{p}-\frac{1}{q}}(f,t_1,t_2)_p\right]^{q^*}\frac{dt_1}{t_1}\frac{dt_2}{t_2}\right)^{\frac{1}{q^*}}$$

$$+\delta_1^{\alpha_1}\left(\int_{\delta_1}^{1}\int_0^{\delta_2}\left[t_1^{-\left(\alpha_1+\frac{1}{p}-\frac{1}{q}\right)}t_2^{-\frac{1}{p}+\frac{1}{q}}\omega_{\gamma_1+\frac{1}{p}-\frac{1}{q},\gamma_2+\frac{1}{p}-\frac{1}{q}}(f,t_1,t_2)_p\right]^{q^*}\frac{dt_1}{t_1}\frac{dt_2}{t_2}\right)^{\frac{1}{q^*}}$$

$$+\delta_2^{\alpha_2}\left(\int_0^{\delta_1}\int_{\delta_2}^{1}\left[t_1^{-\frac{1}{p}+\frac{1}{q}}t_2^{-\left(\alpha_2+\frac{1}{p}-\frac{1}{q}\right)}\omega_{\gamma_1+\frac{1}{p}-\frac{1}{q},\gamma_2+\frac{1}{p}-\frac{1}{q}}(f,t_1,t_2)_p\right]^{q^*}\frac{dt_1}{t_1}\frac{dt_2}{t_2}\right)^{\frac{1}{q^*}}$$

$$+\delta_1^{\alpha_1}\delta_2^{\alpha_2}\left(\int_{\delta_1}^{1}\int_{\delta_2}^{1}\left[t_1^{-\left(\alpha_1+\frac{1}{p}-\frac{1}{q}\right)}t_2^{-\left(\alpha_2+\frac{1}{p}-\frac{1}{q}\right)}\omega_{\gamma_1+\frac{1}{p}-\frac{1}{q},\gamma_2+\frac{1}{p}-\frac{1}{q}}(f,t_1,t_2)_p\right]^{q^*}\frac{dt_1}{t_1}\frac{dt_2}{t_2}\right)^{\frac{1}{q^*}}$$

$$\equiv D_1 + D_2 + D_3 + D_4.$$

Using Lemma 1.4 (item d) and taking into account, that $\gamma_i < \alpha_i$, we get

$$D_2 = \delta_1^{\alpha_1}\left(\int_{\delta_1}^{1}\int_0^{\delta_2}\left[\frac{\omega_{\gamma_1+\frac{1}{p}-\frac{1}{q},\gamma_2+\frac{1}{p}-\frac{1}{q}}(f,t_1,t_2)_p}{t_1^{\left(\gamma_1+\frac{1}{p}-\frac{1}{q}\right)}t_2^{\left(\gamma_2+\frac{1}{p}-\frac{1}{q}\right)}}t_1^{\gamma_1-\alpha_1}t_2^{\gamma_2}\right]^{q^*}\frac{dt_1}{t_1}\frac{dt_2}{t_2}\right)^{\frac{1}{q^*}}$$

$$\ll \delta_1^{\alpha_1}\left(\int_0^{\delta_2}\left[\frac{\omega_{\gamma_1+\frac{1}{p}-\frac{1}{q},\gamma_2+\frac{1}{p}-\frac{1}{q}}(f,\delta_1,t_2)_p}{\delta_1^{\left(\gamma_1+\frac{1}{p}-\frac{1}{q}\right)}t_2^{\left(\gamma_2+\frac{1}{p}-\frac{1}{q}\right)}}\right]^{q^*}t_2^{\gamma_2q^*}\int_{\delta_1}^{1}t_1^{(\gamma_1-\alpha_1)q^*}\frac{dt_1}{t_1}\frac{dt_2}{t_2}\right)^{\frac{1}{q^*}}$$

$$\ll \left(\int_0^{\delta_2}\left[\frac{\omega_{\gamma_1+\frac{1}{p}-\frac{1}{q},\gamma_2+\frac{1}{p}-\frac{1}{q}}(f,\delta_1,t_2)_p}{\delta_1^{\left(\gamma_1+\frac{1}{p}-\frac{1}{q}\right)}t_2^{\left(\gamma_2+\frac{1}{p}-\frac{1}{q}\right)}}\right]^{q^*}t_2^{\gamma_2q^*}\delta_1^{\gamma_1q^*}\frac{dt_2}{t_2}\right)^{\frac{1}{q^*}}$$

$$\ll \left(\int_0^{\delta_2}\left[\frac{\omega_{\gamma_1+\frac{1}{p}-\frac{1}{q},\gamma_2+\frac{1}{p}-\frac{1}{q}}(f,\delta_1,t_2)_p}{\delta_1^{\left(\gamma_1+\frac{1}{p}-\frac{1}{q}\right)}t_2^{\left(\gamma_2+\frac{1}{p}-\frac{1}{q}\right)}}\right]^{q^*}t_2^{\gamma_2q^*}\int_0^{\delta_1}t_1^{\gamma_1q^*}\frac{dt_1}{t_1}\frac{dt_2}{t_2}\right)^{\frac{1}{q^*}}$$

$$\ll \left(\int_0^{\delta_1} \int_0^{\delta_2} \left[\frac{\omega_{\gamma_1+\frac{1}{p}-\frac{1}{q},\gamma_2+\frac{1}{p}-\frac{1}{q}}(f,t_1,t_2)_p}{t_1^{\left(\gamma_1+\frac{1}{p}-\frac{1}{q}\right)} t_2^{\left(\gamma_2+\frac{1}{p}-\frac{1}{q}\right)}} \right]^{q^*} t_2^{\gamma_2 q^*} t_1^{\gamma_1 q^*} \frac{dt_1}{t_1} \frac{dt_2}{t_2} \right)^{\frac{1}{q^*}}$$

$$= \left(\int_0^{\delta_1} \int_0^{\delta_2} \left[(t_1 t_2)^{-\frac{1}{p}+\frac{1}{q}} \omega_{\gamma_1+\frac{1}{p}-\frac{1}{q},\gamma_2+\frac{1}{p}-\frac{1}{q}}(f,t_1,t_2)_p \right]^{q^*} \frac{dt_1}{t_1} \frac{dt_2}{t_2} \right)^{\frac{1}{q^*}} \equiv D_1.$$

Similarly we can show, that $D_3 \ll D_1$ and $D_4 \ll D_1$. Thus, we have proved, that $I \ll D_1$. The proof of Theorem 3.2 is complete. □

Let us show now that if we replace even one of γ_i by α_i, the received inequality will be false. For example we show that for any function $f \in L_p^0$ and any $\delta_1 \in (0,1)$, $\delta_2 \in (0,1)$ for $1 = p < q < \infty$, or $1 < p < q = \infty$ it is impossible to find a constant C, independent of f, δ_1 and δ_2, such that inequality

$$\omega_{\alpha_1,\alpha_2}\big(f,\delta_1,\delta_2\big)_q \le C\left(\int_0^{\delta_1} \int_0^{\delta_2} \left[(t_1 t_2)^{-\frac{1}{p}+\frac{1}{q}} \omega_{\alpha_1+\frac{1}{p}-\frac{1}{q},\gamma_2+\frac{1}{p}-\frac{1}{q}}(f,t_1,t_2)_p \right]^{q^*} \frac{dt_1}{t_1} \frac{dt_2}{t_2} \right)^{\frac{1}{q^*}} \tag{5}$$

holds true.

First of all, let us consider a case of $1 < p < q = \infty$. Let us define the function $f_1(x,y) = \varphi_1(x)\psi_1(y)$, where

$$\varphi_1(x) = \sum_{k=1}^{\infty} \frac{\cos kx}{k^{\alpha_1+1}\big(\ln(k+1)\big)^{\frac{1}{p}}}, \quad \text{for} \quad \alpha_1 \ne 2l-1, l \in \mathbb{N},$$

$$\varphi_1(x) = \sum_{k=1}^{\infty} \frac{\sin kx}{k^{\alpha_1+1}\big(\ln(k+1)\big)^{\frac{1}{p}}}, \quad \text{for} \quad \alpha_1 = 2l-1, l \in \mathbb{N},$$

$$\psi_1(y) = \sin y.$$

For the function $\varphi_1(x)$ it was proved in [10] that

$$\omega_{\alpha_1}(\varphi_1,\delta_1)_\infty \gg \delta_1^{\alpha_1} \left(\ln \frac{1}{\delta_1} \right)^{1-\frac{1}{p}},$$

$$\int_0^{\delta_1} t_1^{-\frac{1}{p}} \omega_{\alpha_1+\frac{1}{p}}(\varphi_1,t_1)_p \frac{dt_1}{t_1} \ll \delta_1^{\alpha_1} \left(\ln\ln \frac{3}{\delta_1} \right)^{\frac{1}{p}}.$$

It is proved in [12] that for the function $\psi_1(y)$ and any $\beta > 0$, $p \in [1,\infty)$ the following relation holds

$$\omega_\beta(\psi_1,\delta_2)_p \asymp \delta_2^\beta.$$

This implies that

$$\omega_{\alpha_2}(\psi_1, \delta_2)_\infty \gg \delta_2^{\alpha_2},$$

$$\int_0^{\delta_2} t_2^{-\frac{1}{p}} \omega_{\gamma_2+\frac{1}{p}}(\psi_1, t_2)_p \frac{dt_1}{t_1} \ll \delta_2^{\gamma_2}.$$

Since $j_1 = \omega_{\alpha_1,\alpha_2}(\varphi_1, \delta_1, \delta_2)_\infty = \omega_{\alpha_1}(\varphi_1, \delta_1)_\infty \cdot \omega_{\alpha_1}(\psi_1, \delta_2)_\infty$, then $j_1 \gg \delta_1^{\alpha_1}\left(\ln \frac{1}{\delta_1}\right)^{1-\frac{1}{p}} \delta_2^{\alpha_2}$. Since

$$j_2 = \int_0^{\delta_1}\int_0^{\delta_2} (t_1 t_2)^{-\frac{1}{p}} \omega_{\alpha_1+\frac{1}{p}\gamma_2+\frac{1}{p}}(f_1, t_1, t_2)_p \frac{dt_1}{t_1}\frac{dt_2}{t_2} =$$

$$= \int_0^{\delta_1} t_1^{-\frac{1}{p}} \omega_{\alpha_1+\frac{1}{p}}(\varphi_1, t_1)_p \frac{dt_1}{t_1} \cdot \int_0^{\delta_2} t_2^{-\frac{1}{p}} \omega_{\gamma_2+\frac{1}{p}}(\psi_1, t_2)_p \frac{dt_2}{t_2},$$

then $j_2 \ll \delta_1^{\alpha_1}\left(\ln\ln \frac{3}{\delta_1}\right)^{\frac{1}{p}} \delta_2^{\gamma_2}$.

For any fixed $\delta_2 \in (0, 1)$ and sufficiently small δ_1, i.e. for $\delta_1 \in (0, \delta_1^{(0)})$, there exists a constant $C_1 > 0$ dependent only on δ_2 such that the inequality

$$\left(\ln\ln \frac{3}{\delta_1}\right)^{\frac{1}{p}} \le C_1 \delta_2^{\alpha_2-\gamma_2}\left(\ln \frac{1}{\delta_1}\right)^{1-\frac{1}{p}}$$

holds.

This yields that for any fixed $\delta_2 \in (0, 1)$ there is no constant $C_2 > 0$ independent of δ_1 and δ_2 such that $j_1 \le C_2 \cdot j_2$ for any $\delta_1 \in (0, 1)$. Hence inequality (5) does not hold for the function $f_1(x, y)$.

Now we consider a case of $1 = p < q < \infty$. Let us define the function $f_2(x, y) = \varphi_2(x)\psi_2(y)$, where $\varphi_2(x)$ is the function such that $\varphi_2^{(\alpha_2+1-\frac{1}{q})} = \sum\limits_{k=1}^{\infty} a_k \cos kx$, the sequence $\{a_k\}$ is such that $a_k - 2a_{k+1} + a_{k+2} \ge 0$ and $a_k \gg \frac{1}{\left(\ln(k+1)\right)^{\frac{1}{q}}}$ for any $k \in \mathbb{N}$; $\psi_2(y) = \sin y$. The following inequalities were proved in [10] and [12]:

$$\omega_{\alpha_1}(\varphi_2, \delta_1)_q \gg \delta_1^{\alpha_1}\left(\ln\ln \frac{3}{\delta_1}\right)^{\frac{1}{q}},$$

$$\left(\int_0^{\delta_1} \left[t_1^{-1+\frac{1}{q}} \omega_{\alpha_1+1-\frac{1}{q}}(\varphi_2, t_1)_1\right]^q \frac{dt_1}{t_1}\right)^{\frac{1}{q}} \ll \delta_1^{\alpha_1},$$

$$\omega_{\alpha_2}(\psi_2, \delta_2)_q \gg \delta_2^{\alpha_2},$$

$$\left(\int_0^{\delta_2} \left[t_2^{-1+\frac{1}{q}} \omega_{\gamma_2+1-\frac{1}{q}}(\psi_2, t_2)_1 \right]^q \frac{dt_2}{t_2} \right)^{\frac{1}{q}} \ll \delta_2^{\gamma_2}.$$

Since $j_3 = \omega_{\alpha_1\alpha_2}(f_2, \delta_1, \delta_2)_q = \omega_{\alpha_1}(\varphi_2, \delta_1)_q \cdot \omega_{\alpha_2}(\psi_2, \delta_2)_q$, then $j_3 \gg \delta_1^{\alpha_1}\left(\ln\ln \frac{3}{\delta_1} \right)^{\frac{1}{q}} \delta_2^{\alpha_2}$.

Since

$$j_4 = \left(\int_0^{\delta_1}\int_0^{\delta_2} \left[(t_1t_2)^{-1+\frac{1}{q}} \omega_{\alpha_1+1-\frac{1}{q},\gamma_2+1-\frac{1}{q}}(f_2, t_1, t_2)_p \right]^q \frac{dt_1}{t_1}\frac{dt_2}{t_2} \right)^{\frac{1}{q}} =$$

$$= \left(\int_0^{\delta_1} \left[t_1^{-1+\frac{1}{q}} \omega_{\alpha_1+1-\frac{1}{q}}(\varphi_2, t_1)_p \right]^q \frac{dt_1}{t_1} \right)^{\frac{1}{q}} \cdot \left(\int_0^{\delta_2} \left[t_2^{-1+\frac{1}{q}} \omega_{\gamma_2+1-\frac{1}{q}}(\psi_2, t_2)_p \right]^q \frac{dt_2}{t_2} \right)^{\frac{1}{q}},$$

then $j_4 \ll \delta_1^{\alpha_1}\delta_2^{\gamma_2}$.

For any fixed $\delta_2 \in (0,1)$ and sufficiently small δ_1, i.e. for $\delta_1 \in (0, \delta_1^{(0)})$, there exists the constant $C_3 > 0$, dependent only on δ_2, such that inequality

$$\delta_2^{\alpha_2-\gamma_2} \le C_3\left(\ln\ln \frac{3}{\delta_1} \right)^{\frac{1}{q}}$$

holds.

This yields that for any fixed $\delta_2 \in (0,1)$ it is impossible to find constant $C_4 > 0$, independent of δ_1 and δ_2, such that $j_3 \le C_4 \cdot j_4$ for any $\delta_1 \in (0,1)$. Hence inequality (5) does not hold for the function $f_2(x,y)$.

Acknowledgements This research was supported by the RFFI (grant N 16-01-00350) and program of support of leading Scientific schools (grant NSH-3862-2014-1).

References

1. N.K. Bary, *A Treatise on Trigonometric Series* (MacMillan, New York, 1964)
2. O.V. Besov, V.P. Il'in, S.M. Nikol'skii, *Integral Representation of Function and Imbedding Theorems* (Wiley, New York, 1978/1979; translated from Russian: Nauka, Moscow, 1975)
3. M.K. Potapov, *On "angular" Approximation, Proc. Conf. Constructive Functions Theory (Budapest, 1969)* (Acad. Kiado, Budapest, 1971), pp. 371–399
4. M.K. Potapov, About one imbedding theorem. Mathematica (Clui) **14**(37), 123–146 (1972)
5. M.K. Potapov, Imbedding of classes of function with a dominating mixed modulus of smoothness. Trudy Mat. Inst. Steklov. **131**, 199–210 (1974)

6. M.K. Potapov, B.V. Simonov, Interrelations between the moduli of smoothness in metrics L_p and C. Vestn. Mosk. Univ. (to be printed)
7. M.K. Potapov, B.V. Simonov, S. Yu. Tikhonov, Relations between the mixed moduli of smoothness and embedding theorems for Nikol'skii classes. Proc. Steklov Inst. Math. **269**, 197–207 (2010) (translation from Russian: Trudy Matem. Inst. V.A. Steklova **269**, 204–214 (2010))
8. M.K. Potapov, B.V. Simonov, S. Yu. Tikhonov, Constructive characteristics of mixed moduli of smoothness of positive orders, in *Proceedings of the* 8th *Congress of the International Society for Analysis, its Applications, and Computation (22–27 August 2011)*, vol. 2, 314–325 (2012)
9. M.K. Potapov, B.V. Simonov, S. Yu. Tikhonov, Mixed moduli of smoothness in $L_p, 1 < p < \infty$: a survey. Surv. Approx. Theor. **8**, 1–57 (2013)
10. B. Simonov, S. Tikhonov, Sharp Ul'yanov-type inequalities using fractional smoothness. J. Approx. Theor. **162**, 1654–1684 (2010)
11. S. Tikhonov, Weak type inequalities for moduli of smoothness: the case of limit value parameters. J. Fourier Anal. Appl. **16**(4), 590–608 (2010)
12. S. Tikhonov, On moduli of smoothness of fractional order. Real Anal. Exch. **30**(2), 507–518 (2005)
13. S. Tikhonov, W. Trebels, Ulyanov inequalities and generalized Liouville derivatives. Proc. Roy. Soc. Edinburgh Sec. A **141**(1), 205–224 (2011)

Reconstruction Operator of Functions from the Sobolev Space

N.T. Tleukhanova

Abstract For function classes with dominant mixed derivative we study the reconstruction operator of functions by their values at a given number of nodes. We prove that the error by the order coincides with the corresponding orthogonal width.

Keywords Fourier series • Hyperbolic cross • Pitt inequality • Quadrature formulas • Reconstruction operator

Mathematics Subject Classification (2000). Primary 31C15, Secondary 44A35, 46E30

1 Introduction

In this work we study the reconstruction problem of periodic functions from the spaces W_p^α with dominant mixed derivative by values of a function at a given number of nodes. Let (X, Y) be a couple of function spaces of 1-periodic functions and $X \subset Y$. We study the following problem: to find nodes $\{t_k\}_{k=1}^M \subset [0, 1]^n$ and functions $\{\phi_k(x)\}_{k=1}^M$ such that the errors

$$\delta_M(X, Y) = \sup_{\|f\|_X=1} \|f(\cdot) - \sum_{k=1}^{M} f(t_k)\phi_k(\cdot)\|_Y$$

are minimal with respect to the decay order with increasing M. In what follows we consider $X = W_p^\alpha[0, 1]^n$ and $Y = L_p[0, 1]^n$. This problem was considered earlier in [6, 7, 15–17, 20, 23–25]. In particular, in these papers there were considered the reconstruction operators such that the decay orders of $\delta_M(X, Y)$ differ from the decay

N.T. Tleukhanova (✉)
L.N. Gumilyov Eurasian National University, Munatpasova str. 7, 010010 Astana, Kazakhstan
e-mail: tleukhanova@rambler.ru

M. Ruzhansky, S. Tikhonov (eds.), *Methods of Fourier Analysis and Approximation Theory*, Applied and Numerical Harmonic Analysis,
DOI 10.1007/978-3-319-27466-9_12

order of orthogonal diameter given by

$$d_M^{\perp}(X,Y) = \inf_{\{g_j\}_{j=1}^M} \sup_{\|f\|_X=1} \|f - \sum_{j=1}^{M}(f,g_j)g_j\|_Y,$$

where the infimum is taken over all orthogonal systems $\{g_j\}_{j=1}^M$ from $L_\infty[0,1]^n$. The notion of the orthogonal diameter was introduced by Temlyakov in [19], see also [3].

The main goal of this paper is to find the reconstruction operator such that the decay orders of $\delta_M(W_p^\alpha[0,1]^n, L_q[0,1]^n)$ and $d_M^{\perp}(W_p^\alpha[0,1]^n, L_q[0,1]^n)$ coincides, i.e., to find $\{t_k\}_{k=1}^M$ and $\{\phi_k(x)\}_{k=1}^M$ such that

$$\delta_M(W_p^\alpha[0,1]^n, L_q[0,1]^n) \sim d_M^{\perp}(W_p^\alpha[0,1]^n, L_q[0,1]^n),$$

in other words

$$C_1\delta_M(W_p^\alpha[0,1]^n, L_q[0,1]^n) \le d_M^{\perp}(W_p^\alpha[0,1]^n, L_q[0,1]^n) \le C_2\delta_M(W_p^\alpha[0,1]^n, L_q[0,1]^n).$$

Let $1 < p < \infty$, f be an 1-periodic function from $L_p[0,1]^n$, and the series $\sum_{k\in\mathbb{Z}^n}\hat{f}(k)e^{2\pi ikx}$ be the Fourier series of f. We say that $f \in W_p^\alpha[0,1]^n$, $\alpha > 0$, if there exists $f^{(\alpha)} \in L_p[0,1]^n$, the Fourier series of which coincides with the series

$$\sum_{k\in\mathbb{Z}^n} \bar{k}^\alpha \hat{f}(k)e^{2\pi ikx},$$

where $\bar{k} = \prod_{j=1}^{n} \max\{2\pi|k_j|, 2\pi\}$. Moreover,

$$\|f\|_{W_p^\alpha[0,1]^n} := \|f^{(\alpha)}\|_{L_p[0,1]^n}.$$

Furthermore, $\mu x = \sum_{j=1}^n \mu_j x_j, \quad |k| = k_1 + \ldots + k_n$,

$$\rho(\nu) = \{\mu = (\mu_1,\ldots,\mu_n) \in \mathbb{N}^n : \left[2^{\nu_j-2}\right] \le |\mu_j| < 2^{\nu_j-1}\},$$

where $[x]$ is integer part of number x.

Let us define the following transform for a function $f \in C[0,1]^n$:

$$F_m(f;x) = \sum_{\substack{|k|=m \\ k\in\mathbb{N}^n}} \frac{1}{2^m} \sum_{0\le r<2^k} f(\frac{r_1}{2^{k_1}},\ldots,\frac{r_n}{2^{k_n}})\phi_{k,r}(x_1+\frac{r_1}{2^{k_1}},\ldots,x_n+\frac{r_n}{2^{k_n}}), \quad (1)$$

where

$$\phi_{k,r}(x) = \sum_{0\le\nu\le k} (-1)^{\sum_{j=1}^{n-1}(r_j+1)sgn(k_j-\nu_j)} \sum_{\mu\in\rho(\nu)} e^{2\pi i\mu x}. \tag{2}$$

Note that the operator $F_m(f)$ is constructed by values of the function f in M points, where $M \sim 2^m(\ln m)^{n-1}$.

The idea of the construction of the operator F_m is based on the fact that for functions from the spaces with dominating mixed derivative, in particular for $W_p^\alpha[0,1]^n$, the best possible approximation is given by trigonometric polynomials with spectrum from step hyperbolic crosses (see [1, 21, 26]). The operator (1)–(2) is constructed so that it reconstructs trigonometric polynomials with spectrum from step hyperbolic crosses of order m:

$$G_m = \bigcup_{|\nu|\le m} \{\mu\in\mathbb{Z}^n;\ [2^{\nu_j-2}] \le |\mu_j| < 2^{\nu_j-1}\} = \bigcup_{|\nu|\le m} \rho(\nu).$$

Let us formulate the main result of the paper.

Theorem 1 *Let $m \ge n$, $F_m(f)$ be given by (1)–(2), and M be the number of nodes in the definition of $F_m(f,x)$. If $1 < p \le 2 \le q \le \infty$, $\alpha > \frac{1}{p}$, then*

$$\sup_{\|f\|_{W_p^\alpha}=1} \|f - F_m(f)\|_{L_q} \sim d_M^\perp(W_p^\alpha, L_q). \tag{3}$$

2 Reconstruction of the Fourier Coefficients

Let us denote through $A[0,1]^n$ the space of absolutely convergent trigonometric series $f = \sum_{\mu\in\mathbb{Z}^n} a_\mu e^{2\pi i\mu x}$. Let $\nu\in\mathbb{N}^n$, $\mu\in\rho(\nu)$, $m\ge|\nu|$. Let us define the functional

$$P_m(f;\mu) =$$

$$= \sum_{\substack{|k+\nu|=m\\ k_j\ge 0}} \frac{1}{2^m} \sum_{0\le r<2^{k+\nu}} (-1)^{\sum_{j=1}^{n-1}(r_j+1)sgnk_j} f\left(\frac{r_1}{2^{k_1+\nu_1}},\dots,\frac{r_n}{2^{k_n+\nu_n}}\right) e^{2\pi i\sum_{i=1}^n \mu_i \frac{r_i}{2^{k_i+\nu_i}}}.$$

Lemma 1 ([9]) *Let $d\in\mathbb{N}^n$, $B = B_1\times\ldots\times B_n$ be parallelepiped from $\mathbb{Z}^n$, $d_j > |B_j|$, $j = 1,\dots,n$, $I = \bigcup_{r\in\mathbb{Z}^n}((B_1+r_1d_1)\times\ldots\times(B_n+r_nd_n))$. If $f\in C[0,1]^n$, $f \sim \sum_{\mu\in\mathbb{Z}^n} \hat f(\mu)e^{2\pi i\mu x}$, then the series $\sum_{\mu\in I}\hat f(\mu)e^{2\pi i\mu x}$ is the Fourier series of the function*

$$\frac{1}{d_1\ldots d_n}\sum_{r_1=0}^{d_1-1}\cdots\sum_{r_n=0}^{d_n-1} f\left(x_1+\frac{r_1}{d_1},\dots,x_n+\frac{r_n}{d_n}\right) D_B\left(\frac{r_1}{d_1},\dots,\frac{r_n}{d_n}\right),$$

where $D_B(x) = \sum_{\mu\in B} e^{2\pi i\mu x}$ *is the Dirichlet kernel corresponding to the parallelepiped B from* $\mathbb{Z}^n$.

Lemma 2 *Let* $b = \{b_s\}_{s\in\mathbb{Z}} \in l_1, m \in \mathbb{N}$, *then for any* $0 \le \nu \le m$ *the following representation holds*

$$\sum_{k=\nu}^{m}(-1)^{sgn(k-\nu)}\sum_{r\in\mathbb{Z}} b_{2^{k-1}(2r+sgn(k-\nu))} = \sum_{r\in\mathbb{Z}} b_{2^m r}.$$

Proof

$$\sum_{k=\nu}^{m}(-1)^{sgnk}\sum_{r\in\mathbb{Z}} b_{2^{k-1}(2r+sgn(k-\nu))} = \sum_{r\in\mathbb{Z}} b_{2^\nu r} - \sum_{k=\nu+1}^{m}\sum_{r\in\mathbb{Z}} b_{2^{k-1}(2r+1)}$$

$$= b_0 + \sum_{k=\nu+1}^{\infty}\sum_{r\in\mathbb{Z}} b_{2^{k-1}(2r+1)} - \sum_{k=\nu+1}^{m}\sum_{r\in\mathbb{Z}} b_{2^{k-1}(2r+1)}$$

$$= b_0 + \sum_{k=m+1}^{\infty}\sum_{r\in\mathbb{Z}} b_{2^{k-1}(2r+1)} = b_0 + \sum_{\substack{r\in\mathbb{Z}\\ r\neq 0}} b_{2^m r} = \sum_{r\in\mathbb{Z}} b_{2^m r}.$$

□

Theorem 2 *Let* $\nu \in \mathbb{N}^n$, $|\nu| \le m$, $\mu \in \rho(\nu)$, *then for* $f \in A[0,1]^n$ *the following equality holds*

$$P_m(f;\mu) = \hat{f}(\mu) + \sum_{l=1}^{n}\sum_{\substack{k_1+\ldots+k_l=m-|\nu|\\ k_j\ge 0}}(-1)^{\sum_{j=1}^{l-1} sgnk_j}\times$$

$$\times\sum_{\substack{r\in\mathbb{Z}^l\\ r_l\neq 0}}\hat{f}(2^{k_1+\nu_1-1}(2r_1+sgnk_1)+\mu_1,\ldots,r_l 2^{k_l+\nu_l}+\mu_l,\mu_{l+1},\ldots,\mu_n). \quad (4)$$

In the case of $\mu = (0,\ldots,0)$, formula (4) is the quadrature formula that was considered in the papers [12] and [13].

Proof The proof of Theorem 2 follows the same lines as the proof of Theorem 1 in [13] using Lemmas 1 and 2. □

Theorem 3 *Let* $\nu \in \mathbb{N}^n, |\nu| \le m$, $\alpha > \frac{1}{\bar{p}} = \max(\frac{1}{p}, \frac{1}{2})$. *Then for* $\mu \in \rho(\nu)$ *we have*

$$\sup_{\|f\|_{W_p^\alpha}=1} |\hat{f}(\mu) - P_m(f,\mu)| \le C\frac{(m+1-|\nu|)^{\frac{n-1}{\bar{p}}}}{2^{m\alpha}}.$$

Proof Let $f \in W_p^\alpha[0,1]^n$ and $\mu \in \rho(\nu) = \{\mu \in \mathbb{N}^n : [2^{\nu_i-2}] \le \mu_i < 2^{\nu_i-1}\}$. Using Theorem 2 and Hölder's inequality, we get

$$|\hat{f}(\mu) - P_m(f,\mu)| \le \left(\sum_{s\in M_\mu}\left(\bar{s}^\alpha|\hat{f}(s)|\right)^{\tilde{p}'}\right)^{1/\tilde{p}'}\left(\sum_{s\in M_\mu}\left(\frac{1}{\bar{s}^\alpha}\right)^{\tilde{p}}\right)^{1/\tilde{p}}$$

$$\le \left(\sum_{s\in M_\mu}\left(|\widehat{f^{(\alpha)}}(s)|\right)^{\tilde{p}'}\right)^{1/\tilde{p}'}\left(\sum_{s\in M_\mu}\left(\frac{1}{\bar{s}^\alpha}\right)^{\tilde{p}}\right)^{1/\tilde{p}}$$

$$\le C\|f^{(\alpha)}\|_{L_{\tilde{p}}}\left(\sum_{s\in M_\mu}\left(\frac{1}{\bar{s}^\alpha}\right)^{\tilde{p}}\right)^{1/\tilde{p}} = C\|f\|_{W_p^\alpha}\left(\sum_{s\in M_\mu}\left(\frac{1}{\bar{s}^\alpha}\right)^{\tilde{p}}\right)^{1/\tilde{p}},$$

where $M_\mu = \bigcup_{l=1}^n \bigcup_{k_1+\ldots+k_l=m-|\nu|}\{(2^{k_1+\nu_1-1}(2r_1 + sgnk_1) + \mu_1, \ldots, 2^{k_l+\nu_l}r_l + \mu_l, \mu_{l+1}, \ldots, \mu_n) : r \in Z^l, r_l \neq 0\}$. Taking into account that for $\alpha > 1/p$

$$\left(\sum_{s\in M_\mu}\left(\frac{1}{\bar{s}^\alpha}\right)^{\tilde{p}}\right)^{1/\tilde{p}} \le C\frac{(m+1-|\nu|)^{\frac{n-1}{\tilde{p}}}}{2^{m\alpha}} \tag{5}$$

we obtain the statement of the theorem. □

3 The Pitt Inequality

In this section we present some inequalities for anisotropic Lorentz spaces $L_{\mathbf{p},\mathbf{q}}$, where $\mathbf{p} = (p_1, \ldots, p_n)$, $\mathbf{q} = (q_1, \ldots, q_n)$; see [8] and [11] for the definitions of these spaces. We will use these estimates in the proof of Theorem 1.

Let $* = (j_1, \ldots, j_n)$ be some rearrangement of sequence of numbers $\{1, 2, \ldots, n\}$. By $f^*(t) = f^{*j_1\cdots *j_n}(t_1, \ldots, t_n)$ let us denote the function obtained by applying non-increasing rearrangements with respect to $*$ as follows: first with respect to the variable x_{j_1} of $x_{j_1}, \ldots, x_{j_n}$ in $[0,1]^n$ with fixed other variables, then sequentially with respect to the second variable with fixed others and etc. The space $L^*_{\mathbf{p},\mathbf{q}}([0,1]^n)$ is defined as the space of functions for which

$$\left(\int_0^1 \cdots \left(\int_0^1 \left|t_1^{\frac{1}{p_1}} \ldots t_n^{\frac{1}{p_n}} f^{*j_1\cdots *j_n}(t_1, \ldots, t_n)\right|^{q_1} \frac{dt_1}{t_1}\right)^{\frac{q_2}{q_1}} \cdots \frac{dt_n}{t_n}\right)^{\frac{1}{q_n}} < \infty,$$

where the expression $\left(\int_0^\infty (F(t))^q \frac{dt}{t}\right)^{1/q}$ for $q = \infty$ we understand as $\sup_{t>0} F(t)$.

If $* = (1, 2, \dots, n)$, then the space $L^*_{\mathbf{p},\mathbf{q}}([0,1]^n)$ is denoted by $L_{\mathbf{p},\mathbf{q}}([0,1]^n)$.

Remark 1 In contrast to the scalar case, when $\mathbf{p} = \mathbf{q}$ space $L_{\mathbf{p},\mathbf{q}}$ does not coincide with the Lebesgue space with mixed norm $L_{\mathbf{p}} = L_{p_1,\dots,p_n}$.

Similarly we define the discrete spaces $l^*_{\mathbf{p},\mathbf{q}} = \{a = \{a_s\}_{s\in\mathbb{Z}^n}$:

$$\|a\|_{l_{\mathbf{p},\mathbf{q}}} = \left(\sum_{k_n=1}^{\infty} k_n^{\frac{q_n}{p_n}-1} \dots \left(\sum_{k_1=1}^{\infty} k_1^{\frac{q_1}{p_1}-1} (a^{*j_1\dots *j_n}_{k_1\dots k_n})^{q_1}\right)^{q_2/q_1}\right)^{1/q_n} < \infty \Bigg\}.$$

Let us formulate the Paley type inequalities for trigonometric series.

Lemma 3 ([11, Theorem 4]) *Let* $f \sim \sum_{k\in\mathbb{Z}^n} \hat{f}(k) e^{2\pi ikx}$, $* = (j_1, \dots, j_n)$ *be some rearrangement of the sequence* $(1, 2, \dots, n)$, *and* $*' = (j_n, j_{n-1} \dots, j_1)$. *Then for* $\mathbf{2} < \mathbf{p} = (p_1, \dots, p_n) < \infty$, $\mathbf{0} < \mathbf{q} = (q_1, \dots, q_n) \le \infty$, $\mathbf{p}' = \mathbf{p}/(\mathbf{p}-\mathbf{1})$

$$\|f\|_{L^*_{\mathbf{p},\mathbf{q}}[0,1]^n} \le c\|\hat{f}\|_{l^{*'}_{\mathbf{p}',\mathbf{q}}},$$

for $\mathbf{1} < \mathbf{p} = (p_1, \dots, p_n) < \mathbf{2}$, $\mathbf{0} < \mathbf{q} = (q_1, \dots, q_n) \le \infty$, $\mathbf{p}' = \mathbf{p}/(\mathbf{p}-\mathbf{1})$

$$\|\hat{f}\|_{l^{*'}_{\mathbf{p}',\mathbf{q}}} \le c\|f\|_{L^*_{\mathbf{p},\mathbf{q}}[0,1]^n}.$$

Lemma 4 (Pitt Type Inequality)

a) If $\mathbf{1} < \mathbf{p} < \mathbf{2}$, $0 < q \le \min_j p_j'$, $\mathbf{q} = (q, \dots, q)$, *then*

$$\left(\sum_{k\in\mathbb{Z}^n} \overline{k}^{\frac{\mathbf{q}}{\mathbf{p}'}-1} |\hat{f}(k)|^q\right)^{\frac{1}{q}} \le c\|f\|_{L_{\mathbf{p},\mathbf{q}}}. \tag{6}$$

b) If $\mathbf{2} < \mathbf{p} < \infty$, $\max_{1\le j\le n} p_j' \le q \le \infty$, $\mathbf{q} = (q, \dots, q)$, *then*

$$\|f\|_{L_{\mathbf{p},\mathbf{q}}} \le c\left(\sum_{k\in\mathbb{Z}^n} \overline{k}^{\frac{\mathbf{q}}{\mathbf{p}'}-1} |\hat{f}(k)|^q\right)^{\frac{1}{q}}. \tag{7}$$

Proof Let $f \in L_{\mathbf{p},\mathbf{q}}([0,1]^n)$, where $\mathbf{1} < \mathbf{p} < \mathbf{2}$, $0 < q \le \min_j p_j'$, $\mathbf{q} = (q, \dots, q)$. Without loss of generality we can assume that $f = \sum_{k\in\mathbb{N}^n} \hat{f}(k) e^{2\pi ikx}$. By Lemma 3 we have

$$\sum_{k_n=1}^{\infty} \dots \sum_{k_1=1}^{\infty} k_1^{\frac{q}{p_i'}-1} \dots k_n^{\frac{q}{p_i'}-1} \left(\hat{f}^{*_n\dots *_1}(k_1, \dots, k_n)\right)^q \le c\|f\|^q_{L_{\mathbf{p},\mathbf{q}}}.$$

Since $\mathbf{q} \leq \mathbf{p}'$, then

$$\sum_{k_n=1}^{\infty} k_n^{\frac{q}{p_n}-1} \dots \sum_{k_1=1}^{\infty} k^{\frac{q}{p_1}-1} \left(\hat{f}^{*_n \dots *_1}(k_1, \dots, k_n)\right)^q$$

$$\geq \sum_{k_n=1}^{\infty} k_n^{\frac{q}{p_n}-1} \dots \sum_{m_1=1}^{\infty} m_1^{\frac{q}{p_1}-1} \left(\hat{f}^{*_n \dots *_2}(m_1, k_2, \dots, k_n)\right)^q$$

$$\geq \sum_{m \in \mathbb{N}^n} m_1^{\frac{q}{p_1}-1} \dots m_n^{\frac{q}{p_n}-1} \left|\hat{f}(m_1, \dots, m_n)\right|^q.$$

Thus, part a) of Lemma is proved. Part b) may be shown similarly using the second statement of Lemma 3. □

Different variants of Pitt's inequalities of type (6)–(7) are contained in [4]; see also the papers [2, 5, 8, 11, 14, 18, 22] for the previous results related to Pitt's inequalities for periodic functions.

4 Proof of Theorem 1

Let $f \in L_1[0,1]^n$ and $f^{(\alpha)} \sim \sum_{k \in \mathbb{Z}^n} \overline{k}^{\alpha} \hat{f}(k) e^{2\pi i k x}$. Let us first introduce the anisotropic Sobolev and Besov spaces as follows:

$$W^{\alpha}_{\mathbf{p},\mathbf{q}}[0,1]^n = \{f : \|f^{(\alpha)}\|_{L_{\mathbf{p},\mathbf{q}}} < \infty\}$$

and

$$B^{\alpha}_{p,q}[0,1]^n = \left\{ f \in L_p : \left(\sum_{k \in \mathbb{Z}^n_+} \left(2^{\alpha \sum_{j=1}^n k_j} \left\| \sum_{\mu \in \rho(k)} \hat{f}(\mu) e^{2\pi i \mu x} \right\|_{L_p} \right)^q \right)^{1/q} < \infty \right\}.$$

Second, we note that the functionals $P_m(f, \mu)$ and the operators $F_m(f; x)$ are connected by

$$F_m(f;x) = \sum_{\mu \in G_m} P_m(f,\mu) e^{2\pi i \mu x} = \sum_{\substack{|\nu| \leq m \\ \nu \in \mathbb{N}^n}} \sum_{\mu \in \rho(\nu)} P_m(f,\mu) e^{2\pi i \mu x}. \tag{8}$$

From [21] we get

$$\sup_{\|f\|_{W_p^\alpha}=1} \|f - S_{G_m}(f)\|_{L_q} \sim d_M^{\perp}(W_p^\alpha, L_q).$$

Therefore, it is enough to prove that

$$\sup_{\|f\|_{W_p^\alpha}=1} \|S_{G_m}(f) - F_m(f)\|_{L_q} \le c d_M^{\perp}(W_p^\alpha, L_q).$$

Let $\alpha > 1/2$ and $q > 2$. From Pitt's inequality (see Lemma 4), we have

$$\|S_{G_m}(f) - F_m(f)\|_{L_q} \le \left(\sum_{\mu \in G_m} \bar{\mu}^{\frac{2}{q'}-1} |\hat{f}(\mu) - P_m(f, \mu)|^2 \right)^{1/2}. \tag{9}$$

By Theorem 2, we have

$$|\hat{f}(\mu) - P_m(f;\mu)| \le \sum_{l=1}^{n} \sum_{\substack{k_1+\ldots+k_l=m-|\nu| \\ k_i \ge 0}} \sum_{r \in D_{kl}} |\hat{f}(r)| = \sum_{r \in M_\mu} |\hat{f}(r)|.$$

Then, using Hölder inequality and taking into account that $\alpha > 1/2$ and inequality (5), we obtain that

$$|\hat{f}(\mu) - P_m(f;\mu)| \le \left(\sum_{r \in M_\mu} |\bar{r}^\alpha \hat{f}(r)|^2 \right)^{1/2} \left(\sum_{r \in M_\mu} \bar{r}^{-2\alpha} \right)^{1/2}$$

$$\le C \frac{(m+1-|\nu|)^{(n-1)/2}}{2^{m\alpha}} \left(\sum_{r \in M_\mu} |\bar{r}^\alpha \hat{f}(r)|^2 \right)^{1/2}. \tag{10}$$

Hence, from (9) we have

$$\|S_{G_m}(f) - F_m(f)\|_{L_q}^2 \le \frac{C}{2^{2\alpha m}} \sum_{s=n}^{m} (m+1-s)^{n-1} \sum_{|k|=m} \sum_{\mu \in \rho(k)} m\bar{u}^{q-2} \sum_{r \in M_\mu} \left| \bar{r}^\alpha \hat{f}(r) \right|^2$$

$$\le C \frac{1}{2^{2(\alpha + \frac{1}{q'} - \frac{1}{2})m}} \|f\|_{W_2^\alpha}^2 = C \frac{1}{2^{2(\alpha - \frac{1}{2} + \frac{1}{q})m}} \|f\|_{W_2^\alpha}^2.$$

Thus, for $q > 2$

$$\sup_{\|f\|_{W_2^\alpha}=1} \|S_{G_m}(f) - F_m(f)\|_{L_q} \le C\frac{1}{2^{(\alpha-\frac{1}{2}+\frac{1}{q})m}} \sim d_M^{\perp}(W_2^\alpha, L_q).$$

Let $1 < p < 2 < q < \infty$. Applying Pitt's inequality and (5), we get

$$\|S_{G_m}(f) - F_m(f)\|_{L_q} \le \left(\sum_{\mu\in G_m} \overline{\mu}^{q-2} \left|\hat{f}(\mu) - P_m(f,\mu)\right|^q\right)^{1/q}$$

$$\le \left(\sum_{\mu\in G_m} \overline{\mu}^{q-2} \left[\left(\sum_{r\in M_\mu} \bar{r}^{\frac{q}{p'}-1} |\bar{r}^\alpha \hat{f}(r)|^q\right)^{1/q} \left(\sum_{r\in M_\mu} \left(\frac{1}{\bar{r}^{\alpha+\frac{1}{p'}-\frac{1}{q}}}\right)^{q'}\right)^{1/q'}\right]^q\right)^{1/q}$$

$$\le C\frac{1}{2^{(\alpha+\frac{1}{p'}-\frac{1}{q})m}} \left(\sum_{s=0}^{\infty} (m-s+1)^{\frac{(n-1)q}{q'}} \sum_{|k|=s} \sum_{\mu\in\rho(k)} \mu^{q-2} \sum_{r\in M_\mu} \bar{r}^{\frac{q}{p'}-1} \left|\bar{r}^\alpha \hat{f}(r)\right|^q\right)^{1/q}$$

$$\le C\frac{1}{2^{(\alpha-\frac{1}{p}+\frac{1}{q})m}} \left(\sum_{r\in\mathbb{Z}^n} \bar{r}^{\frac{q}{p'}-1} |\bar{r}^\alpha \hat{f}(r)|^q\right)^{1/q}.$$

Therefore,

$$\|S_{G_m}(f) - F_m(f)\|_{L_q} \le C\frac{1}{2^{(\alpha-\frac{1}{p}+\frac{1}{q})m}} \left(\sum_{r\in\mathbb{Z}^n} \bar{r}^{\frac{q}{p'}-1} |\bar{r}^\alpha \hat{f}(r)|^q\right)^{1/q}. \qquad (11)$$

By Pitt's inequality, we get

$$\|S_{G_m}(f) - F_m(f)\|_{L_q} \le C\frac{1}{2^{(\alpha-\frac{1}{p}+\frac{1}{q})m}} \|f\|_{W^\alpha_{\mathbf{p},\mathbf{q}}}. \qquad (12)$$

Taking into account that $q > p$ and using the embedding $W_p^\alpha = W^\alpha_{\mathbf{p},\mathbf{p}} \hookrightarrow W^\alpha_{\mathbf{p},\mathbf{q}}$, we have

$$\sup_{\|f\|_{W_p^\alpha}=1} \|S_{G_m}(f) - F_m(f)\|_{L_q} \le \frac{C}{2^{(\alpha-\frac{1}{p}+\frac{1}{q})m}} \sim d_M^{\perp}(W_p^\alpha, L_q)$$

The proof is now complete.

Finally, we state the result similar the one given by Theorem 1 but for the Besov spaces.

Theorem 4 *If* $1 < p \leq 2 \leq q \leq \infty, \ \ \alpha > \frac{1}{p}$, *then*

$$\sup_{\|f\|_{B^{\alpha}_{p,q}}=1} \|f - F_m(f)\|_{L_q} \sim d^{\perp}_{M}(B^{\alpha}_{p,q}, L_q).$$

Proof The statement follows from inequality (12) and the following

Lemma 5 ([10]) *Let* $p = (p,\ldots,p), q = (q,\ldots,q), \ r = (r,\ldots,r), \alpha - \frac{1}{p} = \gamma - \frac{1}{r}, \ \beta > \alpha$, *then*

$$B^{\gamma}_{r,q}[0,1]^n \hookrightarrow W^{\alpha}_{\mathbf{p},\mathbf{q}}[0,1]^n.$$

□

Acknowledgements This research was partially supported by the Ministry of Education and Science of the Republic of Kazakhstan (3311/GF4, 4080/GF4).

References

1. K.I. Babenko, Approximation by trigonometric polynomials in a certain class of periodic functions of several variables. Sov. Math. Dokl. **1**, 513–516 (1960)
2. J.J. Benedetto, H.P. Heinig, Weighted Fourier inequalities: new proofs and generalizations. J. Fourier Anal. Appl. **9**, 1–37 (2003)
3. E.M. Galeev, Orders of the orthoprojection widths of classes of periodic functions of one and of several variables. Mat. Zametki **43**(2), 197–211 (1988)
4. D. Gorbachev, S. Tikhonov, Moduli of smoothness and growth properties of Fourier transforms: two-sided estimates. J. Approx. Theor. **164**(9), 1283–1312 (2012)
5. H.P. Heinig, Weighted norm inequalities for classes of operators. Indiana Univ. Math. J. **33**, 573–582 (1984)
6. L.K. Hua, Y. Wang, *Applications of Number Theory to Numerical Analysis* (Springer, Berlin, Heidelberg, New York, 1981)
7. N.M. Korobov, *Number-Theoretic Methods in Approximate Analysis*. 2nd edn., p. 285 (Tsentr Nepreryvnogo Matematicheskogo Obrazovaniya, Moscow, 2004)
8. E.D. Nursultanov, Interpolation properties of some anisotropic spaces and Hardy-Littlewood type inequalities. East J. Approx. **4**(2), 243–275 (1998)
9. E.D. Nursultanov, On multipliers of Fourier series in a trigonometric system. Math. Notes **63**(1–2), 205–214 (1998)
10. E.D. Nursultanov, Interpolation theorems for anisotropic function spaces and their applications. Dokl. Akad. Nauk **394**(1), 22–25 (2004) [Russian]
11. E.D. Nursultanov, On the application of interpolation methods in the study of the properties of functions of several variables. Mat. Zametki **75**(3), 372–383 (2004) [Russian]; translation in Math. Notes **75**(3–4), 341–351 (2004)
12. E.D. Nursultanov, N.T. Tleukhanova, On the approximate computation of integrals for functions in the spaces $W_p[0,1]n$. Math. Surv. **55**(6), 1165–1167 (2000)
13. E.D. Nursultanov, N.T. Tleukhanova, Quadrature formulae for classes of functions of low smoothness. Sb. Math. **194**(9–10), 1559–1584 (2003)
14. H.R. Pitt, Theorems on Fourier series and dower series. Duke Math. J. **3**, 747–755 (1937)
15. V.S. Rjabenki, Tables and interpolation of a certain class of functions. Sov. Math. Dokl. **1**, 382–384 (1960)

16. S.A. Smolyak, Interpolation and quadrature formulas for classes W_s^α and E_s^α. Dokl. Akad. Nauk SSSR **131**(5), 1028–1031 (1960)
17. S.A. Smolyak, Quadrature and interpolation formulas for tensor products of certain classes of functions. Dokl. Akad. Nauk SSSR **148**(9), 1042–1045 (1963) [Russian]
18. E.M. Stein, Interpolation of linear operators, Trans. Am. Math. Soc. **83**, 482–492 (1956)
19. V.N. Temlyakov, Widths of some classes of functions of several variables. Sov. Math. Dokl. **26**, 619–622 (1982)
20. V.N. Temlyakov, On the approximate reconstruction of periodic functions of several variables. Sov. Math. Dokl. **31**, 246–249 (1985)
21. V.N. Temlyakov, Approximations of functions with bounded mixed derivative. Proc. Steklov Inst. Math. **178**, 1–121 (1989)
22. S. Tikhonov, Trigonometric series with general monotone coefficients. J. Math. Anal. Appl. **326** (1), 721–735 (2007)
23. N.T. Tleukhanova, On the approximate computation of multiplicative transformations of functions from the Korobov and Sobolev classes. Mat. Zh. **2**(3), 79–88 (2002) [Russian]
24. N.T. Tleukhanova, An interpolation formula for functions of several variables. Math. Notes **74**(1), 151–153 (2003)
25. N.T. Tleukhanova, An interpolation formula for multiplicative transformations of functions of several variables. Dokl. Akad. Nauk **390**(2), 169–171 (2003) [Russian]
26. R.M. Trigub, E.S. Belinsky, *Fourier Analysis and Approximation of Functions* (Kluwer Acad. Publ. Dordrecht, 2004)

Part IV
Optimization Theory and Related Topics

Laplace–Borel Transformation of Functions Holomorphic in the Torus and Equivalent to Entire Functions

L.S. Maergoiz

Abstract The object under study of this paper is the class $\mathcal{A}(\mathbb{T}_n)$ of functions holomorphic in the torus $\mathbb{T}_n$ and equivalent to entire functions. We present an approach to constructing some growth theory of this class with the use of the growth theory of entire functions of several variables. This approach is illustrated by investigations of Laplace–Borel transformation of functions of $\mathcal{A}(\mathbb{T}_n)$.

Keywords Entire function • Growth characteristics (order, type, indicator) • Laplace–Borel transformation • Monomial mapping • Multiple Laurent series

Mathematics Subject Classification (2010). Primary 32A22, 32A15

1 Introduction

Let $\mathbb{T}^n = \mathbb{C}_*^n$, where $\mathbb{C}_* = \mathbb{C}\backslash\{0\}$, $n > 1$, be the multidimensional torus, and $H(\mathbb{T}^n)$ be the class of functions, holomorphic in $\mathbb{T}^n$. This class is the natural extension of the class $H(\mathbb{C}^n)$ of entire functions of n complex variables.

The first problem is to find in $H(\mathbb{T}^n)$ a subclass of functions "equivalent" to $H(\mathbb{C}^n)$. The natural analog of entire functions of one variable is given by the analytic functions on the Riemann sphere without a point. We cannot expect such an analog for several variables: the analytic functions of several variables have no isolated singularities. Some possible approach to finding a multidimensional analog of this result is suggested by the following version of the Hadamard–Valiron theorem:

L.S. Maergoiz (✉)
Siberian Federal University, pr. Svobodny 79, 660041 Krasnoyarsk, Russia
e-mail: bear.lion@mail.ru

M. Ruzhansky, S. Tikhonov (eds.), *Methods of Fourier Analysis and Approximation Theory*, Applied and Numerical Harmonic Analysis,
DOI 10.1007/978-3-319-27466-9_13

Theorem A *Let $f(z) = f(z_1, \ldots, z_n)$ be a function, holomorphic in $\mathbb{T}^n$, while*

$$M_f(r) = \max\{|f(z)| : |z_k| = r_k,\ k = 1, \ldots, n\}, \quad r \in \mathbb{R}_0^n,$$

where $\mathbb{R}_0 = \{r > 0 : r \in \mathbb{R}\}$, be the maximum modulus of f. Then $M_f(r)$ and $\ln M_f(r)$ are convex functions of $\ln r_1, \ldots, \ln r_n$.

In the case that f is the trace of an entire function on $\mathbb{T}^n$, note that $M_f(r)$ is an increasing function in each variable in contrast, for example, to the Zhukovskiĭ function $f_0(z) = \frac{1}{2}(z + 1/z)$ whose maximum modulus attains the (global) minimum at some point $r_0 > 0$. Therefore,

$$V_f(u) = \ln M_f(e^{u_1}, \ldots, e^{u_n}), \quad u \in \mathbb{R}^n \tag{1}$$

is a convex function whose directions of decrease comprise the convex *decrease direction cone* K_V which includes $\mathbb{R}_-^n$, where $\mathbb{R}_- = \{u \in \mathbb{R} : u \leq 0\}$.

On the other hand, *if for a given convex function $W(v)$, $v \in \mathbb{R}^n$, different from a constant, K_W satisfies the condition* $\dim K_W = n$ *then there exists a nondegenerate linear mapping $\mathcal{F}_W : \mathbb{R}^n \to \mathbb{R}^n$ with a property*

$$V(u) := [W \circ \mathcal{F}_W](u),\ u \in \mathbb{R}^n$$

is a convex function increasing in each variable. This fact generates the conjecture of the existence of a proper subclass $\mathcal{A}(\mathbb{T}_n)$ in the class $H(\mathbb{T}_n)$ of the holomorphic functions in $\mathbb{T}^n$ that are equivalent to entire functions in the following sense: a function $g \in H(\mathbb{T}_n)$, $n > 1$, belongs to $\mathcal{A}(\mathbb{T}_n)$ if there exists a monomial mapping

$$\mathcal{F} : \mathbb{T}^n \to \mathbb{T}^n; \quad z = \mathcal{F}(w),\ z_j = \prod_{i=1}^{n} w_i^{s_{ij}},\ j = 1, \ldots, n, \tag{2}$$

where $B = \|s_{ij}\|$ is an integer nondegenerate square $n \times n$-matrix,[1] such that the function $f(w) = [g \circ \mathcal{F}](w)$ admits an analytical continuation $F(w)$ to $\mathbb{C}^n$ (i.e. F is an entire function).

Main motivation of this paper is to present an approach to constructing some growth theory for the extension $\mathcal{A}(\mathbb{T}_n)$, with the use of the growth theory of entire functions of several variables [1, 2]. This approach is illustrated by investigations of Laplace–Borel transformation of functions of $\mathcal{A}(\mathbb{T}_n)$.

[1]For fixed $i \in \{1, \ldots, n\}$ i–tuple $(s_{i1}, \ldots, s_{in})$ of B consists of exponents of w_i in (2).

2 Description of a Multiple Laurent Series Equivalent to a Power Series

Main results of this section were announced in [3].

Consider the multiple Laurent series

$$g(z) = \sum_{k \in \mathbb{Z}^n} a_k z^k, \quad z^k = z_1^{k_1} \ldots z_n^{k_n}, \tag{3}$$

with nonempty domain D, of absolute convergence which we conventionally assume to be situated in $\mathbb{T}^n$ (in contrast to multiple power series). The set $S_g = \{k \in \mathbb{Z}^n : a_k \neq 0\}$ is called the *support* of g.

Definition 1 Call a multiple Laurent series $g(z)$ of the form (3) *equivalent to a power series* if there exists a monomial mapping of the form (2) where $B = \|s_{ij}\|_{n \times n}$ is an integer nongenerate square $n \times n$ matrix such that

$$f(w) = [g \circ \mathcal{F}](w) = \sum_{m \in B[S_g]} c_m w^m \tag{4}$$

is a convergent power series. In this case we write $g \sim f$. In particular, if $g \in H(\mathbb{T}^n)$ then, under the condition (2), the function $g(z)$ is said to be *equivalent to an entire function*.

Here the notion of equivalence means that, after a change of variables of the form (2), the function $f(w)$ extends analytically to a neighborhood of the origin (in particular, in $\mathbb{C}^n$).

Remark The inverse mapping $\mathfrak{F}^{-1}$ is a monomial mapping with fractional exponents, consequently, it is multi-valued one. In spite of it, $g(z) = [f \circ \mathcal{F}^{-1}](z)$ is the same Laurent series as $S_g = B^{-1}[S_f]$ (see (3), (4)). Therefore, let us agree to say that the power series f is *equivalent to the Laurent series* g, i.e. $f \sim g$.

Example Let $g(z) = az_1z_2^2 + bz_1^{-3}z_2^{-1} + cz_2/z_1$ be a Laurent polynomial in $\mathbb{T}^2$, where a, b, and c are nonzero complex numbers. Under the mapping

$$z = \mathcal{F}w, \; w \in \mathbb{T}^2, \quad z_1 = w_1^{-1}w_2^{-2}, \; z_2 = w_1^3 w_2 \tag{5}$$

it goes to the polynomial $f(w_1, w_2) = aw_1^5 + bw_2^5 + cw_1^4w_2^3$. The determinant of the exponents of w_1, w_2 (see (5)) is nonzero. Therefore, the polynomials g and f are equivalent by Definition 1. Let K be the closed convex cone in $\mathbb{R}^2$ presenting an obtuse angle with the sides

$$\Gamma_1 = \{u \in \mathbb{R}^2 : u_1 = s, \; u_2 = 2s, \; s \geq 0\},$$

$$\Gamma_2 = \{u \in \mathbb{R}^2 : u_1 = -3t, \; u_2 = -t, \; t \geq 0\}.$$

Since $(1,2) \in \Gamma_1$, $(-3,-1) \in \Gamma_2$, $(-1,1) = 0.8(1,2) + 0.6(-3,-1) \in K$, therefore, $S_g \subset K$.[2]

For description of a multiple Laurent series equivalent to a power series we use of the concept of a *simplicial cone*. This is the cone generated by the simplex, where the apex of the cone coincides with one of the vertices of the simplex. A simplicial cone with the apex at $0 \in \mathbb{R}^n$ is called a *simplicial rational cone* if there is a simplex generating this cone such that all its vertices have only rational coordinates. Note that this conception together with monomial mappings of the form (2) are used in describing the structure of toric varieties [4, 5].

A key property of a simplicial cone is as follows.

Theorem 1 ([3, Theorem 7]) *Let $g(z)$ be a n-dimensional Laurent series of the form* (3). *In the notations of formulas* (4) *the series g is equivalent to the power series $f = g \circ \mathcal{F}$* (see Definition 1) *if and only if there exists a simplicial rational cone K (with the apex at $0 \in \mathbb{R}^n$) that contains the support S_g of g. Moreover, the support S_f of f lies in the sublattice*

$$L_K = \{m \in \mathbb{Z}_+^n : m = Bk, \ k \in K \cap \mathbb{Z}^n\}$$

of $\mathbb{Z}_+^n$, where B is the matrix associated with $\mathcal{F}$.

Proof NECESSITY. Let $g \sim f$. In the notations of Definition 1 the matrix $B := \|s_{ij}\|_{n\times n}$ defines some bijective linear mapping $B : \mathbb{R}^n \to \mathbb{R}^n$. Then (see (4)) $S_f = B[S_g] \subset \mathbb{R}_+^n$, $S_g \subset K$, where S_g and S_f are the supports of the series g and f, respectively, and $K = B^{-1}[\mathbb{R}_+^n]$. Here B^{-1} is the inverse matrix of B. Let e_j, $j = 1, \dots, n$ be the standard basis in $\mathbb{R}_+^n$. Combining this with a structure of B, we deduce that vectors $B^{-1}(e_j)$, $j = 1, \dots, n$ have only rational coordinates and make up the basis of K. Therefore, K is the simplicial rational cone, containing S_g.

SUFFICIENCY. Suppose that K is a simplicial rational cone such that $S_g \subset K$. We may assume without loss of generality that $\dim K = n$. Denote by

$$c_j = (c_{1j}, \dots, c_{nj}), \ j = 1, \dots, n \tag{6}$$

direction vectors of the cone generators with properties: they have only integer coordinates, and, moreover, their sequence defines the positive orientation of $\mathbb{R}^n$. This means that $\Delta := \det A > 0$, where $A = \|c_{ij}\|_{n\times n}$ (the vectors $\{c_j\}_1^n$ are the columns of A (see (6)). Look for a basis in K in the form $b_j = tc_j$, $j = 1, \dots, n$, for some $t > 0$. Each element $k = (k_1, \dots, k_n) \in S_g \subset K$ is representable as

$$k = m_1 b_1 + \dots + m_n b_n = tAm, \ \ m = (m_1, \dots, m_n) \in \mathbb{R}_+^n.$$

[2]There are more complicated examples in [3].

Hence, using the inverse matrix A^{-1} of A, we obtain, $tm = A^{-1}k$. Choosing $t = 1/\Delta$, we have:

$$m = A^*k,\ m_j = A_{1j}k_1 + \cdots + A_{nj}k_n, \quad j = 1, \ldots, n, \tag{7}$$

where A^* is the adjoint matrix of A, whose elements are the cofactors to the elements of A. Consequently, $m \in \mathbb{Z}^n_+$.

In the notations of (7) we consider the mapping

$$\mathcal{F} : \mathbb{T}^n \to \mathbb{T}^n; \quad z = \mathcal{F}(w),\ z_i = \prod_{j=1}^{n} w_j^{A_{ij}},\ i = 1, \ldots, n.$$

Putting $g_k(z) := z^k,\ k \in S_g$, we find

$$f_k(w) := [g_k \circ \mathcal{F}](w) = w^m = w_1^{m_1} \ldots w_n^{m_n},$$

with the exponent of the monomial w^m (see (7)). Therefore, $f(w) = [g \circ \mathcal{F}](w)$ is a power series, and, moreover, in the notations of Definition 1 we have $B = A^*$, i.e. $g \sim f$. □

Remark Let in the notations of Theorem 1 X_g be the closed convex hull of the support S_g of g with property $\dim X_g = m < n = \dim K$. Then it is possible to choose the cone K and the mapping $\mathcal{F}$ of the form (2) such that $f = g \circ \mathcal{F}$ is a power series of m variables (see Example 1 and Theorem 7 in [3]).

Corollary 1 *In the notations of Theorem* 1 *and its proof it is possible to choose the cone K and the matrix B, associated with the mapping of the form* (2)*, such that*

$$B(b_j) = e_j,\ j = 1, \ldots, n, \quad B(K) = \mathbb{R}^n_+, \quad K = B^{-1}[\mathbb{R}^n_+]. \tag{8}$$

It is known that the domain D of absolute convergence of a multiple Laurent series g (see (2)) has the following property:

$$\ln D = \{u = \ln|z| := (\ln|z_1|, \ldots, \ln|z_n|) \in \mathbb{R}^n : z \in D \subset \mathbb{T}^n\}$$

is a convex set in $\mathbb{R}^n$. Moreover, it is valid

Multidimensional Analogue of the Abel's Lemma (see, e.g., [6]) *Let g be a Laurent series with nonempty domain D, of absolute convergence, while K is an arbitrary cone containing the closed convex hull of the support S_g of g, and a is its apex. Then every point $z_0 \in D$ satisfies the condition*

$$\ln|z_0| + K_a^\circ \subset \ln D,\ K_a^\circ = \{v \in \mathbb{R}^n :\ \langle u, v\rangle\ \leq 0\ \ \forall\ u \in K_a\}$$

where K_a° is the polar of the cone $K_a = K - a$ with the apex at $0 \in \mathbb{R}^n$.

If the series g equivalent to a power series, then its domain of absolute convergence has the following similar geometric property.

Corollary 2 (cf. Theorem 2 in [3]) *Let g be a multiple Laurent series g with a nonempty domain $D \subset \mathbb{T}^n$ of absolute convergence. Suppose that g is equivalent to a power series. Denote by K any fixed simplicial rational cone (with the apex at $0 \in \mathbb{R}^n$), containing the support S_g of g. Then every point $z_0 \in D$ satisfies the condition* $\ln |z_0| + K^\circ \subset \ln D$, *where K° is the polar of the cone K.*

3 Growth Characteristics of Functions of $H(\mathbb{T}^n)$.

Remind the notions of growth characteristics of functions of $H(\mathbb{T}^n)$ which were defined in [3] by analogy with the asymptotic characteristics of entire functions of several variables (see [2], Chaps. 6–8).

The growth of each function $g \in H(\mathbb{T}^n)$ was compared to its maximum modulus $M_g(r),\ r \in \mathbb{R}_0^n$ (see Theorem A). This function is in general not increasing in each variable.

Definition 2 (cf. [2], Definition 6.2.4; [3]) *The order function* of a function $g \in H(\mathbb{T}^n)$ is defined in $\mathbb{R}^n$ as

$$\rho_g(u) = \overline{\lim_{t\to\infty}}\, (\ln t)^{-1} \ln^+ \ln^+ M_g(t^u),\ u \in \mathbb{R}^n; \quad t^u = (t^{u_1}, \dots, t^{u_n}).$$

Call g a function *of finite order* if its order function ρ_g is a finite function in $\mathbb{R}^n$. For fixed $x \in \mathbb{R}^n \setminus \{0\}$ the value $\rho_g(x)$ is called the *x-order* of g.

If $g \in H(\mathbb{T}^n)$ is a function of finite order (see Definition 2), then $D_\rho = \{u \in \mathbb{R}^n : \rho_g(u) > 0\}$ is the cone of all parabolic directions of its normal growth order. This set is defined by its section

$$T_g = \{u \in \mathbb{R}_0^n : \rho_g(u) = 1\} \tag{9}$$

which we, by analogy with the case of entire functions(see [2], Definition 6.2.16), call *the order hypersurface* of g.

Definition 3 (cf. [2], Definition 6.3.1; [3]) Let $g \in H(\mathbb{T}_n)$, and let ρ_g be the order function of g. Assume that $\rho_g(x) \in (0, \infty)$ for fixed $x \in \mathbb{R}^n \setminus \{0\}$. Call

$$\sigma_g(r; x) = \overline{\lim_{t\to\infty}}\, t^{-\rho_g(x)} \ln^+ M_g(rt^x), \quad r \in \mathbb{R}_0^n,$$

the *x-type function* for g and refer to $\sigma_g(x) := \sigma_g(\mathbb{I}; x)$ as its *type in the direction x*.

Therefore, without loss of generality, under the assumptions of Definition 3, it suffices to require the fulfillment of the condition $x \in T_g$ (see (9)).

Since

$$\rho_g(\alpha u) = \alpha \rho_g(u) \quad \forall\, \alpha > 0, \quad u \in \mathbb{R}^n, \tag{10}$$

where ρ_g is the order function of g (see Definition 2), we have by Definition 3 that

$$\sigma_g(r\lambda^x; x) = \sigma_g(r; x)\lambda^{\rho_g(x)} \quad \forall\, r \in \mathbb{R}_0^n,\ \lambda > 0; \tag{11}$$

$$\sigma_g(\cdot; x) = \sigma_g(\cdot; x\tau) \quad \forall\, \tau > 0.$$

Therefore, without loss of generality, under the assumptions of Definition 3, it is suffices to require the fulfillment of the condition $x \in T_g$ (see (9)). Such a property can be used in defining the x-indicator of a function g.

Definition 4 (cf. [2], Chap. 8; [3]) Let $g \in H(\mathbb{T}^n)$ be a function of finite order. Assume that $T_g \neq \emptyset$, and $x \in T_g$. The *x-indicator* of g is defined in $\mathbb{T}^n$ as

$$h_g(z; x) = \overline{\lim_{\zeta \to z}}\ \overline{\lim_{r \to \infty}}\, t^{-1} \cdot \ln |g(\zeta t^x)|, \quad z \in \mathbb{T}^n;\ \zeta t^x = (\zeta_1 t^{x_1}, \ldots, \zeta_n t^{x_n}).$$

Study some properties of these asymptotic characteristics of functions of the class $\mathcal{A}(\mathbb{T}^n)$.

Theorem 2 *Let $g \in \mathcal{A}(\mathbb{T}^n)$, and K be a simplicial rational cone, containing the support S_g of g such that* $\dim K = n$. *Assume that $f = g \circ \mathcal{F}$ is an entire function of n variables, where $\mathcal{F}$ is a mapping of the form* (2), *satisfying condition* (8). *Put $\check{K} = -K^\circ$,*[3] *where K° is the polar of K. Suppose $x \in \mathrm{int}\check{K}$. Then the following assertions are true* (see Definitions 2, 3, 4):

1) *the order function ρ_g is finite, if $\rho_g(x) < \infty$;*
2) *the x-type function $\sigma_g(\cdot; x)$ of g is finite, if $0 < \rho_g(x) < \infty$, and $\sigma_g(x) < \infty$, where $\sigma_g(x)$ is the x-type of g;*
3) *under the conditions of* 2), *the $\hat{x}$-indicator $h_g(z; \hat{x})$ of g is a finite plurisubharmonic function in $\mathbb{T}^n$; moreover,*

$$h_g(z\lambda^{\hat{x}}; \hat{x}) = \lambda h_g(z; \hat{x}), \quad z \in \mathbb{T}^n,\ \lambda > 0,\ \hat{x} = x/\rho_g(x); \tag{12}$$

$$\sup_{|\varphi_j| \le \pi} h_g[(r_1 e^{\varphi_1}, \ldots, r_n e^{\varphi_n}; \hat{x}] = \sigma_g(r; \hat{x}) \quad \forall\, r \in \mathbb{R}_0^n. \tag{13}$$

To prove Theorem 2, we need an auxiliary assertion.

[3]This set is called the *dual* cone.

Lemma 1 *Let, under the conditions of Theorem 2, $K_V(g)$ be the cone of decrease directions of $V_g(u) = \ln M_g(e^u)$, where M_g is the maximum modulus of g* (see (1)). *In the notations of Theorem 2 we have*

$$[B']^{-1}[\check{K}] = \mathbb{R}^n_+, \quad K^\circ \subset K_V(g), \tag{14}$$

where $B := \|s_{ij}\|_{n\times n}$ is associated with $\mathcal{F}$ the matrix satisfying equalities (8), *and B' is the transpose of B.*

Proof By Corollary 14.6.1 in [7], we have $\dim K^\circ = n$, where the cone K° does not contains lines. Using (8), we find

$$\check{K} = \{x \in \mathbb{R}^n : \langle x, B^{-1}u\rangle \geq 0 \;\forall\, u \in \mathbb{R}^n_+\}$$

Hence, taking into consideration that B^{-1} is a linear operator in $\mathbb{R}^n$, we deduce: $[B^{-1}]'\check{K} = \mathbb{R}^n_+$. Since $[B^{-1}]' = [B']^{-1}$, the equality in (14) follows. Therefore,

$$[B']^{-1}[K^\circ] = \mathbb{R}^n_- \tag{15}$$

On the other hand, consider the trace of $\mathcal{F}$ (see (2)) in $\mathbb{R}^n_0$:

$$r = \mathcal{F}(p),\; p \in \mathbb{R}^n_0; \quad r_j = \prod_{i=1}^{n} p_i^{s_{ij}},\; j = 1, \ldots, n, \tag{16}$$

Denote by M_f the maximum modulus of f. Under the conditions of Theorem 3, the functions $f(w)$ and $M_f(r)$ are functions of n variables. Then

$$M_f(p) = [M_g \circ \mathcal{F}](p),\; p \in \mathbb{R}^n_0. \tag{17}$$

Taking the logarithm of both sides of Eq. (16), we find (see (1))

$$V_f(v) = V_g[B'(v)], v \in \mathbb{R}^n;\; V_g(u) - [V_f \circ (B')^{-1}](u), u \in \mathbb{R}^n. \tag{18}$$

Here B' is the transpose of B, and $(B')^{-1}$ is the inverse matrix of B'. But $M_f(p)$ and $V_f(v)$ are increasing functions in each variable, therefore, $\mathbb{R}^n_- \subset K_V(f)$, where $K_V(f)$ is the cone of decrease directions of $V_f(v)$. Combining this with (15), (18), we obtain

$$B'[\mathbb{R}^n_-] = K^\circ \subset K_V(g).$$

□

Proof of Theorem 2

1. By Definition 2 and (18), we obtain that

$$\rho_g(u) \equiv [\rho_f \circ (B')^{-1}](u), \ u \in \mathbb{R}^n, \tag{19}$$

where ρ_g, ρ_f are the order functions of g and f, respectively, Since $x \in int\check{K}$, we find by Lemma 1 (see (14)): $y := [B']^{-1}(x) \in \mathbb{R}^n_0$. Under the condition of Theorem 1 (see 1)), together with (19), (10), we receive:

$$\rho_g(tx) = \rho_f(ty) = t\rho_g(x) < \infty \quad \forall\, t > 0. \tag{20}$$

Combining this with (19), we deduce that the order function $\rho_f(v)$, increasing in each variable, is finite. By (19), g is a function of finite order, according to Definition 2. Therefore, the statement 1) of Theorem 1 is valid.

2. The conclusions of item 1 have a complete analog for the x-type function of g. In the above notations, together with (16) and (17), we find

$$\mathcal{F}(pt^y) = t^x\mathcal{F}(p), \ M_g(rt^x) = M_g[t^x\mathcal{F}(p)] = M_f(pt^y), p \in \mathbb{R}^n_0, \tag{21}$$

where $x = B'(y), \ y \in \mathbb{R}^n_0$. Now by Definition 3, we have

$$\sigma_g(r; x) = \sigma_g[\mathcal{F}(p); B'(y)] \equiv \sigma_f(p; y), \ p \in \mathbb{R}^n_0.$$

Specifically, taking into consideration (11), we obtain

$$\sigma_g(\lambda^x; x) = \sigma_g(x)\lambda^{\rho_g(x)} = \sigma_f(y)\lambda^{\rho_f(y)} = \sigma_f(\lambda^y; y) < \infty \quad \forall\, \lambda > 0.$$

Here, the quantities $\sigma_g(x), \sigma_f(y)$ are the x-types of g and f, respectively. Next, using the same arguments as in item 1, we conclude that the statement 2) of Theorem 2 is true.

3. Note that the element $\hat{x} := x/\rho_g(x) \in T_g$ (see (9)). At first, formula (12) follows from Definition 4. In the notations of item 1 put $\hat{y} := y/\rho_f(y)$. Taking into account (10) and the equality $\rho_g(x) = \rho_f(y)$ (see (19)), we have $\hat{x} = B'(\hat{y}), \ \hat{y} \in T_f$. Combining Definition 4 with (2) and (21), we obtain

$$g[t^{\hat{x}}\mathcal{F}(w)] = g[\mathcal{F}(wt^{\hat{y}})] = f(wt^{\hat{y}}); \ h_g(z; \hat{x})] \equiv h_f(w; \hat{y}), \ w \in \mathbb{T}^n. \tag{22}$$

Growth characteristics of f, its $\hat{y}$-type function

$$\sigma_f(p; \hat{y}) = \varlimsup_{b \to r} \varlimsup_{r \to \infty} t^{-1} \ln^+ M_g(bt^{\hat{y}}) \tag{23}$$

(see [2], Definition 6.3.1, Proposition 6.3.2) and its $\hat{y}$-indicator $h_f(w;\hat{y})$ are defined in $\mathbb{R}^n_+$ and $\mathbb{C}^n$, respectively. Moreover, by item 1 and 2 of the proof,

$$\rho_f(\hat{y}) = \rho_g(\hat{x}) = 1, \quad \sigma_f(\hat{y}) = \sigma_g(\hat{x}) < \infty. \tag{24}$$

Therefore, properties 8.1.1 and 8.1.3 in [2] (see also Chap. 3, §5 in [1]) imply that $\hat{y}$-indicator $h_f(w;\hat{y})$ of f is a finite plurisubharmonic function in $\mathbb{C}^n$ and satisfies the formula of the form (13), where the function g and the vector $\hat{x}$ are replaced by f and $\hat{y}$, respectively. Now the statement 3) of Theorem 2 follows from (22). □

Remark Theorem 2 is true also in the case if $f = g \circ \mathcal{F}$ is an entire function of m variables, where $m < n$. Then there exist more complicated relations between growth characteristics of functions g and f.

Corollary 3 *If under the conditions of the statement* 2) *of Theorem* 2, *specifically,* $\sigma_g(x) = 0$, *then* $h_g(z;x) \equiv 0,\ z \in \mathbb{T}^n$.

Proof By analyze of item 2 of the proof (see (21), (23)), we verify, in the notations of Theorem 2, that

$$\sigma_g(r;x) \equiv 0,\ r \in \mathbb{R}^n_0; \quad \sigma_f(p;y) \equiv 0,\ p \in \mathbb{R}^n_+.$$

Then $h_f(w;y) \le 0,\ w \in \mathbb{C}^n$, and $h_f(w_0;y) = 0$ for some $w_0 \in \mathbb{C}^n$. Together with Theorem 2.1.4 in [1], we deduce that $h_f(w;y) \equiv 0,\ w \in \mathbb{C}^n$. Now Corollary 3 follows from (22). □

4 Laplace–Borel Transformation of Functions Equivalent to Entire Functions with Supports in a Simplicial Cone

The object under study at the first part of this section are geometrical growth characteristics of functions of the class $H(\mathbb{T}^n)$. Introduce a natural generalization of the geometrical image in $\mathbb{C}^n$, $n > 1$ of the nonnegative x-indicator of an entire function in the case $x \in \mathbb{R}^n_0$ (see [8]; Definition 8.1.5 in [2]).

Definition 5 Let $g \in H(\mathbb{T}^n)$, and $x \in \mathbb{R}^n \setminus \{0\}$. Put

$$h(z) = h_g^+(z;x) := \max\{h_g(z;x), 0\},\ z \in \mathbb{T}^n, \tag{25}$$

where $h_g(\cdot;x)$ is the x-indicator of g (see Definition 4). The set

$$\Omega_h := \Omega_h(g;x) = \bigcup \Pi_\tau | \Re\tau > 0,\ \Pi_\tau = \{z \in \mathbb{T}^n : h(\tau^x z) < \mathrm{Re}\tau\}, \tag{26}$$

where $\tau^{x_k} = |\tau|^{x_k} \exp\{ix_k \arg\tau\},\ k = 1, \dots, n$, is called the *x-indicator diagram* of g.

The geometrical structure of Ω_h (see (26)) is similar to the above case (see Sect. 8.2.1 in [2]). Consider the disk

$$K = \{\tau \in \mathbb{C} \setminus \{0\} : \operatorname{Re} \tau^{-1} \geq 1\}.$$

For $w \in \mathbb{T}^n$, we put

$$O_x(w) = \{\tau^x w = (\tau^{x_1}, \dots, \tau^{x_n}) \in \mathbb{T}^n : \tau \in K\}.$$

Definition 6 A set Ω in $\mathbb{T}^n$ is said to be *x-circular* if, together with every point z, Ω contains the set $O_x(w)$ with some $w = w(z) \in \mathbb{T}^n$. The set

$$\omega = \{w \in \mathbb{T}^n : O_x(w) \subset \Omega\}$$

is called the *kernel* of Ω.

The set $\{(z_1, z_2) \in \mathbb{T}^2 : |z_1 z_2| > 1\}$ is the example of the $(-1, -1)$-circular domain. By Definition 6, in the notations of Definition 5 the x-indicator diagram Ω_h of $g \in H(\mathbb{T}^n)$ is the x-circular set; moreover, the set Π_1 (see (26)) is the kernel of Ω_h. The set Ω_h and its kernel Π_1 are x-parabolically star-like sets, i.e., if, for example, $z \in \Omega_h$, then $zt^x \in \Omega_h \ \forall\, t \in (0, 1)$ (see (12)). Besides, the function h is the *x-parabolic Minkowski functional* of Ω_h, i. e.

$$h(z) = \inf\{s > 0 : zs^{-x} \in \Omega_h\}, \quad z \in \mathbb{T}^n$$

(see [9], Definition 3; [2], Chap. 8).

Fix a simplicial rational cone $K \subset \mathbb{R}^n$ such that $\dim K = n > 1$, and associated with K a mapping $\mathcal{F}$ of the form (2), satisfying the condition (8). We consider two classes of functions.

The first class is the subclass $\mathcal{A}_K(\mathbb{T}^n) = \{g\}$ of $\mathcal{A}(\mathbb{T}^n)$ with properties $S_g \subset K$, where S_g is the support of Laurent series expansion of g, and $f = g \circ \mathcal{F}$ is the entire function of n variables (see Theorem 2). Next, using the notations of Theorem 2, fix $x \in \operatorname{int}\check{K}$, and suppose that

$$\rho_g(x) = 1, \; 0 < \sigma_g(x) < \infty \quad \forall\, g \in \mathcal{A}_K(\mathbb{T}^n). \tag{27}$$

Denote by $P_n(x) = \{p = h_g^+(\cdot; x) : g \in \mathcal{A}_K(\mathbb{T}^n)\}$ the class of nonnegative functions of the form (25). By Theorem 2, every element $p \in P_n(x)$ is a plurisubharmonic function in $\mathbb{T}^n$ such that

$$0 \leq p(z) < \infty, \quad p(zt^x) = tp(z), \quad z \in \mathbb{T}^n, \; t > 0. \tag{28}$$

On the other hand, we consider the following class of entire functions of n variables. Let in the above notations $y := [B']^{-1}(x) \in \mathbb{R}_0^n$ (see (14)), where B is the matrix, associated with $\mathcal{F}$ and satisfying the equalities (8), and $[B^{-1}]'$ is the transpose of B^{-1}. Denote by $\mathcal{D}_K(\mathbb{C}^n) = \{f\}$ the subclass of $H(\mathbb{C}^n)$, consisting of entire functions with the property (see Theorem 1)

$$S_f \subset L_K, \quad L_K = \{m \in \mathbb{Z}^n_+ : m = Bk, \ k \in K \cap \mathbb{Z}^n\}. \tag{29}$$

Here S_f is the support of the power series which represents f. Suppose that

$$\rho_f(y) = 1, \ 0 < \sigma_f(y) < \infty \quad \forall f \in \mathcal{D}_K(\mathbb{C}^n), \tag{30}$$

where $\rho_f(y)$, $\sigma_f(y)$ are the y-order and the y-type of f, respectively. Let $Q_n(y)$ be the class of the nonnegative truncations of the y-indicators for functions of the class $\mathcal{D}_K(\mathbb{C}^n)$. This is the class of plurisubharmonic functions in $\mathbb{C}^n$ with the properties

$$0 \le q(w) < \infty, \quad q(wt^y) = tq(w), \quad w \in \mathbb{C}^n, \ t \ge 0. \tag{31}$$

Besides, taking into account Remark to Definition 1 and (29), we conclude that every element $q \in Q_n(y)$ possesses the following property: $q \circ \mathfrak{F}^{-1}$ is an univalent function in $\mathbb{T}^n$. Combining this with the proof of Theorem 2 and Definition 5, we find relations between growth characteristics of the functions $g \in \mathcal{A}_K(\mathbb{T}^n)$ and $f = g \circ \mathcal{F} \in \mathcal{D}_K(\mathbb{C}^n)$.

Proposition 1 *In the above notations consider the operator*

$$\mathcal{T} : \mathcal{A}_K(\mathbb{T}^n) \to \mathcal{D}_K(\mathbb{C}^n), \ \mathcal{T}(g) = g \circ \mathcal{F} = f. \tag{32}$$

This mapping is an isomorphism with the following properties (see (24), (26), (22), (32))*:*

$$\rho_g(x) = 1 = \rho_f(y), \quad 0 < \sigma_g(x) = \sigma_f(y) < \infty;$$

$$h_g[\mathcal{F}w; x] \equiv h_f(w; y), w \in \mathbb{T}^n; \ h_f[\mathcal{F}^{-1}z; y)] \equiv h_g(z; x), z \in \mathbb{T}^n;$$

$$\Omega_p = \mathcal{F}[\Omega_q \cap \mathbb{T}^n], \ \Omega_q \cap \mathbb{T}^n = \mathcal{F}^{-1}[\Omega_p],$$

where $p = h_g^+(\cdot; x)$, $q = h_f^+(\cdot; y)$ are the nonnegative truncations of the x-indicator $h_g(\cdot; x)$ and the y-indicator $h_f(\cdot; y)$, respectively.

Now we study some properties of Laplace–Borel transformation of functions of the class $\mathcal{A}_K(\mathbb{T}^n)$, based on investigations in the case $K = \mathbb{R}^n_+$, $x \in \mathbb{R}^n_0$ (see [2], Chap. 8).

Definition 7 Let $K \subset \mathbb{R}^n$ be a simplicial rational cone such that $\dim K = n > 1$, and $\check{K}$ be its dual cone (see the footnote to Theorem 2). Put $g \in \mathcal{A}_K(\mathbb{T}^n)$. *The Laplace–Borel transformation* of g at fixed $x \in \operatorname{int}\check{K}$ is called the system of integrals

$$G_\theta(z) = \int_0^{\infty(\arg \tau=\theta)} g(z\tau^x)e^{-\tau}d\tau\,, \quad |\theta| < \pi/2. \tag{33}$$

Fix any function $p \in P_n(x)$ (see (28)) and introduce the subclass

$$\mathcal{A}_K(p;x) = \{g \in \mathcal{A}_K(\mathbb{T}^n) : \ h_g(z;x) \le p(z) \ \forall\ z \in \mathbb{T}^n\}, \tag{34}$$

where $h_g(\cdot;x)$ is the x-indicator of g (see Definition 4).

Proposition 2 *Let* $g \in \mathcal{A}_K(p;x)$, *and* $S_g \subset K$ *be the support of its expansion with the Laurent series* (3). *The system of integrals* $\{G_\theta,\ \theta \in \Delta := (-\pi/2, \pi/2)\}$ (see (33)) *is an analytic extension of the Laurent series*

$$G(z) = \sum_{k\in S_g} a_k\Gamma(\langle k, x\rangle + 1)z^k \tag{35}$$

to the x-circular domain (see (26), Definition 6)

$$\Omega_p = \cup\Pi(\theta)|\theta \in \Delta, \quad \Pi(\theta) = \{z \in \mathbb{T}^n : \ p[z(e^{i\theta})^x] < \cos\theta\}. \tag{36}$$

In particular, integral G_θ *converges in* $\Pi(\theta)$ *for any* $\theta \in \Delta$.

Proof Take the mapping $\mathcal{F}$ of the form (2), fixed in this section. By (22), we obtain after elementary manipulations of the integral G_θ for any $\theta \in \Delta$ (see (33), (14)):

$$F_\theta(w) := [G_\theta \circ \mathcal{F}](w) = \int_0^{\infty(\arg \tau=\theta)} f(w\tau^y)e^{-\tau}d\tau\ ,\ x = B'(y), y \in \mathbb{R}^n_0, \tag{37}$$

where $f = g \circ \mathcal{F}$. Put $q = p \circ \mathcal{F}$, and (see (29))

$$\mathcal{D}_K(q;y) = \{g \in \mathcal{D}_K(\mathbb{C}^n) : \ h_f(w;y) \le q(w) \ \forall\ w \in \mathbb{C}^n\}. \tag{38}$$

According to Proposition 1, the operator $\mathcal{T} : \mathcal{A}_K(p;x) \to \mathcal{D}_K(q;y)$ is an isomorphism (see (32)). Therefore, $f \in \mathcal{D}_K(q;y)$. Consider the expansion of f with the Laurent series (4) in $\mathbb{T}^n$. Together with Lemma 8.3.2 in [2], the power series

$$F(w) = \sum_{m\in B[S_g]} c_m\Gamma(\langle m, y\rangle + 1)w^m \tag{39}$$

is holomorphic in a neighborhood of the origin in $\mathbb{C}^n$, and this series extends analytically to the domain

$$\Omega_q = \cup\Pi(\theta)|\theta \in \Delta, \quad \Pi(\theta) = \{w \in \mathbb{C}^n : q[w(e^{i\theta})^y] < \cos\theta\} \tag{40}$$

by the system of integrals $\{F_\theta\}$ (see (37)). Then the series $[F \circ \mathcal{F}^{-1}](z)$ has a nonempty domain of absolute convergence in $\mathbb{T}^n$. But

$$\langle m, y\rangle = \langle Bk, y\rangle = \langle k, B'(y)\rangle = \langle k, x\rangle, \ \ m \in S_f = B[S_g], \ k \in S_g.$$

Consequently, in notations of (35) $[F \circ \mathcal{F}^{-1}](z) \equiv G(z)$. Recalling that $p = q \circ \mathcal{F}^{-1}$, $\Omega_p = \mathcal{F}[\Omega_q \cap \mathbb{T}^n]$ (see Proposition 1), we deduce the statement of Proposition 2 by (37), (39), (40). □

Remark In the above notations the set Ω_p (36) is called *the generalized Borel x-polygon of the series G*. Put $H(\Omega_p)$ the class of holomorphic functions in Ω_p. It is possible to consider the Laplace–Borel transform as the operator

$$\mathcal{L}_x(p) : \mathcal{A}_K(p; x) \to H(\Omega_p), \ \mathcal{L}_x(g) = G \tag{41}$$

(see Definition 7, (34), (35)). In similar form it is used in [10] for description of the domain of absolute convergence of the integral G_0 (see (33)) in the case $K = \mathbb{R}^n_+$, $x \in \mathbb{R}^n_0$.

Now we formulate the following analog of the Polya–Martineau–Ehrenprice theorem (cp. [2], Sect. 8.3).

Theorem 3 *Let* $p \in P_n(x)$ (see (28)). *In the notations of Definition* 7, (34), (3) *the operator* $\mathcal{L}_x$ *is an isomorphism between* $\mathcal{A}_K(p; x)$ *and* $H(\Omega_p)$. *Moreover, the following statements are equivalent:*

1) *for a function* $g \in \mathcal{A}_K(p; x)$, *its x- indicator diagram is equal to* Ω_p, *or the nonnegative trunkation of its x-indicator* $h_g^+(\cdot; x) = p$;
2) *the largest x-circular domain* (see Definition 6) *to which the function* $G = \mathcal{L}_x(g)$ *extends analytically coincides with* Ω_p.

Proof A scheme of the proof is the same as in the case of Proposition 2: the statement of Theorem 3 is true if (and only if) it is true the same assertion for the Laplace–Borel transform

$$\mathcal{L}_y : \mathcal{D}_K(q; y) \to H(\Omega_q), \ \mathcal{L}_y(f) = F,$$

where $q = p \circ \mathcal{F} \in Q_n(y)$, $H(\Omega_q)$ is the class of holomorphic functions in Ω_q (see (31), (38), (40)). By Theorem 8.3.1 in [2], this statement is valid for more wide class of entire functions $\{f\}$: relation (29) is replaced by the condition $S_f \in \mathbb{R}^n_+$. The proof of Theorem 3 is the same in the present case. □

Remark As was noted by the author in [11], Kiselman [8] proved a multidimensional analog of the Polya theorem. In that analog, the geometrical image Ω_h of the radial indicator h of an entire function (i.e. $x = I := (1, \dots, 1)$) lies in the projective space $\mathbb{P}^n$:

$$\omega_h = \bigcup \Pi_h(\tau)|\tau \in \mathbb{C};\ \Pi_h(\tau) = \{(z_1, \dots, z_n, t) \in \mathbb{P}^n :\ h(\tau z) < \Re t\tau\}.$$

This construction generates the indicator diagram Ω_h in $\mathbb{C}^n$ in the case $h \geq 0,\ x = I$ (see Definition 5). A review of near to Theorem 3 results is contained in [12].

References

1. L.I. Ronkin, *Introduction to the Theory of Entire Functions of Several Variables* (Nauka, Moscow, 1971; American Mathematical Society, Providence, RI, 1974)
2. L.S. Maergoiz, *Asymptotic Characteristics of Entire Functions and Their Applications in Mathematics and Biophysics* (Nauka, Novosibirsk, 1991; Kluwer Academic Publishers, Dordrecht, 2003)
3. L.S. Maergoiz, Multidimensional analogue of the laurent series expansion of a holomorphic function and related issues. Dokl. Math. **88**(2), 569–572 (2013)
4. A.G. Chovanskii, The Newton polytopes (solution of singularities), in *Results of Science and Technology. Modern Mathematical Problems*, pp. 207–239 (VINITY, Moscow, 1983) [in Russian]
5. W. Fulton, *Introduction to Toric Varieties* (Princeton University Press, Princeton, NJ, 1993)
6. M. Passare, T. Sadykov, A. Tsikh, Singularities of hypergeometric functions in several variables. Compos. Math. **141**, 787–810 (2005)
7. R. Rockafellar, *Convex Analysis* (Princeton University Press, Princeton, 1970; Mir, Moscow, 1973)
8. C.O. Kiselman, On entire functions of exponential type and indicators of analytic functionals. Acta Math. **117**, 1–35 (1967)
9. L.S. Maergoiz, Functions of types of an entire function of of several variables in the directions of its growth. Sibirsk. Math. Zh. **14**(5), 1037–1056 (1973)
10. E.I. Yakovlev, On summation of multiple power series in a parabolic Mittag-Lefler star, No. 1, pp. 112–118 (Krasnoyarsk State University, Vestnik, 2006) [in Russian]
11. L.S. Maergoiz, On analogs of the theorem Polya for entire functions of many variables, in *Holomorphic Functions of Several Complex Variables*, pp. 193–199 (IF SO AN SSSR, Krasnoyarsk, 1976) [in Russian]
12. L.S. Maergoiz, The Borel–Laplace transforms and analytic continuation of n-multiple Laurent Sertes, in *Mathematical Forum. Investigations on Mathematical Analysis*, vol. 3, pp. 129–142 (CSC RAN, Vladikaukasus, 2009) [in Russian]

Optimization Control Problems for Systems Described by Elliptic Variational Inequalities with State Constraints

Simon Serovajsky

Abstract The control system described by variational inequality is considered. It is approximated by the system described by a nonlinear equation with using the penalty method. The convergence of the approximate method is proved. The necessary conditions of optimality for approximate optimization control problem are obtained. The optimal control for the approximate optimization problem is chosen as an approximate solution of the initial problem.

Keywords Necessary conditions of optimality • Optimization • Penalty method • Variational inequality

Mathematics Subject Classification (2000). Primary 49K20, Secondary 35J85

1 Introduction

Many mathematical physics problems are described by variational inequalities (see, for example, [1–5]). The mathematical theory of these problems is well known (see [2–6]). So optimization control problems for these systems are interesting enough. A lot of results for optimization control problems of systems described by variational inequalities are known (see, for example, [7–15] for elliptic case, [7, 9, 16–18] for parabolic case, and [9] for hyperbolic case). The control systems for variational inequalities with state constraints are analyzed in [8, 10, 11, 15].

We consider the control system with state constraint in the form of the general inclusion. The analysis is based on the Warga's concept of the search of minimizing sequences, but not optimal controls [19] (see also [20, 21]). Besides we will use a double regularization of the optimization control problem. At first the variational inequality, which defines the state of the system, is approximated by a nonlinear

S. Serovajsky PhD (✉)
Professor, al-Farabi Kazakh national university, 71 al Farabi avenue, 050078 Almaty, Kazakhstan
e-mail: serovajskys@mail.ru

M. Ruzhansky, S. Tikhonov (eds.), *Methods of Fourier Analysis and Approximation Theory*, Applied and Numerical Harmonic Analysis,
DOI 10.1007/978-3-319-27466-9_14

equation with using the penalty method. The analogical technique was used for the classical theory of variational inequalities (see [6]) and optimization control theory (see [7, 9, 11]). Hence we obtain the optimization control problem for a nonlinear elliptic equation with state constraint. It is approximated by the minimization problem for a penalty functional on the set of admissible pairs "state-control". This method was used in [22] for the analysis of the distributed singular systems without state constraints. However our system is regular and we have state constraints. Besides the penalty method was used in [22] for obtaining necessary conditions of optimality for the initial optimization problem (see also [7, 9, 11]). We apply it for finding minimizing sequences with using the idea of Warga [19] (see also [20, 21]). But this means was used for the extension of optimization controls problems in the case of its insolvability there. However we prove the solvability of our problem, and this method is applied for finding an optimal control.

Thus an approximate solution of the initial optimization control problem is chosen as the optimal control for approximate problem for large enough step of the algorithm. The necessary conditions of optimality for the approximate optimization control problem are obtained in the standard form.

2 Problem Statement

Let Ω be an open bounded n-dimensional set, where $n \leq 3$. Define the space $H_0^1(\Omega)$ and its subset Z that consists all functions with non-negative values. We consider the control system described by the variational inequality

$$\int_{\Omega} (\Delta y + v)(z - y)dx \leq 0 \ \forall z \in Z, \tag{1}$$

where v is the control, and y is the state function.

For any control v from the space $L_2(\Omega)$ the problem (1) is solvable on the set Z (see [6], Sect. 3, Example 5.1). The inequality (1) was approximated in [6] by the homogeneous Dirichlet problem for the nonlinear elliptic equation

$$-\Delta y + \frac{1}{\varepsilon_k} a(y) = v, \tag{2}$$

where $\varepsilon_k > 0$ and $\varepsilon_k \to 0$ as $k \to \infty$, $a(y) = 0$ for $y \geq 0$ and $a(y) = y^3$ if $y < 0$. By monotony method (see [6], Sect. 2, Theorem 2.1) for any $v \in L_2(\Omega)$ the Eq. (2) has a unique solution $y = y_k[v]$ from the space $H_0^1(\Omega) \cap H^2(\Omega)$, and the mapping

$$y_k[\cdot] : L_2(\Omega) \to H_0^1(\Omega)$$

is weakly continuous. Besides $y_k[v] \to y$ weakly in $H_0^1(\Omega)$ after extracting a subsequence by Theorem 5.2 (see [6], Chap. 3), where y is a solution of the variational inequality (1) for this control. Note that the norm of the solution of Eq. (2) is estimated by the norm of the absolute term by this theorem. Then the mentioned convergence is uniform with respect to v from any bounded subset of $L_2(\Omega)$.

Consider convex closed bounded subsets V of $L_2(\Omega)$ and Y of $H_0^1(\Omega)$. The pair (v, y) from the set $V \times Y$ is called admissible if it satisfies the inequality (1) (see [22]). By U denote the set of all admissible pairs. Suppose this set is non-empty for nontriviality of the problem. Consider the functional

$$I(v,y) = \frac{1}{2}\int_{\Omega}\left[(y-y_{\partial})^2 + \chi v^2\right]dx,$$

where y_{∂} is a given function from $H_0^1(\Omega)$, $\chi > 0$. We have the following optimization control problem.

Problem P1 *Minimize the functional I on the set U.*

Prove the weak continuity of the solution of the variational inequality (1) with respect to the control. By $y[v]$ denote its solution for the control v.

Lemma 2.1 *If $\{v_s\} \subset V$ and $v_s \to v$ weakly in $L_2(\Omega)$, then $y[v_s] \to y[v]$ weakly in $H_0^1(\Omega)$ after extracting a subsequence.*

Proof We have

$$y[v_s] - y[v] = \big(y[v_s] - y_k[v_s]\big) + \big(y_k[v_s] - y_k[v]\big) + \big(y_k[v] - y[v]\big).$$

Then $y_k[w] \to y[w]$ weakly in $H_0^1(\Omega)$ uniformly with respect to $w \in V$ after extracting a subsequence as $k \to \infty$. So $y_k[v] \to y[v]$ and $(y[v_s] - y_k[v_s]) \to 0$ weakly in $H_0^1(\Omega)$. Besides $y_k[v_s] \to y_k[v]$ weakly in $H_0^1(\Omega)$ for all k as $s \to \infty$. Hence the assertions of the lemma follow from the last equality. □

Theorem 2.2 *Problem P1 is solvable.*

Proof Let the sequence of pairs $\{(v_s, y_s)\}$ be minimizing. So we have the inclusions $v_s \in V$, $y_s \in Y$, the variational inequality

$$\int_{\Omega}(\Delta y_s + v_s)(z - y_s)dx \le 0 \;\forall z \in Z,$$

and the convergence $I(v_s) \to \inf I(U)$. The sequence $\{v_s\}$ is bounded in $L_2(\Omega)$ by the boundedness of V. Then $v_s \to v$ weakly in $L_2(\Omega)$ after extracting a subsequence. Using Lemma 2.1, we get $y_s \to y[v]$ weakly in $H_0^1(\Omega)$ after extracting a subsequence. So we obtain the inclusions $v \in V$ and $y[v] \in Y$ by the convexity and the closeness of the sets V and Y. Then

$$\big(v, y[v]\big) \in U.$$

Using the lower semicontinuity of the square of the norm for Hilbert space, we have

$$I\big(v, y[v]\big) \leq \varliminf_{s \to \infty} I(v_s, y_s).$$

Thus

$$I\big(v, y[v]\big) \leq I(U).$$

Therefore the pair $\big(v, y[v]\big)$ is a solution of our problem. This completes the proof of the Theorem 2.2. □

Hence the Problem P1 has a solution. Our aim is the development and the substantiation of the method of its resolution.

3 Approximation of the Optimization Control Problem

The optimization control problems for systems described by equations are easier than for systems described by variational inequalities. So we will use the known approximation of the system (1) by the nonlinear elliptic equation (2) for the analysis of Problem P1. Consider the set

$$V_k = \big\{ v \in V \mid \; y_k[v] \in Y \big\}$$

and the functional

$$I_k(v) = \frac{1}{2} \int\limits_{\Omega} \Big[(y_k[v] - y_\partial)^2 + \chi v^2 \Big] dx,$$

Problem P2 *Minimize the functional I_k on the set V_k.*

Prove the non-triviality of the set V_k at first. We supposed that the set U is non-empty. Use now the more strong assumption. Suppose the existence of a point $v \in V$ such that the state $y[v]$ belongs to the interior of the set Y with respect to the weak topology of $H_0^1(\Omega)$.

Lemma 3.1 *Under this supposition the set V_k is non-empty for large enough value k.*

Proof By our assumption the state $y[v]$ belongs to the interior of the set Y for some control $v \in V$. Then there exists a neighborhood O of $y[v]$ such that it is the subset of this set. By convergence $y_k[v] \to y[v]$ weakly in $H_0^1(\Omega)$ the point $y_k[v]$ belongs to O for a large enough k. Then $y_k[v] \in Y$. So the set V_k is non-empty. □

Using the weakly continuity of the map

$$y_k[\cdot] : L_2(\Omega) \to H_0^1(\Omega),$$

we obtain the following result.

Lemma 3.2 *Problem P2 is solvable.*

By v_k denote a solution of Problem P2. Prove the convergence of the approximation method.

Theorem 3.3 *We have the convergence $I\big(v_k, y[v_k]\big) \to \inf I(U)$ as $k \to \infty$ and $v_k \to v_*$ in $L_2(\Omega)$ after extracting a subsequence, where v_* is a solution of Problem P1.*

Proof We have

$$I_k(v_k) = \min I_k(V_k) \le I_k(v_*).$$

Using the definition of the approximate functional, we get

$$I_k(v_*) = \frac{1}{2}\int\limits_{\Omega}\Big\{(y_k[v_*] - y_\partial)^2 + \chi v^2\Big\}dx$$

$$= I(v_*) + \frac{1}{2}\int\limits_{\Omega}\Big\{[(y_k[v_*] - y_\partial)]^2 - [(y[v_*] - y_\partial)]^2\Big\}dx.$$

Then

$$I_k(v_k) \le \inf I(U) + \frac{1}{2}\Big\|y_k[v_*] + y[v_*] - 2y_\partial\Big\|_2\Big\|y_k[v_*] - y[v_*]\Big\|_2,$$

where $\|\cdot\|_p$ is the norm of the space $L_p(\Omega)$. The sequence $\{y_k[v_*]\}$ is bounded in the space $H_0^1(\Omega)$. Besides $y_k[v_*] \to y[v_*]$ weakly in $H_0^1(\Omega)$ and strongly in $L_2(\Omega)$ by Rellich–Kondrashov Theorem. Then we obtain

$$\overline{\lim_{k\to\infty}}\, I_k(v_k) \le \inf I(U). \tag{3}$$

The sequences $\{v_k\}$ and $\{y_k\}$, where $y_k = y_k[v_k]$, are bounded in the spaces $L_2(\Omega)$ and $H_0^1(\Omega)$ because of the boundedness of the set V and Y. Then we get $v_k \to v$ weakly in $L_2(\Omega)$ and $y_k \to y$ weakly in $H_0^1(\Omega)$ after extracting subsequences. Using convexity and closeness of the set V and Y, we get $v \in V$ and $y \in Y$. We have the equality

$$-\Delta y_k + \frac{1}{\varepsilon_k} a(y_k) = v_k. \tag{4}$$

Then

$$a(y_k) = \varepsilon_k (v_k + \Delta y_k).$$

By boundedness of the sequence $\{y_k\}$ in $H_0^1(\Omega)$ the sequence $\{\Delta y_k\}$ is bounded in $H^{-1}(\Omega)$. Using the convergence $y_k \to y$ weakly in $H_0^1(\Omega)$, we have $\Delta y_k \to \Delta y$ weakly in $H^{-1}(\Omega)$. After passing to the limit in the last equality, we get $a(y_k) \to 0$ weakly in $H^{-1}(\Omega)$.

By Sobolev Theorem we have the continuous embedding $H_0^1(\Omega) \subset L_4(\Omega)$ and $L_{4/3}(\Omega) \subset H^{-1}(\Omega)$. Then the sequence $\{y_k\}$ is bounded in the space $L_4(\Omega)$. Using the definition of the function , we obtain

$$\|a(y_k)\|_{4/3}^{4/3} = \int_\Omega |a(y_k)|^{4/3} dx = \int_{\Omega_k} |y_k|^4 dx \leq \|y_k\|_4^4,$$

where

$$\Omega_k = \{x \in \Omega \,|\, y_k(x) \leq 0\}.$$

Then the sequence $\{a(y_k)\}$ is bounded in the space $L_{4/3}(\Omega)$. By Rellich–Kondrashov Theorem we have the convergence $y_k \to y$ strongly in $L_2(\Omega)$ and a.e. in Ω after extracting a subsequence. So $a(y_k) \to a(y)$ a.e. in Ω . Using Lemma 1.3 [6, Chap. 1], we have $a(y_k) \to a(y)$ weakly in $L_{4/3}(\Omega)$ and in $H^{-1}(\Omega)$ too. Then $a(y) = 0$, so $y \geq 0$ on Ω. Hence the inclusion $y \in Z$ is true.

Using the equality (1), we have

$$\int_\Omega (\Delta y_k + v_k)(z - y_k)dx = \frac{1}{\varepsilon_k}\int_\Omega a(y_k)(z - y_k)dx = -\frac{1}{\varepsilon_k}\int_\Omega \big[a(z) - a(y_k)\big](z - y_k)dx$$

$$= -\frac{1}{\varepsilon_k}\int_{\Omega_k} \big[z^3 - (y_k)^3\big](z - y_k)dx \;\forall z \in Z. \tag{5}$$

Besides we get

$$\int_\Omega \Delta y_k(y_k - y)dx = -\int_\Omega v_k(y_k - y)dx + \frac{1}{\varepsilon_k}\int_\Omega a(y_k)(y_k - y)dx = -\int_\Omega a(y_k)(y_k - y)dx$$

$$= -\int_\Omega v_k(y_k - y)dx + \frac{1}{\varepsilon_k}\int_\Omega \big[a(y_k) - a(y)\big](y_k - y)dx \geq -\int_\Omega v_k(y_k - y)dx.$$

Then

$$\overline{\lim_{k\to\infty}} \int_\Omega \Delta y_k(y_k - y)dx \geq -\overline{\lim_{k\to\infty}} \int_\Omega v_k(y_k - y)dx = 0. \tag{6}$$

By inequalities (5) and (1) we have

$$\int_\Omega (\Delta y + v)(z - y)dx = \lim_{k\to\infty} \int_\Omega \Big[\Delta y_k(z - y) + v_k(z - y)\Big]dx$$

$$= \lim_{k\to\infty} \int_\Omega \big[\Delta y_k(z - y_k) + v_k(z - y_k) + \Delta y_k(y_k - y)\big]dx$$

$$\leq \overline{\lim_{k\to\infty}} \int_\Omega (\Delta y_k + v_k)(z - y_k)dx + \overline{\lim_{k\to\infty}} \int_\Omega \Delta y_k(y_k - y)dx \leq 0 \ \forall z \in Z.$$

So $y = y[v]$, then $(v, y) \in U$.

Using the convergence $v_k \to v$ weakly in $L_2(\Omega)$ and $y_k \to y$ weakly in $H_0^1(\Omega)$, we get

$$\|v\|_2 \leq \inf \lim_{k\to\infty} \|v_k\|_2, \quad \|y - y_\partial\|_2 \leq \inf \lim_{k\to\infty} \|y_k - y_\partial\|_2.$$

Then $I_k(v_k) \to \inf I(U)$.

We have the inequality

$$\Big|I_k(v_k) - I(v_k, y_k)\Big| \leq \frac{1}{2}\int_\Omega \Big|\big(y_k[v_k] - y_\partial\big)^2 - \big(y[v_k] - y_\partial\big)^2\Big|dx$$

$$\leq \frac{1}{2}\big\|y_k[v_k] - y[v_k]\big\|_2 \big\|y_k[v_k] + y[v_k] - 2y_\partial\big\|_2$$

$$\leq \frac{1}{2}\Big\{\big\|y_k] - y\big\|_2 + \big\|y[v_k] - y[v]\big\|_2\Big\}\big\|y_k[v_k] + y[v_k] - 2y_\partial\big\|_2.$$

By the convergence $v_k \to v$ weakly in $L_2(\Omega)$ we get $y[v_k] \to y[v]$ weakly in $H_0^1(\Omega)$. Using the convergence $y_k \to y$ weakly in $H_0^1(\Omega)$, we obtain $y[v_k] \to y[v]$ and $y_k \to y$ strongly in $L_2(\Omega)$. Using the last inequality, we have

$$\lim_{k\to\infty} \left| I_k(v_k) - I(v_k) \right| \to 0,$$

so $I(v_k, y_k) \to \inf I(U)$.

We proved that a subsequence of solutions of Problem P2 is minimizing for the Problem P1. Suppose the existence of a subsequence of $\{I(v_k, y_k)\}$ such that it does not have $\inf I(U)$ as a limit point. Using considered technique, extract its subsequence that convergences to $\inf I(U)$. So the whole sequence $\{I(v_k, y_k)\}$ converges to $\inf I(U)$.

By the convergence $v_k \to v$ weakly in $L_2(\Omega)$ and $y_k \to y$ strongly in $L_2(\Omega)$ we have

$$\|v\|_2 \leq \underline{\lim}_{k\to\infty} \|v_k\|_2, \quad \|y[v] - y_\partial\|_2 = \lim_{k\to\infty} \|y_k[v] - y_\partial\|_2.$$

Then

$$I(v, y) \leq \underline{\lim}_{k\to\infty} I(v_k, y_k) = \inf I(U).$$

Using the inclusion $(v, y) \in U$, we prove that v is a solution of Problem P1.

By $\{v_k\}$ denote the subsequence, which correspond the lower limit of last inequalities. Suppose the strong inequality

$$\|v\|_2 < \inf \lim_{k\to\infty} \|v_k\|_2.$$

Then we obtain the strong inequality

$$I(v, y) < \inf I(U).$$

This contradiction prove the convergence $\|v_k\|_2 \to \|v\|_2$. Using the convergence $v_k \to v$ weakly in $L_2(\Omega)$, we prove that $v_k \to v$ strongly in $L_2(\Omega)$. This completes the proof of Theorem 3.3. □

Remark 3.4 Problem P1 can have many solutions. In this case different subsequences of $\{v_k\}$ can converge to different solutions of this problem. However our conclusions are true for all its convergent subsequence. Therefore the set of limit points of $\{v_k\}$ consists of solutions of Problem P1. However it is possible that some solution does not belong to this set.

The known results the optimization control problems for systems described by variational inequalities include as a rule the justification of the necessary conditions of optimality (see, for example, [7–14]). However we solve optimization control problems only approximately. The known necessary conditions of optimality are difficult enough. So it is more naturally to find the approximate solution of the problem, rather than necessary conditions of optimality. This idea was used in [19–21] in the case of insolvability of extremum problems. By Theorem 2 we can choose the optimal control for Problem P2 for large enough value of k as an approximate solution of Problem P1. So we will solve the solution of Problem P2. It is easier than Problem P1 because the system is described by equation, rather than variational inequality.

4 Second Approximation of the Problem

The general difficulty of Problem P2 is the state constraint. We cannot to use the standard variational method for this case because we do not know how we can change the control for saving the state constraint. We could apply results of the general extremum theory (Lagrange principle and some other methods, see, for example, [23–25]). But it uses very difficult properties of the linearized operator and the state constraint. However some results for optimization control problems for nonlinear elliptic equations with state constraints are known (see, for example, [26–32]). Our aim is the search of minimizing sequences in contrast to these results. Then we transform our problem to an easier one. Using the penalty method [22], we change our optimization problem by the minimization problem for the penalty functional on the set of admissible "control-state" pairs. Note that this technique was used in [22] for the case of the absence of the state constraint. The unique solvability of the state equation was not guarantee there. However our boundary problem is well-posed, but we have the state constraint.

Define the functional

$$I_k^m(v,y) = \frac{1}{2}\int\limits_{\Omega}\Big\{(y-y_\partial)^2 + \chi v^2 + \frac{1}{\delta_m}\big[\Delta y + \varepsilon_k^{-1}a(y) + v\big]^2\Big\}dx,$$

where $\delta_m > 0$ and $\delta_m \to 0$ as $m \to \infty$. Define the space

$$W = L_2(\Omega) \times H_0^1(\Omega)$$

and the set $U_\partial = V \times Y$. We have the following problem.

Problem P3 *Minimize the functional I_k^m on the set U_∂.*

Lemma 4.1 *Problem P3 is solvable.*

Proof Let $\{u_s\} = \big\{v_s, y_s\big\}$ be a minimizing sequence for the Problem P3, so $u_s \in U_\partial$ and $I_k^m \to \inf I_k^m(U_\partial)$ as $s \to \infty$. Using the boundedness of the set U_∂, we prove that the sequence $\{u_s\}$ is bounded in the space W. By definition of the functional we have the equality

$$-\Delta y_s = \varepsilon_k^{-1} a(y_s) + v_s + f_s,$$

where the sequence $\{f_s\}$ is bounded in the space $L_2(\Omega)$. Using the boundedness of the sequence $\{y_s\}$ in $H_0^1(\Omega)$ and in $L_6(\Omega)$ too because of Sobolev Embedding Theorem, we prove the boundedness of the sequence $\{a(y_s)\}$ in the space $L_2(\Omega)$. Then the term in the right side of the last equality is bounded in the space $L_2(\Omega)$. So $\{\Delta y_s\}$ is bounded in $L_2(\Omega)$. Hence we get $v_s \to v$ weakly in $L_2(\Omega)$, $y_s \to y$ weakly in $H_0^1(\Omega)$, $a(y_s) \to \varphi$ weakly in $L_2(\Omega)$, $\Delta y_s \to \Delta y$ weakly in $L_2(\Omega)$ after extracting subsequences. Using the convexity and the closeness of the sets V and Y, we have the inclusions $v \in V$ and $y \in Y$, then $u \in U_\partial$, where $u = (v, y)$. By Rellich–Kondrashov Theorem we get $y_s \to y$ strongly in $L_2(\Omega)$ and a.e. on Ω, then $a(y_s) \to a(y)$ a.e. on Ω. Using Lemma 1.3 (see [6], Chap. 1), we obtain $a(y_s) \to a(y)$ weakly in $L_2(\Omega)$, so $\varphi = a(y)$. By the weak lower semicontinuous of the norm in Hilbert spaces we get

$$I_k^m(u) \leq \inf I_k^m(U_\partial),$$

so u is a solution of Problem P3. This completes the proof of Lemma 4.1. □

Let $u_k^m = \left(v_k^m, y_k^m\right)$ be a solution of Problem P3.

Theorem 4.2 *For any k $I_k(v_k^m) \to \inf I_k(V_k)$ as $m \to \infty$, besides $v_k^m \to v_k$ in $L_2(\Omega)$ after extracting a subsequence.*

Proof We have the inequality

$$I_k^m(u_k^m) = \min I_k^m(U_\partial) \leq I_k^m\left(v_k, y_k[v_k]\right) = I_k(v_k). \tag{7}$$

By boundedness of the set U_∂ the sequence $\{u_k^m\}$ is bounded in the space W. Using the inequality (7) and the definition of the functional I_k^m, we get

$$-\Delta y_k^m = \varepsilon_k^{-1} a(y_k^m) + v_k^m + \sqrt{\delta_m} f_k^m, \tag{8}$$

where the sequence $\{f_k^m\}$ is bounded in $L_2(\Omega)$. Then (see the proof of Lemma 4.1), the sequence $\{\Delta y_k^m\}$ is bounded in $L_2(\Omega)$. Then $v_k^m \to v_k^*$ weakly in $L_2(\Omega)$, $y_k^m \to y_k^*$ weakly in $H_0^1(\Omega)$, $f_k^m \to f_k$ weakly in $L_2(\Omega)$, and $\Delta y_k^m \to \Delta y_k^*$ weakly in $L_2(\Omega)$ as $m \to \infty$ after extracting subsequences. Using the technique from the proof of Lemma 4.1, we obtain $v_k^* \in V$, $y_k^* \in Y$, and $a(y_k^m) \to a(y_k^*)$ weakly in $L_2(\Omega)$. After passing to the limit in the equality (8) we get $y_k^* = y_k[v_k^*]$.

By definition of the functional I_k^m we have

$$I_k^m(u_k^m) \geq \frac{1}{2} \int_\Omega \left[\left(y_k^m - y_\partial\right)^2 + \chi\left(v_k^m\right)^2\right] dx.$$

Hence

$$\min I_k(V_k) = I_k(v_k) \le I_k(v_k^*) = \frac{1}{2}\int_\Omega \Big\{\big(y_k[v_k^*] - y_\partial\big)^2 + \chi\big(v_k^*\big)^2\Big\}dx$$

$$\le \frac{1}{2}\varliminf_{m\to\infty}\int_\Omega \Big[\big(y_k^m - y_\partial\big)^2 + \chi\big(v_k^m\big)^2\Big]dx \le \varliminf_{m\to\infty} I_k^m(u_k^m).$$

Using (7), we obtain $I_k^m(u_k^m) \to \min I_k(V_k)$. By inequalities

$$I_k(v_k) \le I_k(v_k^m) \le I_k^m(u_k^m)$$

we have $I_k(v_k^m) \to \inf I_k(V_k)$. The proof is ended with using the technique from Theorem 4.2 . □

Remark 4.3 All assertions of Remark 3.4 are true in this case.

By proved theorem a sequence of solutions of Problem P1 minimizes the functional I_k on the set V_k. So the value v_k^m for large enough m can be chosen as an approximate solution of Problem P2. Then the control v_k^m for large enough value m and k can be chosen as an approximate solution of Problem P1. Our next step is finding of this control. We will prove that the obtained result is sufficient for the analysis of the given optimization problem without any constraints.

5 Necessary Conditions of Optimality

We have the minimization problem for an integral functional on a convex set. The necessary condition of the minimum at the point u of Gateaux differentiable functional J on a convex set W is the variational inequality

$$\langle J'(u), w-u\rangle \ge 0 \;\forall w \in W, \tag{9}$$

where $\langle \varphi, \lambda\rangle$ is the value of a linear continuous functional φ at a point λ. We prove the differentiability of the functional I_k^m for using this result in our case.

Lemma 5.1 *The functional I_k^m has the partial derivatives*

$$I_{kv}^m(v,y) = \chi v + p_k^m(v,y),\; I_{ky}^m(v,y) = y - y_\partial + \Delta p_k^m(v,y) + \varepsilon_k^{-1}a'(y)p_k^m(v,y), \tag{10}$$

at the arbitrary point (v, y), where

$$p_k^m(v,y) = \frac{1}{\delta_m}\big[\Delta y + \varepsilon_k^{-1}a(y) + v\big]. \tag{11}$$

Proof For any function $h \in L_2(\Omega)$ and the value σ we have the equality

$$I_k^m(v+\sigma h, y) - I_k^m(v,y) = \frac{\chi}{2}\int_\Omega \Big[(v+\sigma h)^2 - v^2\Big]dx$$

$$+\frac{1}{2\delta_m}\int_\Omega \Big\{\big[\Delta y + \varepsilon_k^{-1}a(y) + v + \sigma h\big]^2 - \big[\Delta y + \varepsilon_k^{-1}a(y) + v\big]^2\Big\}dx$$

$$= \sigma \int_\Omega \big[\chi v + p_k^m(v,y)\big]h dx + \frac{\sigma}{2}\int_\Omega \Big\{\chi + \delta_m\big[p_k^m(v,y)\big]^2\Big\}h^2 dx.$$

So the first equality (10) is true. For any function $h \in H_0^1(\Omega)$ and the value σ we get

$$I_k^m(v, y+\sigma h) - I_k^m(v,y) = \frac{1}{2}\int_\Omega \big[(y - y_\partial + \sigma h)^2 - (y-y_\partial)^2\big]dx$$

$$+\frac{1}{2\delta_m}\int_\Omega \Big\{\big[\Delta(y+\sigma h) + \varepsilon_k^{-1}a(y+\sigma h) + v\big]^2 - \big[\Delta y + \varepsilon_k^{-1}a(y) + v\big]^2\Big\}dx$$

$$= \sigma \int_\Omega \Big\{(y-y_\partial)h + p_k^m(v,y)\big[\Delta h + \varepsilon_k^{-1}a'(y)h\big]\Big\}dx + \eta(\sigma)$$

$$+\sigma \int_\Omega \Big\{(y-y_\partial) + \big[\Delta p_k^m(v,y) + \varepsilon_k^{-1}a'(y)p_k^m(v,y)\big]\Big\}h dx + \eta(\sigma),$$

where $a'(y) = 0$ for $y \geq 0$, $a'(y) = 3y^2$ for $y < 0$ and $\eta(\sigma) \to 0$ as $\sigma \to 0$. So the first equality (10) is true. This completes the proof of Lemma 5.1. □

Thus by the inequality (9) we get a necessary condition of optimality.

Theorem 5.2 *The solution* $\big(v_k^m, y_k^m\big)$ *of Problem 3 satisfies the following system*

$$\int_\Omega \big(\chi v_k^m + p_k^m\big)\big(v - v_k^m\big)dx \geq 0 \ \forall v \in V, \tag{12}$$

$$\int_\Omega \Big[y_k^m - y_\partial + \Delta p_k^m + \varepsilon_k^{-1}a'(y_k^m)p_k^m\Big]\big(y - y_k^m\big)dx \geq 0 \ \forall y \in Y, \tag{13}$$

$$\Delta y_k^m + \varepsilon_k^{-1}a(y_k^m) + v_k^m = \delta_m p_k^m. \tag{14}$$

We obtain the standard necessary condition of optimality. It can be solved with using an iterative method (see, for example, [33–35]). Then the control v_k^m can be chosen as an approximate solution of the initial optimization problem for large enough values of k and m.

Remark 5.3 This system is simplified in the case of the absence of the state constraint. The variational inequality (13) can be transformed to the standard adjoint equation

$$\Delta p_k^m + \varepsilon_k^{-1} a'(y_k^m) p_k^m = y_\partial - y_k^m$$

in this case. Hence necessary conditions of optimality include the state equation (14), this adjoint equation and classical variational inequality (12). If we do not have any constraints, then we can find the control $v_k^m = -\chi p_k^m$ from (12). Then we obtain two elliptic equations

$$\Delta p_k^m + \varepsilon_k^{-1} a'(y_k^m) p_k^m = y_\partial - y_k^m,$$

$$\Delta y_k^m + \varepsilon_k^{-1} a(y_k^m) + v_k^m = \delta_m p_k^m.$$

After solving this system we can find v_k^m by the obtained formula.

Analogical results could be obtained for controls systems described by parabolic and hyperbolic variational inequalities. Laplace operator can be substituted by general linear elliptic operators and some nonlinear elliptic operators. We could consider also a general integral functional with corresponding assumptions.

References

1. G. Fichera, Sul problema elastostatico di Signorini con ambigue condizioni al contorno. Rendiconti della Accademia Nazionale dei Lincei, Classe di Scienze Fisiche, Matematiche e Naturali **834**(2), 138–142 (1963)
2. G. Duvaut, J.-L. Lions, *Les inéquations en mécanique et en physique* (Dunod, Paris, 1972)
3. D. Kinderlehrer, G. Stampacchia, *An Introduction to Variational Inequalities and Their Applications* (Academic, New York, 1980)
4. R. Glowinski, J.-L. Lions, R. Trémolier, *Numerical Analysis of Variational Inequalities* (North Holland, Amsterdam, 1981)
5. C. Baiocchi, A. Capelo, *Variational and Quasivariational Inequalities: Applications to Free-Boundary Problems* (Wiley, New York, 1984)
6. J.-L. Lions, *Quelques Méthods de resolution des problèmes aux limites Non linèaires* (Dunod, Gautier-Villars, Paris, 1969)
7. V. Barbu, *Optimal Control of Variational Inequalities*. Research Notes in Mathematics, vol. 100 (Pitman, Boston, 1984)
8. Z.X. He, State constrained control problems governed by variational inequalities. SIAM J. Control Optim. **25**, 1119–1145 (1987)
9. D. Tiba, *Optimal Control of Nonsmooth Distributed Parameter Systems*. Lecture Notes in Mathematics, vol. 1459 (Springer, Berlin, 1990)

10. J. Bonnans, E. Casas, An extension of Pontryagin's principle for state-constrained optimal control of semilinear elliptic equations and variational inequalities. SIAM J. Control Optim. **33**, 274–298 (1995)
11. G. Wang, Y. Zhao, W. Li, Some optimal control problems governed by elliptic variational inequalities with control and state constraint on the boundary. J. Optim. Theory Appl. **106**, 627–655 (2000)
12. Y. Ye, Q. Chen, Optimal control of the obstacle in a quasilinear elliptic variational inequality. J. Math. Anal. Appl. **294**, 258–272 (2004)
13. Q. Chen, D. Chu, R.C.E. Tan, Optimal control of obstacle for quasi-linear elliptic variational bilateral problems. SIAM J. Control Optim. **44**, 1067–1080 (2005)
14. Y.Y. Zhou, X.Q. Yang, K.L. Teo, The existence results for optimal control problems governed by a variational inequality. J. Math. Anal. Appl. **321**, 595–608 (2006)
15. S. Serovajsky, State-constrained optimal control of nonlinear elliptic variational inequalities. Optim. Lett. **8**(7), 2041–2051 (2014)
16. P. Neittaanmaki, D. Tiba, *Optimal Control of Nonlinear Parabolic Systems: Theory, Algorithms, and Applications* (Marcel Dekker, New York, 1994)
17. D.R. Adams, S. Lenhart, Optimal control of the obstacle for a parabolic variational inequality. J. Math. Anal. Appl. **268**, 602–614 (2002)
18. Q. Chen, Optimal obstacle control problem for semilinear evolutionary bilateral variational inequalities. J. Math. Anal. Appl. **307**, 677–690 (2005)
19. J. Warga, *Optimal Control of Differential and Functional Equations* (Academic, New York, 1972)
20. V. F. Krotov, *Global Methods in Optimal Control Theory* (Marcel Dekker, New York, 1996)
21. M.I. Sumin, Suboptimal control of a semilinear elliptic equation with state constraints. Izv. Vuzov **6**, 33–44 (2000)
22. J.-L. Lions, *Contrôle Optimal de systèmes Distributes Singuliers* (Gautier-Villars, Paris, 1983)
23. A.D. Ioffe, V.M. Tihomirov, *Extremal Problems Theory* (Nauka, Moscow, 1974)
24. A.S. Matveev, V.A. Yakubivich, *Abstract Theory of Optimal Control* (St. Petersburg University, St. Petersburg, 1994)
25. K. Madani, G. Mastroeni, A. Moldovan, Constrained extremum problems with infinite-dimensional image: selection and necessary conditions. J. Optim. Theory Appl. **135**, 37–53 (2007)
26. P. Michel, Necessary conditions for optimality of elliptic systems with positivity constraints on the state. SIAM J. Control Optim. **18**, 91–97 (1980)
27. E. Casas, Boundary control of semilinear elliptic equations with pointwise state constraints. SIAM J. Control Optim. **33**, 993–1006 (1993)
28. N. Arada, J.P. Raymond, State-constrained relaxed problems for semilinear elliptic equations J. Math. Anal. Appl. **223**, 248–271 (1998)
29. J.F. Bonnans, A. Hermant, Conditions d'optimalité du second ordre nécessaires ou suffisantes pour les problèmes de commande optimale avec une contrainte sur l'état et une commande scalaires. C. R. Acad. Sci. Paris Ser. I. **343**, 473–478 (2006)
30. A. Rösch, F. Tröltzsch. Sufficient second-order optimality conditions for an elliptic optimal control problem with pointwise control-state constraints. SIAM J. Optim. **17**, 776–794 (2006)
31. A. Rösch, F. Tröltzsch, On regularity of solutions and Lagrange multipliers of optimal control problems for semilinear equations with mixed pointwise control-state constraints. SIAM J. Control Optim. **46**, 1098–1115 (2007)
32. E. Casas, J.C. Reyes, F. Tröltzsch, Sufficient second-order optimality conditions for semilinear control problems with pointwise state constraints. SIAM J. Optim. **19**, 616–643 (2008)
33. R. Fletcher, *Practical Methods of Optimization* (Wiley, New York, 2000)
34. J.F. Bonnans, J.C. Gilbert, C. Lemaréchal, C. Sagastizábal, *Numerical Optimization: Theoretical and Practical Aspects*. Universitext (Springer, Berlin, 2006)
35. C. Floudas, P. Pardalos (eds.), *Encyclopedia of Optimization* (Springer, Berlin, 2008)

Two Approximation Methods of the Functional Gradient for a Distributed Optimization Control Problem

Ilyas Shakenov

Abstract An optimization control problem for one-dimensional parabolic equation is considered. Given a value for Gateaux derivative of the target functional. Obtained problem is solved numerically by considering approximation of functional gradient and gradient of functional approximation. Some experiments are done to compare the effects of two methods on resolvability of the problem on final time.

Keywords Approximation • Control • Gateaux derivative • Gradient • Optimization, Parabolic equation

Mathematics Subject Classification (2000). 49K20

1 Introduction

Problem setting. We consider a problem for parabolic equation in one dimension. Mathematical statement looks as following:

$$\partial_t u(t,x) = \partial_x^2 u(t,x) + f(x,t), \quad 0 < x < L, \quad 0 < t < T, \tag{1}$$

$$u(0,x) = \varphi(x), \quad 0 < x < L, \tag{2}$$

$$\partial_x u(t,0) = b(t), \quad 0 < t < T, \tag{3}$$

$$\partial_x u(t,L) = y(t), \quad 0 < t < T, \tag{4}$$

Function $y(t)$ is unknown and must be determined. For that purposes we use an additional information $u(t,0) = a(t)$. We transform this problem to optimization

I. Shakenov (✉)
al-Farabi Kazakh National University, 71 al-Farabi Avenue, 050040 Almaty, Kazakhstan
e-mail: ilias.shakenov@gmail.com

M. Ruzhansky, S. Tikhonov (eds.), *Methods of Fourier Analysis and Approximation Theory*, Applied and Numerical Harmonic Analysis,
DOI 10.1007/978-3-319-27466-9_15

problem which requires to minimize a functional

$$I = \int_0^T \Big(u\big(t, 0; y(t)\big) - a(t)\Big)^2 dt.$$

If the functional gets its minimum value then $u(t, 0)$ is the most close to $a(t)$ and additional information is fulfilled.

The most commonly used method for solving such problems is gradient method. For that, we construct a sequence

$$y_{n+1}(t) = y_n(t) - \alpha_n I'\big(y_n(t)\big),$$

where $\alpha_n > 0$ with an appropriate initial value for $y_0(t)$.

2 Gradient in Continuous Form

Question arises, what is $I'\big(y_n(t)\big)$? Expression for Gateaux derivative is given by the following

Theorem 1 *Gateaux derivative of the functional I at the point y is determined by the formula*

$$I'\big(y(t)\big) = \psi(t, L),$$

where $\psi(t, x)$ is the solution of the adjoint system:

$$\partial_t \psi(t, x) + \partial_x^2 \psi(t, x) = 0, \quad 0 < x < L, \quad 0 < t < T, \tag{5}$$

$$\psi(T, x) = 0, \quad 0 < x < L, \tag{6}$$

$$\partial_x \psi(t, L) = 0, \quad 0 < t < T, \tag{7}$$

$$\partial_x \psi(t, 0) = -2\big(u(t, 0; y(t)) - a(t)\big), \quad 0 < t < T, \tag{8}$$

Proof The proof of this theorem you can see in [1]. □

According to this theorem we can write

$$y_{n+1}(t) = y_n(t) - \alpha_n \psi(t, L). \tag{9}$$

We choose an initial approximation $y_0(t)$ and the process is ready to be started. But at this moment we encounter a huge problem.

3 Problem of Improvement on the Right Side of Time Interval

One of the boundary conditions says that $\psi(T,x) = 0$ and also $\psi(T,L) = 0$. If we put $t = T$ in (9) we obtain that $y_{n+1}(T) = y_n(T)$ and it means that on the right side of time interval we can't get more precise values than the initial approximation has. One of the methods to avoid this problem is to cut time interval and consider a numerical solution only at a part of $0 < t < T$ interval. This approach is described in [1].

4 Numerical Solution

Lets consider an approximation of direct problem (1)–(4)

$$\frac{u_i^{j+1} - u_i^j}{\tau} = \frac{u_{i-1}^{j+1} - 2u_i^{j+1} + u_{i+1}^{j+1}}{h^2} + f_i^j,$$

$$i = 1, 2, \ldots, N-1, \quad j = 0, 1, \ldots, M-1, \tag{10}$$

$$u_i^0 = \varphi_i, \quad i = 0, 1, \ldots, N, \tag{11}$$

$$\frac{u_1^j - u_0^j}{h} = b_j, \quad j = 0, 1, \ldots, M, \tag{12}$$

$$\frac{u_N^j - u_{N-1}^j}{h} = y_j, \quad j = 0, 1, \ldots, M, \tag{13}$$

where $u_i^j = u(j\tau, ih), f_i^j = f(j\tau, ih)$, $\tau = \frac{T}{M}$, $h = \frac{L}{N}$ are steps by time and space variables.

The following is an approximation of adjoint problem (5)–(8) [2, 3].

$$\frac{\psi_i^{j+1} - \psi_i^j}{\tau} + \frac{\psi_{i-1}^j - 2\psi_i^j + \psi_{i+1}^j}{h^2} = 0,$$

$$j = M-1, M-2, \ldots, 0, \quad i = 1, 2, \ldots, N-1, \tag{14}$$

$$\psi_i^M = 0, \quad i = 0, 1, \ldots, N, \tag{15}$$

$$\frac{\psi_N^j - \psi_{N-1}^j}{h} = 0, \quad j = M, M-1, \ldots, 0, \tag{16}$$

$$\frac{\psi_1^j - \psi_0^j}{h} = -2\big(u_0^j - a_j\big), \quad j = M, M-1, \ldots, 0. \tag{17}$$

Solving (10)–(13) and (14)–(17) numerically and using iterative process to determine new improved values of $y(t)$ we can solve the stated problem. As we expected, there is no any movement of the endpoint where $t = T$. Numerical experiments show it very clearly.

It also must be noted that we considered

$$\tilde{I}(y) = \tau \sum_{j=0}^{M-1} \left(u_0^j - a_j\right)^2 \tag{18}$$

as an approximation of functional ("left rectangles"). Here are three experiments that were done with different initial approximations in which the following common input data is used:

1. $\varepsilon = 0.05^2$ (it is how close is I to zero).
2. $T = 1, L = 1$. Both space and time intervals are divided by 100 steps.
3. $\alpha_n = 1$ in iterative process $y_{n+1}(t) = y_n(t) - \alpha_n I'(y_n(t))$.
4. Exact solution is $y(t) = 2e^{2-t}$.

In all figures (–) imply exact solution and (...) is numerical solution and i is a number of iterations.

The results of experiments are shown on Figs. 1, 2, 3.

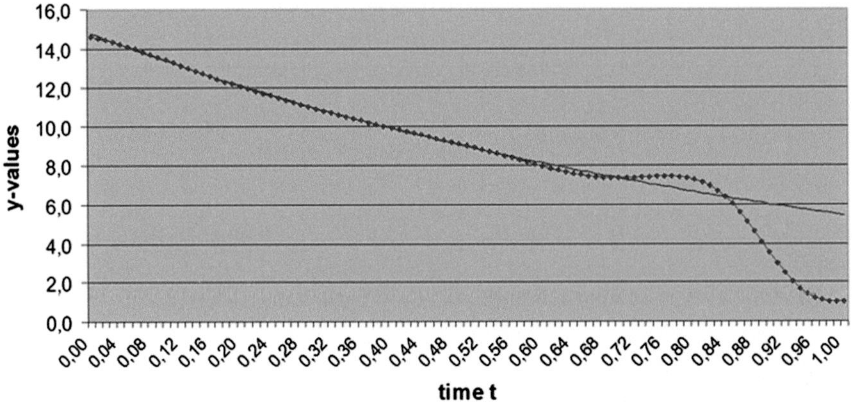

Fig. 1 $y_0 = 1, i = 2767, ||y_n - y_{\text{exact}}||_{\mathbf{H}} = 2.299$

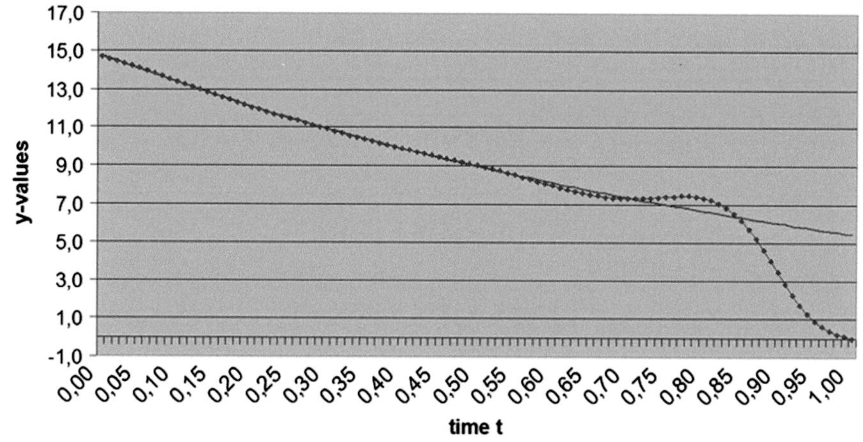

Fig. 2 $y_0 = 10 - 10t$, $i = 2995$, $||y_n - y_{\text{exact}}||_{\mathbf{H}} = 1.453$

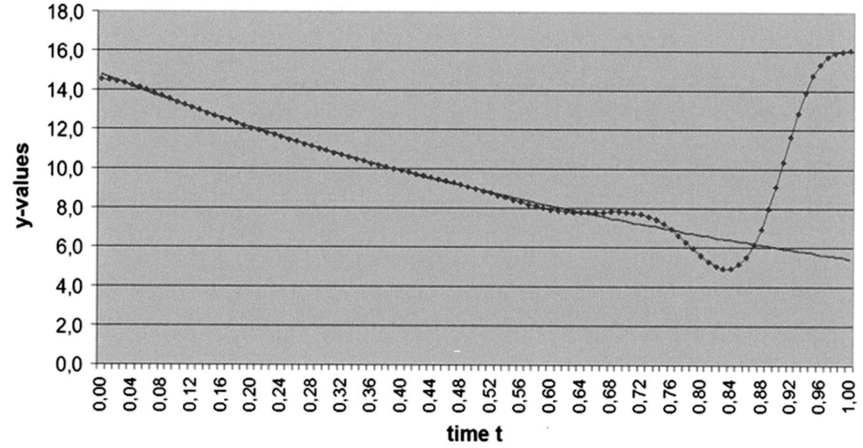

Fig. 3 $y_0 = 16$, $i = 9708$, $||y_n - y_{\text{exact}}||_{\mathbf{H}} = 2.630$

5 Gradient of Approximation Method

One more method is to derive an adjoint problem from (10)–(13), that is, we do not approximate adjoint problem, we derive an adjoint problem from approximation. Usually it is much more tedious and requires a lot of rigorous computations [4]. Let $\tilde{u}_i^j$ is a solution of (10)–(13) when y_j is replaced with $y_j + \sigma H_j$, where H_j are arbitrary values. Lets subtract two Eqs. (10)–(13) with solutions $\tilde{u}_i^j$ and u_i^j from each other and we take $\left(\text{assuming } z_i^j = \dfrac{\tilde{u}_i^j - u_i^j}{\sigma} = \dfrac{\delta u_i^j}{\sigma}\right)$

$$\frac{z_i^{j+1} - z_i^j}{\tau} = \frac{z_{i-1}^{j+1} - 2z_i^{j+1} + z_{i+1}^{j+1}}{h^2}, \quad i = 1, 2, \ldots, N-1, \quad j = 0, 1, \ldots, M-1, \tag{19}$$

$$z_i^0 = 0, \quad i = 0, 1, \ldots, N, \tag{20}$$

$$\frac{z_1^j - z_0^j}{h} = 0, \quad j = 0, 1, \ldots, M, \tag{21}$$

$$\frac{z_N^j - z_{N-1}^j}{h} = H_j, \quad j = 0, 1, \ldots, M. \tag{22}$$

We multiply both sides of (19) by unknown values ψ_i^j and sum by all $i = 1, 2, \ldots, N-1$ and $j = 0, 1, \ldots, M-1$:

$$0 = \sum_{i=1}^{N-1} \sum_{j=0}^{M-1} \left(\frac{z_i^{j+1} - z_i^j}{\tau} \psi_i^j - \frac{z_{i-1}^{j+1} - 2z_i^{j+1} + z_{i+1}^{j+1}}{h^2} \psi_i^j \right).$$

Using summation by parts we obtain:

$$\sum_{i=1}^{N-1} \left(\frac{z_i^M \psi_i^M}{\tau} - \frac{z_i^0 \psi_i^0}{\tau} - \sum_{j=0}^{M-1} z_i^{j+1} \left(\frac{\psi_i^{j+1} - \psi_i^j}{\tau} \right) \right) -$$
$$\sum_{j=0}^{M-1} \left(\frac{z_N^{j+1} \psi_N^j}{h^2} - \frac{z_1^{j+1} \psi_1^j}{h^2} - \frac{z_{N-1}^{j+1} \psi_N^j}{h^2} - \frac{z_0^{j+1} \psi_1^j}{h^2} \right) +$$
$$\sum_{j=0}^{M-1} \sum_{i=1}^{N-1} \left(\frac{z_{i+1}^{j+1} - z_i^{j+1}}{h} \cdot \frac{\psi_{i+1}^j - \psi_i^j}{h} \right).$$

here we use statements (21) and (22) and convert the last expression to the following by extracting one addend for $i = N-1$ from the last expression:

$$\sum_{i=1}^{N-1} \left(\frac{z_i^M \psi_i^M}{\tau} - \sum_{j=0}^{M-1} z_i^{j+1} \left(\frac{\psi_i^{j+1} - \psi_i^j}{\tau} \right) \right) - \sum_{j=0}^{M-1} \frac{\psi_N^j H_j}{h} +$$
$$\sum_{j=0}^{M-1} \left(\frac{z_N^{j+1} - z_{N-1}^{j+1}}{h} \cdot \frac{\psi_N^j - \psi_{N-1}^j}{h} + \frac{z_{N-1}^{j+1}(\psi_N^j - \psi_{N-1}^j)}{h^2} - \frac{z_1^{j+1}(\psi_2^j - \psi_1^j)}{h^2} - \right.$$
$$\left. \sum_{i=1}^{N-2} z_{i+1}^{j+1} \left(\frac{\psi_{i+2}^j - 2\psi_{i+1}^j + \psi_i^j}{h^2} \right) \right).$$

After further rearranging:

$$-\sum_{j=0}^{M-1}\sum_{i=1}^{N-1} z_i^{j+1}\left(\frac{\psi_i^{j+1}-\psi_i^j}{\tau}+\frac{\psi_{i+1}^j-2\psi_i^j+\psi_{i-1}^j}{h^2}\right)+$$
$$\sum_{i=1}^{N-1}\frac{z_i^M\psi_i^M}{\tau}-\sum_{j=0}^{M-1}\frac{\psi_N^jH_j}{h}+\sum_{j=0}^{M-1} z_N^{j+1}\left(\frac{\psi_N^j-\psi_{N-1}^j}{h^2}\right)-\sum_{j=0}^{M-1} z_1^{j+1}\left(\frac{\psi_1^j-\psi_0^j}{h^2}\right)$$

Now we put some restrictions on ψ_i^j to simplify the last expression:

$$\frac{\psi_N^j-\psi_{N-1}^j}{h}=0,\ \ j=0,1,\ldots,M-1,\ \big(\text{corresponds to }\ \partial_x\psi(t,L)=0\big),$$

$$\frac{\psi_i^{j+1}-\psi_i^j}{\tau}+\frac{\psi_{i+1}^j-2\psi_i^j+\psi_{i-1}^j}{h^2}=0,\ \ j=0,1,\ldots,M-1,\ \ i=1,2,\ldots,N-1,$$
$$\big(\partial_t\psi(t,x)+\partial_x^2\psi(t,x)=0\big),$$

$$\psi_i^M=0,\ \ i=1,2,\ldots,N-1,\ \big(\psi(T,x)=0\big),$$

From these we conclude that:

$$-\sum_{j=0}^{M-1}\frac{\psi_N^jH_j}{h}-\sum_{j=0}^{M-1} z_1^{j+1}\left(\frac{\psi_1^j-\psi_0^j}{h^2}\right)=0.$$

One interesting fact follows from (21) and says that $z_1^{j+1}=z_0^{j+1}$.

Our next step is to find variation of functional in discrete form:

$$\frac{\tilde I(y+\sigma H)-\tilde I(y)}{\sigma}=\frac{\tau}{\sigma}\sum_{j=0}^{M-1}\left(\left(\tilde u_0^{j+1}-a_{j+1}\right)^2-\left(u_0^{j+1}-a_{j+1}\right)^2\right)=$$
$$\frac{\tau}{\sigma}\sum_{j=0}^{M-1}2\delta u_0^{j+1}\left(u_0^{j+1}-a_{j+1}\right)+\frac{\tau}{\sigma}\sum_{j=0}^{M-1}\left(\delta u_0^{j+1}\right)^2.$$

Pay attention that we get another method for approximation of functional ("right rectangles"). As $\sigma\to 0$ and using notation of z we get: $\tau\sum_{j=0}^{M-1}2z_0^{j+1}\left(u_0^{j+1}-a_{j+1}\right)$.

This is a time to summarize what we obtained before.

$$\sum_{j=0}^{M-1}\psi_N^jH_j=-\sum_{j=0}^{M-1}z_0^{j+1}\left(\frac{\psi_1^j-\psi_0^j}{h}\right)$$

and we assume $\dfrac{\psi_1^j - \psi_0^j}{h} = -2\left(u_0^{j+1} - a_{j+1}\right)$. It follows that

$$\tau \sum_{j=0}^{M-1} 2z_0^{j+1}\left(u_0^{j+1} - a_{j+1}\right) = \tau \sum_{j=0}^{M-1} \psi_N^j H_j.$$

Reasonings above allows us to formulate a

Theorem 2 *Gateaux derivative of the functional I at the point y is determined by the formula*

$$\tilde{I}'(y) = \psi_N^j,$$

where ψ_N^j *is the solution of the adjoint system:*

$$\frac{\psi_i^{j+1} - \psi_i^j}{\tau} + \frac{\psi_{i+1}^j - 2\psi_i^j + \psi_{i-1}^j}{h^2} = 0,$$

$$j = M-1, M-2, \ldots, 0, \;\; i = 1, 2, \ldots, N-1, \tag{23}$$

$$\psi_i^M = 0, \;\; i = 0, 1, \ldots, N, \tag{24}$$

$$\frac{\psi_N^j - \psi_{N-1}^j}{h} = 0, \;\; j = M, M-1, \ldots, 0, \tag{25}$$

$$\frac{\psi_1^j - \psi_0^j}{h} = -2\left(u_0^{j+1} - a_{j+1}\right), \;\; j = M-1, \ldots, 0. \tag{26}$$

6 Comparing Two Methods

There is a slight difference if you compare (17) and (26). In the right hand side index by time variable has changed to $j+1$ instead of j. That was a hope to avoid a problem that we met in first part of the paper. Only numerical experiments can answer the question does the new method resolve the problem or not.

The numerical experiments are done with the same parameters as were introduced in graphs of Figs. 1, 2, 3. If graphs of Figs. 1, 2 and 3 are compared with graphs of Figs. 4, 5 and 6 respectively, then there is no visible changes. It means that method of obtaining gradient of approximated problem doesn't resolve problem of

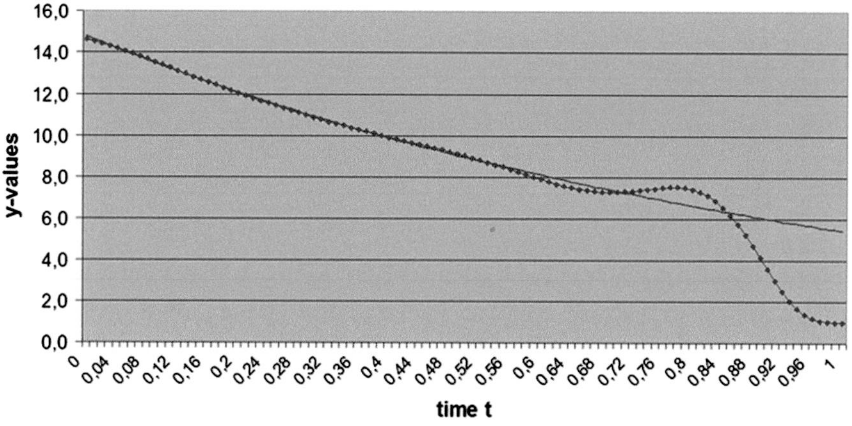

Fig. 4 $y_0 = 1, i = 2472, ||y_n - y_{\text{exact}}||_{\mathbf{H}} = 1.297$

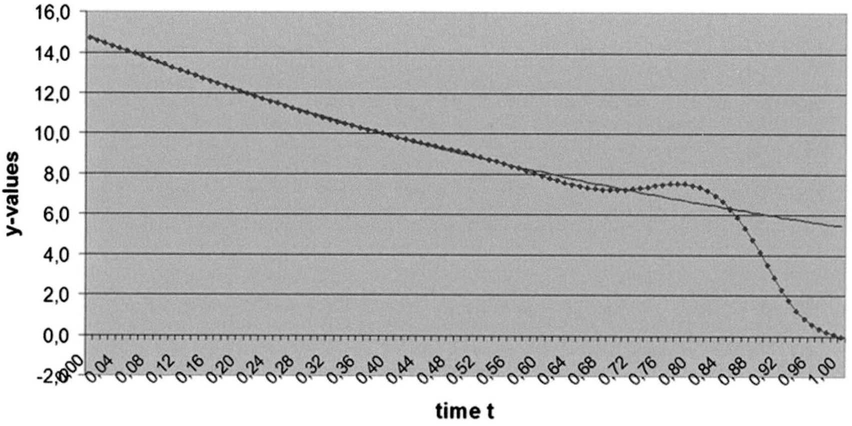

Fig. 5 $y_0 = 10 - 10t, i = 2669, ||y_n - y_{\text{exact}}||_{\mathbf{H}} = 1.450$

improvement on the right side of time interval. But some notable effect presents in a number of iterations. The second method allows to decrease a number of iterations by approximately 10 %.

One more experiment was done to see what happens if time interval becomes larger and now varies as $0 < t < 5$. All other parameters kept the same. See Figs. 7 and 8.

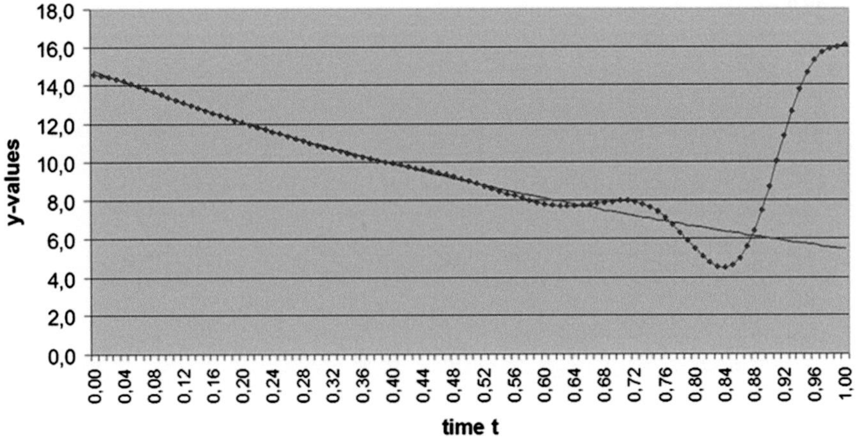

Fig. 6 $y_0 = 16, i = 8942, ||y_n - y_{\text{exact}}||_{\mathbf{H}} = 2.613$

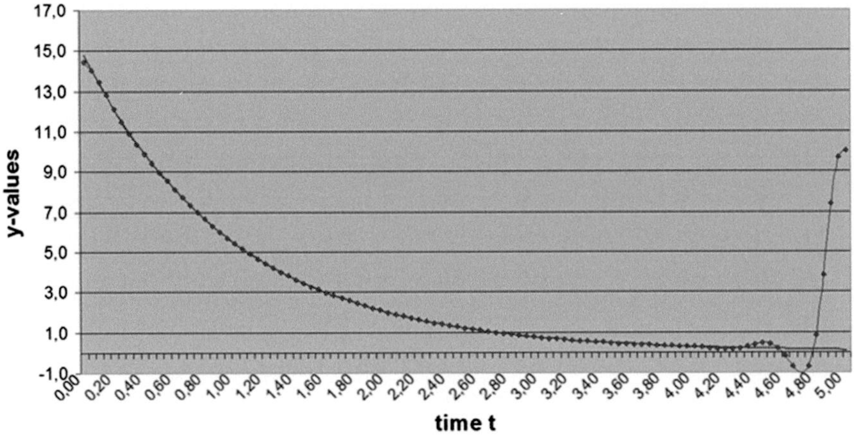

Fig. 7 $y_0 = 10, i = 4620, ||y_n - y_{\text{exact}}||_{\mathbf{H}} = 3.168, \alpha_n = 0.1, \ T = 5$

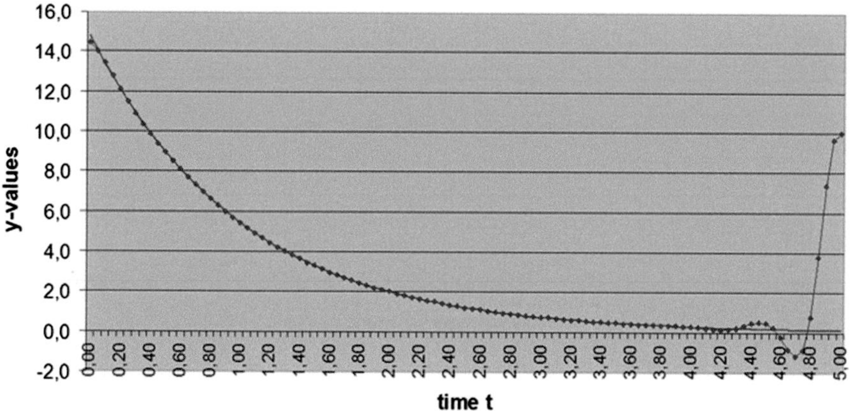

Fig. 8 $y_0 = 10, i = 4134, ||y_n - y_{\text{exact}}||_{\mathbf{H}} = 3.171, \alpha_n = 0.1,\ T = 5$

7 Conclusions

1. Approximation of gradient of the functional I and gradient of approximation of the functional I are not the same. The only difference arises in boundary condition. Obtaining a gradient of approximation of the functional is more difficult.
2. The second method of obtaining the adjoint system (gradient of approximation) doesn't avoid the problem that the error of numerical solution is large when t comes to T.
3. The number of iterations becomes less approximately by 10 % if we use the second method to obtain the adjoint system. That is the new method gives some advantages to reach solution more quickly.
4. If time interval increases (that is $[0, T]$ becomes larger) the situation doesn't change cardinally.

References

1. I.K. Shakenov, Inverse problems for parabolic equations on unlimited horizon of time. Bull. Kazakh Natl. Univ. Ser. Math. Mech. Inform. **70**(3), 36–48 (2011)
2. S.I. Kabanihin, *Inverse and Ill-Conditioned Problem* (Nauka, Novosibirsk, 2009)
3. A.A. Samarsky, *Theory of Difference Schemes* (Nauka, Moscow, 1983)
4. S.Ya. Serovajsky, *Optimization and Differentiation*, vol. 2 (Publisher of Kazakh University, Almaty, 2009)

Numerical Modeling of the Linear Relaxational Filtration by Monte Carlo Methods

Kanat Shakenov

Abstract Four models of linear relaxational filtration is considered. Initial and boundary conditions are set for them (Dirichlet, Neumann and mixed). Obtained problem solved by Monte Carlo methods—"random walk on spheres", "random walk on balls" and "random walk on lattices" of Monte Carlo methods and by probability difference methods.

Keywords Approximation • Difference • Expectation • Filtration • Method • Model • Monte Carlo • Numerical modeling • Probability • Random walk • Relaxation

Mathematics Subject Classification (2000). 65C05

1 Introduction

The linear relaxational filtration is described by the conservation law of pulse of resistance force, by the linearized conservation law of a fluid mass and determining relations for pulse of resistance forces and fluid mass. After exception of a pulse density of resistance forces ($\mathbf{J}$) and $(m\rho)$ this system with respect to pressure (p) and velocity of filtration $(\mathbf{W})$ is

$$\Delta p(x,t) = \frac{F(0)\Phi(0)}{\rho_0}\frac{\partial^2 p(x,t)}{\partial t^2} + \int_0^\infty \left(\frac{F(0)}{\rho_0}\frac{d\Phi(t')}{dt'} + \frac{\Phi(0)}{\rho_0}\frac{dF(t')}{dt'} + \frac{1}{\rho_0}\int_0^{t'} \frac{dF(\tau)}{d\tau}\frac{d\Phi(t'-\tau)}{d(t'-\tau)}d\tau \right) \frac{\partial^2 p(x,t-t')}{\partial (t-t')^2}dt', \tag{1}$$

K. Shakenov (✉)
al-Farabi Kazakh National University, 71 al-Farabi avenue, 050040 Almaty, Kazakhstan
e-mail: shakenov2000@mail.ru

M. Ruzhansky, S. Tikhonov (eds.), *Methods of Fourier Analysis and Approximation Theory*, Applied and Numerical Harmonic Analysis,
DOI 10.1007/978-3-319-27466-9_16

$$- F(0)\frac{\partial \mathbf{W}(x,t)}{\partial t} - \int_0^\infty \frac{dF(t')}{dt'} \frac{\partial \mathbf{W}(x,t-t')}{\partial (t-t')} dt' - \mathrm{grad}_x p(x,t). \tag{2}$$

Here $F(t)$ and $\Phi(t)$ are relaxation kernels of the filtration law and fluid mass [1].

We consider four models of relaxational filtration.

I. A model of classical elastic filtration. This model of filtration relates to the kernels of relaxation $F(t) = \frac{\mu}{\kappa} t\eta(t)$, $\Phi(t) = \rho_0 \beta \eta(t)$, and Eqs. (1) and (2) take form

$$\chi \Delta p(x,t) = \frac{\partial p(x,t)}{\partial t}, \tag{3}$$

$$\mathbf{W}(x,t) = -\frac{\kappa}{\mu} \mathrm{grad}_x p(x,t). \tag{4}$$

II. The simplest model of filtration with a constant speed of disturbance spread. This model is defined with kernels of relaxation: $F(t) = \frac{\mu}{\kappa}(t+\tau)\eta(t)$, $\Phi(t) = \rho_0 \beta \eta(t)$. For the given model the system (1)–(2) has a form:

$$\chi \Delta p(x,t) = \frac{\partial p(x,t)}{\partial t} + \tau \frac{\partial^2 p(x,t)}{\partial t^2}, \tag{5}$$

$$\tau \frac{\partial \mathbf{W}(x,t)}{\partial t} + \mathbf{W}(x,t) = -\frac{\kappa}{\mu} \mathrm{grad}_x p(x,t). \tag{6}$$

III. Filtration model in relaxationaly-compressed porous environment realized by the linear Darcy law. Corresponding kernels are $F(t) = \frac{\mu}{\kappa} t\eta(t)$, $\Phi(t) = \rho_0 \Big(\beta - \frac{\lambda_m - \lambda_p}{\lambda_m} \beta_c \exp\big(-\frac{t}{\lambda_m}\big)\Big)\eta(t)$. System (1)–(2) has a form:

$$\chi \Delta \left(p(x,t) + \lambda_m \frac{\partial p(x,t)}{\partial t} \right) = \frac{\partial}{\partial t} \left(p(x,t) + \lambda'_m \frac{\partial p(x,t)}{\partial t} \right), \tag{7}$$

$$\mathbf{W}(x,t) = -\frac{\kappa}{\mu} \mathrm{grad}_x p(x,t). \tag{8}$$

where $\lambda'_m = \lambda_m \frac{\beta_*}{\beta}$. In a particular case of incompressible fluid, $\beta_f = 0$ and $\lambda_p = 0$, instead of (7)–(8) we have

$$\chi \Delta \left(p(x,t) + \lambda_m \frac{\partial p(x,t)}{\partial t} \right) = \frac{\partial p(x,t)}{\partial t}, \tag{9}$$

$$\mathbf{W}(x,t) = -\frac{\kappa}{\mu} \mathrm{grad}_x p(x,t). \tag{10}$$

Model (9)–(10) describes a filtration of incompressible fluid in relaxationaly-compressed porous environment for $\lambda_p = 0$, and also in fractured-porous environment with infinitesimal elasticity of fractures and conductivity of blocks.

IV. Model of filtration by the simplest unbalanced law in elastic porous environment. Here the kernels of relaxation have form: $F(t) = \frac{\mu}{\kappa}\Big(t - \big(t_W - t_p\big)\Big(1 - \exp\big(-\frac{t}{\tau_p}\big)\Big)\Big)\eta(t)$, $\Phi(t) = \rho_0\beta\eta(t)$. For this model system (1)–(2) lead to form:

$$\chi\Delta\left(p(x,t) + \tau_p\frac{\partial p(x,t)}{\partial t}\right) = \frac{\partial}{\partial t}\left(p(x,t) + \tau_W\frac{\partial p(x,t)}{\partial t}\right), \tag{11}$$

$$\tau_W\frac{\partial \mathbf{W}(x,t)}{\partial t} + \mathbf{W}(x,t) = -\frac{\kappa}{\mu}\,\mathrm{grad}_x\left(p(x,t) + \tau_W\frac{\partial p(x,t)}{\partial t}\right). \tag{12}$$

We describe functions and parameters incoming in four filtration models. τ is time of relaxation, κ is penetrability coefficients, m is a porosity, t is time, β is elasticity capacity coefficient of the layer, $\beta = \beta_c + m_0\beta_f$, β_c is compressibility coefficient of the porous environment, m_0 is a fluid porosity in the unperturbed layer conditions, β_f is compressibility coefficient of the fluid, μ is a fluid viscosity, ρ is a fluid density, $\chi = \frac{\kappa}{\mu\beta}$ is piezoconductivity coefficient of the layer, $\eta(t)$ is Heaviside function, $\eta(t) = 1$ for $t > 0$, $\eta(t) = 1/2$ for $t = 0$, $\eta(t) = 0$ for $t < 0$, τ_W and τ_p nonnegative constants relaxation times of filtration velocity and pressure, $\lambda'_m = \lambda\frac{\beta_*}{\beta}$, $\beta_* = m_0\beta_f + \beta_c\lambda_p/\lambda_m$ is dynamics coefficient of elasticity capacity, λ_m is the relaxation time of porosity under the constant overfull of pressure, λ_p is the relaxation time of pressure under the constant porosity, p_0 is pressure in the unperturbed layer conditions, ρ_0 is density in the unperturbed layer conditions. All parameters are nonnegative given numbers [1].

Mathematical problems for models I–IV. Initial conditions for all four models. First of all in all four models in bounded region of filtration $\Omega \in \mathbb{R}^3$ with boundary $\partial\Omega$ and for $t \in [0, T]$ and for pressure $p(x,t)$ we consider Eqs. (3), (5), (7), (9) and (11). Then we set initial conditions for them. For Eqs. (3) and (9):

$$p(x,t) = a(x), \text{while } t = 0, \tag{13}$$

and for Eqs. (5), (7) and (11) besides condition (13) we give an additional condition

$$\frac{\partial p(x,t)}{\partial t} = b(x), \text{while } t = 0. \tag{14}$$

Boundary conditions for all four models.

Problem 1 (Dirichlet Problem) *In bounded filtration region $\Omega \in \mathbb{R}^3$ with boundary $\partial\Omega$ and for time $t \in [0, T]$, function $p(x, t)$ satisfies the boundary condition*

$$p(x,t) = p_1(x,t) \;\; for \;\; x \in \partial\Omega \times [0,T]. \tag{15}$$

Problem 2 (Neumann Problem) *In bounded filtration region $\Omega \in \mathbb{R}^3$ with boundary $\partial\Omega$ and for time $t \in [0, T]$, function $p(x, t)$ satisfies the boundary condition*

$$\frac{\partial p(x,t)}{\partial \mathbf{n}} = p_2(x,t) \;\; for \;\; x \in \partial\Omega \times [0,T], \tag{16}$$

where **n** *is an internal normal.*

Problem 3 (Mixed Problem) *In bounded filtration region $\Omega \in \mathbb{R}^3$ with boundary $\partial\Omega$ and for time $t \in [0, T]$, function $p(x, t)$ satisfies the boundary condition*

$$\alpha_1 p(x,t) + \beta_1 \frac{\partial p(x,t)}{\partial \mathbf{n}} = p_3(x,t) \;\; for \;\; x \in \partial\Omega \times [0,T], \tag{17}$$

where **n** *is an internal normal.*

An idea for solving by Monte Carlo methods.

Initial-boundary problem with respect to pressure $p(x, t)$ is discretized only by variable t. For that, interval $[0, T]$ split into N equal steps of $\Delta\tau = \frac{T}{N}$, $t_n = n\Delta\tau$, $n = 1, 2, \ldots, N$. As a result we get discrete boundary problem by time variable. Obtained, the problems for elliptic type PDEs (Helmholtz equation) solved by Monte Carlo methods.

2 Solution of the Initial Boundary Value Problems by Monte Carlo Methods

We demonstrate a solution of the initial boundary value problems by Monte Carlo methods on the following model—Filtration in relaxationaly-compressed porous environment realized by the linear Darcy law, that is model III. For this model we have a mathematical problem:

$$\chi\Delta\left(p(x,t) + \lambda_m \frac{\partial p(x,t)}{\partial t}\right) = \frac{\partial}{\partial t}\left(p(x,t) + \lambda'_m \frac{\partial p(x,t)}{\partial t}\right), \tag{18}$$

$$p(x,t) = a(x), \text{while} \;\; t = 0, \tag{19}$$

$$\frac{\partial p(x,t)}{\partial t} = b(x), \text{while} \;\; t = 0, \tag{20}$$

$$p(x,t) = p_1(x,t) \;\; \text{for} \;\; x \in \partial\Omega \times [0,T]. \tag{21}$$

2.1 Solution of the Dirichlet Problem (18)–(21)

Let coefficients $\chi, \ \lambda_m, \ \lambda'_m$ are while positive fixed values. Let us divide interval $t \in [0, T]$ into N equal parts with length $\Delta\tau$. So that $t_n = n \cdot \Delta\tau, \quad n = 0, 1, \dots, N, \quad \Delta\tau = \frac{T}{N}, \quad \Delta\tau > 0$, and we digitize only with respect to t using implicit scheme. In result taking into account λ'_m, we obtain Eq. (18) on time layer t_{n+1}

$$\Delta p^{n+1}(x) - a_1 \cdot p^{n+1}(x) = f^n(x), \tag{22}$$

where $f^n(x) = b_1 \cdot p^n(x) + c_1 \cdot \Delta p^{n-1}(x) + d_1 \cdot p^{n-1}(x)$, $c_1 = \dfrac{\lambda_m}{2\Delta\tau + \lambda_m}$, $a_1 = \dfrac{m_0\beta_f(\Delta\tau + 2\lambda_m) + \beta_c(\Delta\tau + 2\lambda_p)}{\wp}$, $b_1 = -\dfrac{4(m_0\beta_f\lambda_m + \beta_c\lambda_p)}{\wp}$, $d_1 = \dfrac{m_0\beta_f(2\lambda_m - \Delta\tau) + \beta_c(2\lambda_p - \Delta\tau)}{\wp}$, $\quad \wp = \Delta\tau\chi(2\Delta\tau + \lambda_m) \cdot (\beta_c + m\beta_f)$.

The algorithm "Random walk on spheres" of Monte Carlo methods. It is clear that $a_1 > 0$, as parameters $m_0, \ \beta_f, \ \Delta\tau, \ \lambda_m, \ \beta_c, \ \lambda_p, \ \chi$ are positive. Combining the initial condition with (22) we obtain

$$p^0(x) = a(x), \quad x \in \Omega, \quad \frac{p^1(x) - p^0(x)}{\Delta\tau} = b(x), \quad x \in \Omega, \tag{23}$$

which are the difference analogues of the initial data (19) and (20) respectively. For this problem the boundary condition transformed to:

$$p^{n+1}(x) = p_1^{n+1}(x), \quad x \in \partial\Omega. \tag{24}$$

We shall call the boundary $\partial\Omega$ (and $\partial\Omega_\varepsilon$) satisfying the Dirichlet condition as absorbing boundary. It is known that the problem (22)–(24) (Dirichlet problem for the Helmholtz equation of a time layer t_{n+1}), is solved with the help of "random walk on spheres" algorithm of Monte Carlo methods. The constructed ε–displaced estimation of the solution $p^{n+1}(x)$ with the help of "random walk on spheres" algorithm has a uniformly bounded variance by ε [2–7].

The algorithm "Random walk on balls" of Monte Carlo methods. "Random walk on balls" algorithm for solving a Dirichlet problem. This algorithm is similar to algorithm "random walk on spheres". In algorithm "random walk on balls" a "particle" passes from the center of the ball to a random point inside the ball and including a bound of ball (sphere), that is the following state of Markov chain inside the ball and including a bound of ball. It can be proved that Markov's chain converges in the same manner as for "random walk on spheres" algorithm and for finite number of steps to the ε-bound of $\partial\Omega_\varepsilon$. But it is obvious that convergence of Markov chain for "random walk on balls" algorithm is slower than for "random walk on spheres". For that reason, "random walk on balls" algorithm is almost unusable for numerical modeling by Monte Carlo methods. The constructed ε-displaced estimation of the solution $p^{n+1}(x)$ with the help of "random walk on balls" algorithm has a uniformly bounded variance by ε [5].

The algorithm "Random walk on lattices" of Monte Carlo methods. At first we approximate the solution (22)–(24) with the help of finite difference method and construct Markov chain, its transition probabilities are defined with the help of coefficients and parameters of the difference problem (22)–(24). For this purpose we use the following approximation of the second derivative with respect to x, i.e. $p^{n+1}_{x_i x_i}(x) = \dfrac{p^{n+1}(x+e_i h)+p^{n+1}(x-e_i h)-2p^{n+1}(x)}{h^2}$, where h is step along x, e_i is the unit vector along the axis x_i. Obviously $O(h^2)$ is a precision of the such approximation. Let's denote approximation of a domain Ω by ω_h, and boundary $\partial\Omega$—by γ_h. Now by time lowering superscripts $n+1,\ n,\ n-1$ from (22), we obtain the following finite difference equation

$$p(x_i) = \frac{1}{2+a_1 h^2}\cdot p(x_i+e_i h) + \frac{1}{2+a_1 h^2}\cdot p(x_i-e_i h) - \frac{h^2}{2+a_1 h^2}\cdot f(x_i). \qquad (25)$$

It's obvious that

$$\frac{2}{2+a_1 h^2} \longrightarrow 1 \ \text{ for } h\to 0,\ \ \tau\to,\ \ \lambda_m\to 0, \qquad (26)$$

where h is step along x, τ is time step. That is realization of (26) correspond to convergence requirements of a difference schemes and relaxation process. Let's denote $\alpha(x_i, y_i, h, \Delta\tau) = \dfrac{1}{2+a_1 h^2}$. As $\alpha(x_i, y_i, h, \Delta\tau) > 0$ and $\alpha + \alpha \le 1$ on y_i for $\forall x_i$, then $\alpha(x_i, y_i, h, \Delta\tau)$ are transition probabilities of Markov chain. Here $y_i = x_i \pm e_i h$, e_i is unit vector.

Algorithm At first we play a coordinate axis with probability $1/3$ for $\Omega \in \mathbb{R}^3$. Then the "particle" moves (along the direction $-e_i$ or $+e_i$) with identical probability α from node x_i into one of the neighboring node $x_i \pm e_i h$. It is necessary to take into account the "weight" of node, it is proportional to $\dfrac{h^2}{2+a_1 h^2}\cdot f(x_i)$. And so on until the "particle" achieves the discrete boundary γ_h. As soon as the "particle" achieves the boundary γ_h, boundary data $p_1(x_i)$ is add to a counter. Thus a random variable $\xi^h_{N_h}$ is defined along a discrete Markov chain with random length N_h. Then we average it on all trajectories, that is the estimation of the solution $p^{n+1}(x_i)$ in the node x_i is defined from $p^{n+1}(x_i) \approx \dfrac{1}{M}\sum_{i=1}^{M}\left(\xi^h_{N_h}\right)_i$, where M is trajectories amount of Markov chain starting from the node x_i [7–10].

Then we have the following

Theorem 1 *The Neumann–Ulam scheme is applicable to the finite difference problem for (22)–(24).*

Proof Proof of the theorem follows from algorithm of the discrete solution of problem (22)–(24). The complete proof see in [10]. The theorem is proved. □

In this case variance of an estimation of the solution $p^{n+1}(x_i)$ will be bounded, it can be explicitly calculated [5, 8, 10].

Probability difference method. Let's consider the finite difference problem (25) for a time layer $n+1$ with a discrete boundary condition $p(x_i) = p_1(x_i)$ $x_i \in \gamma_h$. Let's denote by $\{\zeta_i^h,\ i = 0, 1, \dots\}$ value of transition chain. Let $p_1(x)$ is the arbitrary continuous function for $x \in \gamma_h$. Let N_h is a moment of the first way out of a discrete domain ω_h: $N_h = \min\{i : \zeta_i^h \notin \omega_h\}$. Combining (25) with a boundary condition we obtain

$$p(x) = \mathbf{E}_x p(\zeta_1^h) + \Delta t^h \alpha f(x), \quad x \in \omega_h, \quad p(x) = p_1(x), \quad x \in \gamma_h \tag{27}$$

If $\mathbf{E}_x N_h < \infty$, then the problem (27) has a unique solution

$$p_h(x) = \mathbf{E}_x\left\{\sum_{i=0}^{N_h-1} f(\zeta_i^h) \cdot \Delta t_i^h + p_1(\zeta_{N_h}^h)\right\}. \tag{28}$$

Here $\Delta t_i^h = \Delta t^h(\zeta_i^h)$ is a process parameter. If $f(x) = 0$, $\mathbf{P}_x\left\{N_h < \infty\right\} = 1$, then (27) has the unique solution

$$p_h(x) = \mathbf{E}_x\left\{p_1(\zeta_{N_h}^h) \cdot I_{\{N_h<\infty\}}\right\}. \tag{29}$$

[7, 11–14].

2.2 *Solution of the Neumann Problem 2*

Let's consider (22) with initial conditions (23) and boundary conditions

$$\frac{\partial p^{n+1}(x)}{\partial \mathbf{n}} = p_2^{n+1}(x), \qquad x \in \partial\Omega. \tag{30}$$

The boundary $\partial\Omega$ (and $\partial\Omega_\varepsilon$), that correspond to the Neumann condition is called the reflecting boundary.

The algorithm "Random walk on spheres" of Monte Carlo methods. Let the solution of the problem (22)–(23), (30) is defined in a point $x_0 \in \Omega_\varepsilon$, where $\Omega_\varepsilon \subset \Omega$ is a domain with the boundary $\partial\Omega_\varepsilon$. $\partial\Omega_\varepsilon$ is ε—vicinity of the boundary Ω. State of Markov chain $\{x_i\}$ is defined with the help of the "random walk by spheres" process, by reaching $\partial\Omega_\varepsilon$—boundary the "particle" is reflected from $\partial\Omega_\varepsilon$—boundary into previous point (chain returns to the state before reflection). The "particle" continues random walk. After reflection the "weight" of boundary proportional to $p_2^{n+1}(x)$ is add to the counter. The chain breaks with the given probability $\zeta(\varepsilon)$, it is "small", $\zeta(\varepsilon) \to 0$ for $\varepsilon \to 0$. Here we shall note, that the "particle" moves to the $\partial\Omega_\varepsilon$—boundary along the normal **n** in "random walk on spheres" algorithm. We obtain ε—displaced estimation of a solution $p^{n+1}(x)$ of the problem (25), (23), (30) in point x by averaging of random variable η_{N_α} constructed along Markov chain of random length N_α. Probability error follows from the central limit theorem. It can be estimated as $\mathbf{P}\{\text{choice error } < \varepsilon\} \cong \text{erf}\left(\frac{|\epsilon|\sqrt{M/2}}{\sigma^2}\right)$, where **P** denote probability the error is no more than $|\varepsilon|$, M is quantity of trajectories, σ^2 is sampling of variance [5, 7, 15–18].

The algorithm "Random walk on balls" of Monte Carlo methods. "Random walk on balls" algorithm for solving a Neumann problem. This algorithm works in the same way as " random walk on balls" algorithm for Dirichlet problem. In this case, when "particle" reaches ε-bound of $\partial\Omega_\varepsilon$, is reflected in previous point and modeling of Markov chain is continued. The chain breaks with the given probability $\zeta(\varepsilon)$, it is "small", $\zeta(\varepsilon) \to 0$ for $\varepsilon \to 0$. Here we shall note, that the "particle" moves to the $\partial\Omega_\varepsilon$-boundary along the normal **n** in "random walk on balls" algorithm. We obtain ε-displaced estimation of a solution $p^{n+1}(x)$ of the problem (25), (23), (30) in point x by averaging of random variable η_{N_α} constructed along Markov chain of random length N_α. Probability error follows from the central limit theorem. It can be estimated as $\mathbf{P}\{\text{choice error } < \varepsilon\} \cong \text{erf}\left(\frac{|\epsilon|\sqrt{M/2}}{\sigma^2}\right)$, where **P** denote probability the error is no more than $|\varepsilon|$, M is quantity of trajectories, σ^2 is sampling of variance [5, 7, 15–18].

The algorithm "Random walk on lattices" of Monte Carlo methods. Just as in a case of Dirichlet problem we get finite difference Neumann problem for three-point difference equation on a time layer $n + 1$, that is (25), (23), (30). Here condition (26) for α is realized, i.e. $\alpha(x_i, y_i, h, \tau)$ are transition probabilities of Markov chain.

Algorithm At first we play coordinate axis with probability $1/3$. Then the "particle" moves (along the direction $-e_i$ or $+e_i$) with equal probability α from the node x_i into one of a neighboring node $x_i \pm e_i h$. It's necessary to take into account the "weight" of node proportional $\frac{h^2}{2 + a_1 h^2} \cdot f(x_i)$. And so on until the "particle" achieves a discrete boundary γ_h. By reaching γ_h—boundary the "particle" is reflected into previous point, and boundary data proportional $p_2(x_i)$ is add to the counter. Near the boundary $\partial\Omega$ a step of grid h^* along the direction to boundary γ_h such that the "particle" gets on discrete ε—boundary γ_h^ε. Random walk process continues. The chain breaks with the given probability $\zeta(\varepsilon)$, it is "small" value, $\zeta(\varepsilon) \to 0$ for $\varepsilon \to 0$. Thus, random variable $\eta_{N_h}^h$ is defined along a discrete Markov chain with random length N_h. Then we average it on all trajectories, that is the estimation of the solution $p^{n+1}(x_i)$ in node x_i is defined from

$p^{n+1}(x_i) \approx \frac{1}{M}\sum_{i=1}^{M}\left(\eta^h_{N_h}\right)_i$, where M is trajectories amount of Markov chain starting from the node x_i [7, 8, 10, 12–14].

Probability difference method. Let's consider the problem (25), (23), (26). Let $p_2(x)$ is a real bounded continuous function on a set $\partial\Omega$. $\partial\Omega$ is reflecting boundary. Approximation of (30) gives

$$\big(d(\partial\Omega)\cdot\nabla\big)p(x) = p_2(x). \tag{31}$$

Let the set $\partial\Omega_h$ approximate $\partial\Omega$ "from within". That is either $x \in \overline{\Omega}\bigcap\mathbb{R}^3_h$ or $x \in \partial\Omega$ or straight line connecting x with one of the nearest node $x_i \pm e_ih,\ x_i \pm e_ih \pm e_jh$ or $x_i \pm e_ih \mp e_jh$ touches $\partial\Omega$. Then α gives transition probabilities of the approximating chain ξ^h_i in Ω_h. The chain breaks with the given probability $\zeta(\varepsilon)$, it is "small", $\zeta(\varepsilon) \to 0$ for $\varepsilon \to 0$. It should be noted that $\mathbf{E}_x\left\{\xi^h_{n+1} - \xi^h_n \middle| \xi^h_n = y_i \in \partial\Omega_h\right\} = \upsilon(y)\,h/|\upsilon(y)|$. It is coordinated that reflection from the point $\partial\Omega_h$ happens along the direction $\upsilon(y)$. $\upsilon(y)$ is direction of hit in interior node. $\upsilon(x) = \sum_{i=1}^{3}|\upsilon_i(x)|$. Transition probabilities on $\partial\Omega^{\mathbb{R}}_h$: $\varrho_h(x, x \pm e_i\,h) = \upsilon^{\pm}_i/|\upsilon(x)|$. Let's define $A^h_n = \prod_{i=0}^{n}\exp\left(-a(\xi^h_i)\cdot\Delta t^h_i\cdot I_{\Omega_h}(\xi^h_i)\right)$, where t^h_i is a discrete time, parameter of ξ^h_i process. For the chain with random length N_h we get unique discrete approximation of solution of the problem (25), (23), (26)

$$p_h(x) = \mathbf{E}_x\left\{\sum_{i=0}^{N_h-1} A^h_i\cdot f(\xi^h_i)\cdot\Delta t^h_i\cdot I_{\Omega_h}(\xi^h_h) + \sum_{i=0}^{N_h-1} A^h_i\cdot p_2(\xi^h_i)\cdot d\phi^h_i\right\} \tag{32}$$

[7, 8, 12, 14].

2.3 *Solution of the Mixed Problem 3*

Let's consider the problem (22), (23). To this problem we'll connect approximation of the mixed boundary condition (17) on a time layer $n+1$

$$\alpha_1 p^{n+1}(x) + \beta_2 p^{n+1}(x) = p^{n+1}_3(x), \quad x \in \partial\Omega_\varepsilon, \tag{33}$$

where $\beta_2 = \beta_1\big(d(\partial\Omega)\cdot\nabla\big)$.

The algorithm "Random walk on spheres" of Monte Carlo methods. As in a case of the Dirichlet problem we construct Markov chain by "random walk on spheres". In general, by reaching $\partial\Omega_\varepsilon$—boundary of a domain Ω the "particle" is absorbed or reflected with equal probability $1/2$. But in our case, if the "particle" is absorbed, then in each point the value of "weight" $p_3(x_i)/\alpha_1$ is added to Markov chain, and if the "particle"reflected then we add $p_3(x_i)/\beta_2$. The chain breaks if a "particle" is absorbed. We get ε—displaced estimation of the solution $p_\varepsilon(x)$ of the problem (25), (23), (33) in the point x by averaging random variable ξ_i constructed along Markov chain with random length N.That is $p_\varepsilon(x) = \frac{1}{M}\sum_{i=1}^{M}\xi_i$ [7, 13, 14, 19, 20].

The algorithm "Random walk on balls" of Monte Carlo methods. "Random walk on balls" for solving a mixed problem. As in a case of the Dirichlet problem we construct Markov chain by "random walk on balls". In general, by reaching $\partial\Omega_\varepsilon$-boundary of a domain Ω the "particle" is absorbed or reflected with equal probability 1/2. But in our case, if the "particle" is absorbed, then in each point the value of "weight" $p_3(x_i)/\alpha_1$ is added to Markov chain, and if the "particle"reflected then we add $p_3(x_i)/\beta_2$. The chain breaks if a "particle" is absorbed. We get ε-displaced estimation of the solution $p_\varepsilon(x)$ of the problem (25), (23), (33) in the point x by averaging random variable ξ_i constructed along Markov chain with random length N. That is $p_\varepsilon(x) = \frac{1}{M}\sum_{i=1}^{M}\xi_i$ [7, 13, 14, 19, 20].

The algorithm "Random walk on lattices" of Monte Carlo methods. Let's consider the following finite difference problem (25), (23)

$$\alpha_1 p(x_i) + \beta_2 p(x_i) = p_3(x_i), \quad x_i \in \gamma_h. \tag{34}$$

The problem (25), (23), (34) is considered on a time layer $n + 1$.

Algorithm At first we play coordinate axis with probability $1/3$ for $\Omega \in \mathbb{R}^3$. Then the "particle" moves (along the direction $-e_i$ or $+e_i$) with equal probability α from the node x_i into one of a neighboring node $x_i \pm e_i h$. It is necessary to take into account the "weight" of node, it proportional $\frac{h^2}{2 + a_1 h^2} \cdot f(x_i)$. And so on until the "particle" achieves the discrete boundary γ_h. In general, on a boundary γ_h the "particle" is absorbed or reflected with equal probability $1/2$. But in our case, the chain breaks if the "particle" is absorbed, and we add to counter a "weight" of absorbing boundary node $p_3(x_i)/\alpha_1$, at reflection—$p_3(x_i)/\beta_2$. Thus, we define a random variable $\xi^h_{N_h}$ along a discrete Markov chain with random length N_h. The estimation of solution $p_h(x_i)$ in a node x_i is defined by $p_h(x_i) \approx \frac{1}{M}\sum_{i=1}^{M}\left(\xi^h_{N_h}\right)_i$, where M is trajectories amount of Markov chain starting from the node x_i [7, 8, 10, 11, 13, 14, 19].

Probability difference method. The problem (25), (23), (34) is considered on a time layer $n + 1$. Let $p_3(x)$ is the real bounded continuous function on a set $\partial\Omega$. Let the set $\partial\Omega^{\mathbb{R}}_h$ approximate $\partial\Omega$ "from within". That is either $x \in \overline{\Omega}\bigcap\mathbb{R}^3_h$ or $x \in \partial\Omega$ or straight line connecting x with one of the nearest node $x_i \pm e_i h$, $x_i \pm e_i h \pm e_j h$ or $x_i \pm e_i h \mp e_j h$ touches $\partial\Omega$. The set is determined in $\overline{\Omega}\bigcap\mathbb{R}^3_h$. Let's define digitization $\Omega_h = \Omega\bigcap\mathbb{R}^3_h - \partial\Omega^{\mathbb{R}}_h$ of interior Ω and digitization of a stopping set

$\partial\Omega_h^A = \mathbb{R}_h^3 - \Omega_h - \partial\Omega_h^{\mathbb{R}}$. Then α gives transitive probabilities of the approximating chain ξ_i^h in Ω_h. The chain breaks at the first contact with $\partial\Omega_h^A$. Let's notice that $\mathbf{E}_x\left\{\xi_{n+1}^h - \xi_n^h | \xi_n^h = y_i \in \partial\Omega_h^{\mathbb{R}}\right\} = \upsilon(y)\, h/|\upsilon(y)|$. It is coordinated that the reflection from the point $\partial\Omega_h^{\mathbb{R}}$ happens along direction $\upsilon(y)$. $\upsilon(y)$ is the direction of hit into interior node. Let's define $A_n^h = \prod_{i=0}^{n} \exp\big(-a(\xi_i^h)\cdot \Delta t_i^h \cdot I_{\Omega_h}(\xi_i^h)\big)$, $C_n^h = \prod_{i=0}^{n} \exp\big(-\beta_1(\xi_i^h)\, d\phi_i^h\big)$, $D_n^h = A_n^h\, C_n^h$. We consider the case $\alpha_1 = \alpha_1(x)$, $\beta_1 = \beta_1(x)$, t_i^h is a discrete time, parameter of the process ξ_i^h, $d\phi^h = h/|\upsilon(x)|$, $d\phi_i^h = d\phi^h(\xi_i^h)\, I_{\partial\Omega_h^R}(\xi_i^h)$. For the chain with random length $N_h = \min\{n : \xi_n^h \in \partial\Omega_h^A\}$ we obtain unique discrete approximation of a solution of the problem (25), (23), (34)

$$p_h(x) = \mathbf{E}_x\Big\{\sum_{i=0}^{N_h-1} D_i^h \cdot f(\xi_i^h)\cdot \Delta t_i^h \cdot I_{\Omega_h}(\xi_i^h) + D_{N_h-1}^h\, \alpha_1(\xi_{N_h}^h) + \sum_{i=0}^{N_h-1} D_i^h \cdot p_3(\xi_i^h)\cdot d\phi_i^h\Big\}. \tag{35}$$

[7, 8, 10–14, 19].

3 Solution of the Initial Boundary Value Problem for the Model (I): Classical Elastic Filtration Model by Monte Carlo Methods

3.1 *Mathematical Setting of Dirichlet Problem*

for this model has the following form:

$$\chi\Delta p(x,t) = \frac{\partial p(x,t)}{\partial t}, \tag{36}$$

$$p(x,t) = a(x), \text{while } \ t = 0, SplitEq \tag{37}$$

$$p(x,t) = p_1(x,t) \ \text{ for } \ x \in \partial\Omega \times [0,T]. \tag{38}$$

After approximation only by time variable Eq. (36) has the form:

$$\Delta p^{n+1}(x) - a_1 p^{n+1}(x) = f^n(x), \tag{39}$$

where $a_1 = \dfrac{1}{\Delta\tau\chi}$, $f^n(x) = -\dfrac{1}{\Delta\tau\chi}p^n(x)$. Initial and boundary conditions (37) and (38):

$$p^0(x) = a(x), \ \ x \in \Omega, \tag{40}$$

$$p^{n+1}(x) = p_1^{n+1}(x), \ \ x \in \partial\Omega. \tag{41}$$

Problem (39)–(41) for a fixed time layer $n = 0, 1, \ldots, N-1$ can be considered as a Dirichlet problem for Helmholtz equation.

3.2 Mathematical Setting of Neumann Problem

for classical elastic filtration model:

$$\chi \Delta p(x,t) = \frac{\partial p(x,t)}{\partial t}, \tag{42}$$

$$p(x,t) = a(x), \text{ while } t = 0, \tag{43}$$

$$\frac{\partial p(x,t)}{\partial \mathbf{n}} = p_2(x,t) \text{ for } x \in \partial\Omega \times [0,T], \tag{44}$$

where $\mathbf{n}$ is an internal normal. After approximation only by time variable t Eq. (42) has the form:

$$\Delta p^{n+1}(x) - a_1 p^{n+1}(x) = f^n(x), \tag{45}$$

where $a_1 = \frac{1}{\Delta\tau\chi}, f^n(x) = -\frac{1}{\Delta\tau\chi} p^n(x)$. Initial condition (43) has the same form as for approximated Dirichlet problem:

$$p^0(x) = a(x), \quad x \in \Omega, \tag{46}$$

and boundary Neumann condition (44):

$$\frac{\partial p^{n+1}(x)}{\partial \mathbf{n}} = p_2^{n+1}(x) \quad x \in \partial\Omega. \tag{47}$$

Problem (45)–(47) for a fixed time layer $n = 0, 1, \ldots, N-1$ can be considered as a Neumann problem for Helmholtz equation.

3.3 Mathematical Setting of Mixed Problem

for classical elastic filtration model:

$$\chi \Delta p(x,t) = \frac{\partial p(x,t)}{\partial t}, \tag{48}$$

$$p(x,t) = a(x), \text{ while } t = 0, \tag{49}$$

$$\alpha_1 p(x,t) + \beta_1 \frac{\partial p(x,t)}{\partial \mathbf{n}} = p_3(x,t) \text{ for } x \in \partial\Omega \times [0,T], \tag{50}$$

where **n** is an internal normal. After approximation only by time variable t Eq. (48) has the form:

$$\Delta p^{n+1}(x) - a_1 p^{n+1}(x) = f^n(x), \tag{51}$$

where $a_1 = \frac{1}{\Delta\tau\chi}, f^n(x) = -\frac{1}{\Delta\tau\chi} p^n(x)$. Initial condition (49) has the same form as for approximated Dirichlet problem:

$$p^0(x) = a(x), \;\; x \in \Omega, \tag{52}$$

and mixed boundary condition (50):

$$\alpha_1 p^{n+1}(x) + \beta_2 p^{n+1}(x) = p_3^{n+1}(x), \quad x \in \partial\Omega_\varepsilon, \tag{53}$$

where $\beta_2 = \beta_1 \big(d(\partial\Omega) \cdot \nabla\big)$. Problem (51)–(53) for a fixed time layer $n = 0, 1, \ldots, N-1$ can be considered as a mixed problem for Helmholtz equation.

Problems described in 3.1, 3.2 and 3.3, that is Dirichlet, Neumann and mixed problems (39)–(41), (45)–(47) and (51)–(53) are solved by Monte Carlo methods algorithms in the same way as problems from 2, as for model **III**—Filtration in relaxationaly-compressed porous environment realized by the linear Darcy law.

4 Solution of the Initial Boundary Value Problem for the Model (II): The Simplest Model of Filtration with a Constant Speed of Disturbance Spread by Monte Carlo Methods

4.1 Mathematical Setting of Dirichlet Problem

for this model has the following form:

$$\chi \Delta p(x,t) = \frac{\partial p(x,t)}{\partial t} + \tau \frac{\partial^2 p(x,t)}{\partial t^2}, \tag{54}$$

$$p(x,t) = a(x), \text{ while } t = 0, \tag{55}$$

$$\frac{\partial p(x,t)}{\partial t} = b(x), \text{ while } t = 0, \tag{56}$$

$$p(x,t) = p_1(x,t) \text{ for } x \in \partial\Omega \times [0,T]. \tag{57}$$

After approximation only by time variable t Eq. (54) has the form:

$$\Delta p^{n+1}(x) - a_1 p^{n+1}(x) = f^n(x), \tag{58}$$

where $a_1 = \dfrac{\Delta\tau + \tau}{\Delta\tau^2\chi}$, $f^n(x) = -\dfrac{\Delta\tau + 2\tau}{\Delta\tau^2\chi}p^n(x) + \dfrac{\tau}{\Delta\tau^2\chi}p^{n-1}(x)$. Initial and boundary conditions (55), (56) and (57):

$$p^0(x) = a(x), \quad x \in \Omega, \tag{59}$$

$$\frac{p^1(x) - p^0(x)}{\Delta\tau} = b(x), \quad x \in \Omega, \tag{60}$$

$$p^{n+1}(x) = p_1^{n+1}(x), \quad x \in \partial\Omega. \tag{61}$$

Problem (58)–(61) for a fixed time layer $n = 0, 1, \ldots, N-1$ can be considered as a Dirichlet problem for Helmholtz equation.

4.2 *Mathematical Setting of Neumann Problem*

for this model has the following form:

$$\chi\Delta p(x,t) = \frac{\partial p(x,t)}{\partial t} + \tau\frac{\partial^2 p(x,t)}{\partial t^2}, \tag{62}$$

$$p(x,t) = a(x), \text{while } t = 0, \tag{63}$$

$$\frac{\partial p(x,t)}{\partial t} = b(x), \text{while } t = 0, \tag{64}$$

$$\frac{\partial p(x,t)}{\partial \mathbf{n}} = p_2(x,t) \text{ for } x \in \partial\Omega \times [0,T], \tag{65}$$

where $\mathbf{n}$ is an internal normal. After approximation only by time variable t Eq. (62) has the form:

$$\Delta p^{n+1}(x) - a_1 p^{n+1}(x) = f^n(x), \tag{66}$$

where $a_1 = \dfrac{\Delta\tau + \tau}{\Delta\tau^2\chi}$, $f^n(x) = -\dfrac{\Delta\tau + 2\tau}{\Delta\tau^2\chi}p^n(x) + \dfrac{\tau}{\Delta\tau^2\chi}p^{n-1}(x)$. Initial and boundary conditions (63), (64) and (65):

$$p^0(x) = a(x), \quad x \in \Omega, \tag{67}$$

$$\frac{p^1(x) - p^0(x)}{\Delta\tau} = b(x), \quad x \in \Omega, \tag{68}$$

$$\frac{\partial p^{n+1}(x)}{\partial \mathbf{n}} = p_2^{n+1}(x), \qquad x \in \partial\Omega. \tag{69}$$

Problem (66)–(69) for a fixed time layer $n = 0, 1, \ldots, N-1$ can be considered as a Neumann problem for Helmholtz equation.

4.3 Mathematical Setting of Mixed Problem

for this model has the following form:

$$\chi \Delta p(x,t) = \frac{\partial p(x,t)}{\partial t} + \tau \frac{\partial^2 p(x,t)}{\partial t^2}, \tag{70}$$

$$p(x,t) = a(x), \text{ while } t = 0, \tag{71}$$

$$\frac{\partial p(x,t)}{\partial t} = b(x), \text{ while } t = 0, \tag{72}$$

$$\alpha_1 p(x,t) + \beta_1 \frac{\partial p(x,t)}{\partial \mathbf{n}} = p_3(x,t) \text{ for } x \in \partial\Omega \times [0,T], \tag{73}$$

where $\mathbf{n}$ is an internal normal. After approximation only by time variable t Eq. (70) has the form:

$$\Delta p^{n+1}(x) - a_1 p^{n+1}(x) = f^n(x), \tag{74}$$

where $a_1 = \frac{\Delta\tau + \tau}{\Delta\tau^2 \chi}$, $f^n(x) = -\frac{\Delta\tau + 2\tau}{\Delta\tau^2 \chi} p^n(x) + \frac{\tau}{\Delta\tau^2 \chi} p^{n-1}(x)$. Initial and boundary conditions (71), (72) and (73):

$$p^0(x) = a(x), \quad x \in \Omega, \tag{75}$$

$$\frac{p^1(x) - p^0(x)}{\Delta\tau} = b(x), \quad x \in \Omega, \tag{76}$$

$$\alpha_1 p^{n+1}(x) + \beta_2 p^{n+1}(x) = p_3^{n+1}(x), \quad x \in \partial\Omega_\varepsilon, \tag{77}$$

where $\beta_2 = \beta_1\big(d(\partial\Omega) \cdot \nabla\big)$. Problem (74)–(77) for a fixed time layer $n = 0, 1, \ldots, N-1$ can be considered as a mixed problem for Helmholtz equation.

Problems described in 4.1, 4.2 and 4.3, that is Dirichlet, Neumann and mixed problems (58)–(61), (66)–(69) and (74)–(77) are solved by Monte Carlo methods algorithms in the same way as problems from 2, as for model **III**—Filtration in relaxationaly-compressed porous environment realized by the linear Darcy law.

5 Solution of the Initial Boundary Value Problem for the Model (IV): Model of Filtration by the Simplest Unbalanced Law in Elastic Porous Environment by Monte Carlo Methods

5.1 Mathematical Setting of Dirichlet Problem

for this model has the following form:

$$\chi\Delta\left(p(x,t)+\tau_p\frac{\partial p(x,t)}{\partial t}\right)=\frac{\partial}{\partial t}\left(p(x,t)+\tau_W\frac{\partial p(x,t)}{\partial t}\right), \tag{78}$$

$$p(x,t)=a(x), \text{while } t=0, \tag{79}$$

$$\frac{\partial p(x,t)}{\partial t}=b(x), \text{while } t=0, \tag{80}$$

$$p(x,t)=p_1(x,t) \text{ for } x\in\partial\Omega\times[0,T]. \tag{81}$$

After approximation only by time variable t Eq. (78) has the form:

$$\Delta p^{n+1}(x)-a_1p^{n+1}(x)=f^n(x), \tag{82}$$

where $a_1=\dfrac{\Delta\tau+\tau_W}{\Delta\tau\chi(\Delta\tau+\tau_p)}$, $f^n(x)=\dfrac{\tau_p}{\Delta\tau+\tau_p}\Delta p^n(x)-\dfrac{\Delta\tau+2\tau_W}{\Delta\tau\chi(\Delta\tau+\tau_p)}p^n(x)+\dfrac{\tau_W}{\Delta\tau\chi(\Delta\tau+\tau_p)}p^{n-1}(x)$. Initial and boundary conditions (79), (80) and (81):

$$p^0(x)=a(x),\quad x\in\Omega, \tag{83}$$

$$\frac{p^1(x)-p^0(x)}{\Delta\tau}=b(x),\quad x\in\Omega, \tag{84}$$

$$p^{n+1}(x)=p_1^{n+1}(x),\quad x\in\partial\Omega. \tag{85}$$

Problem (82)–(85) for a fixed time layer $n=0,1,\ldots,N-1$ can be considered as a Dirichlet problem for Helmholtz equation.

5.2 Mathematical Setting of Neumann Problem

for this model has the following form:

$$\chi\Delta\left(p(x,t)+\tau_p\frac{\partial p(x,t)}{\partial t}\right)=\frac{\partial}{\partial t}\left(p(x,t)+\tau_W\frac{\partial p(x,t)}{\partial t}\right), \tag{86}$$

$$p(x,t) = a(x), \text{while } t = 0, \tag{87}$$

$$\frac{\partial p(x,t)}{\partial t} = b(x), \text{while } t = 0, \tag{88}$$

$$\frac{\partial p(x,t)}{\partial \mathbf{n}} = p_2(x,t) \text{ for } x \in \partial\Omega \times [0,T], \tag{89}$$

where $\mathbf{n}$ is an internal normal. After approximation only by time variable t Eq. (62) has the form:

$$\Delta p^{n+1}(x) - a_1 p^{n+1}(x) = f^n(x), \tag{90}$$

where $a_1 = \dfrac{\Delta\tau + \tau_W}{\Delta\tau\chi(\Delta\tau + \tau_p)}$, $f^n(x) = \dfrac{\tau_p}{\Delta\tau + \tau_p}\Delta p^n(x) - \dfrac{\Delta\tau + 2\tau_W}{\Delta\tau\chi(\Delta\tau + \tau_p)}p^n(x) + \dfrac{\tau_W}{\Delta\tau\chi(\Delta\tau + \tau_p)}p^{n-1}(x)$. Initial and boundary conditions (87), (88) and (89):

$$p^0(x) = a(x), \quad x \in \Omega, \tag{91}$$

$$\frac{p^1(x) - p^0(x)}{\Delta\tau} = b(x), \quad x \in \Omega, \tag{92}$$

$$\frac{\partial p^{n+1}(x)}{\partial \mathbf{n}} = p_2^{n+1}(x), \qquad x \in \partial\Omega. \tag{93}$$

Problem (90)–(93) for a fixed time layer $n = 0, 1, \ldots, N-1$ can be considered as a Neumann problem for Helmholtz equation.

5.3 *Mathematical Setting of Mixed Problem*

for this model has the following form:

$$\chi\Delta\left(p(x,t) + \tau_p\frac{\partial p(x,t)}{\partial t}\right) = \frac{\partial}{\partial t}\left(p(x,t) + \tau_W\frac{\partial p(x,t)}{\partial t}\right), \tag{94}$$

$$p(x,t) = a(x), \text{while } t = 0, \tag{95}$$

$$\frac{\partial p(x,t)}{\partial t} = b(x), \text{while } t = 0. \tag{96}$$

$$\alpha_1 p(x,t) + \beta_1\frac{\partial p(x,t)}{\partial \mathbf{n}} = p_3(x,t) \text{ for } x \in \partial\Omega \times [0,T], \tag{97}$$

where **n** is an internal normal. After approximation only by time variable t Eq. (94) has the form:

$$\Delta p^{n+1}(x) - a_1 p^{n+1}(x) = f^n(x), \tag{98}$$

where $a_1 = \dfrac{\Delta\tau + \tau_W}{\Delta\tau\chi(\Delta\tau + \tau_p)}$, $f^n(x) = \dfrac{\tau_p}{\Delta\tau + \tau_p}\Delta p^n(x) - \dfrac{\Delta\tau + 2\tau_W}{\Delta\tau\chi(\Delta\tau + \tau_p)}p^n(x) + \dfrac{\tau_W}{\Delta\tau\chi(\Delta\tau + \tau_p)}p^{n-1}(x)$. Initial and boundary conditions (95), (96) and (97):

$$p^0(x) = a(x), \quad x \in \Omega, \tag{99}$$

$$\frac{p^1(x) - p^0(x)}{\Delta\tau} = b(x), \quad x \in \Omega, \tag{100}$$

$$\alpha_1 p^{n+1}(x) + \beta_2 p^{n+1}(x) = p_3^{n+1}(x), \quad x \in \partial\Omega_\varepsilon, \tag{101}$$

where $\beta_2 = \beta_1\big(d(\partial\Omega)\cdot\nabla\big)$. Problem (98)–(101) for a fixed time layer $n = 0, 1, \ldots, N-1$ can be considered as a mixed problem for Helmholtz equation.

Problems described in 5.1, 5.2 and 5.3, that is Dirichlet, Neumann and mixed problems (82)–(85), (90)–(93) and (98)–(101) are solved by Monte Carlo methods algorithms in the same way as problems from 2, as for model **III**—Filtration in relaxationaly-compressed porous environment realized by the linear Darcy law.

6 Solution of the Initial Boundary Value Problem for the Model (III): Filtration Model in Relaxationaly-Compressed Porous Environment Realized by the Linear Darcy Law, Part 2 ($\beta_f = 0$ and $\lambda_p = 0$), by Monte Carlo Methods

6.1 Mathematical Setting of Dirichlet Problem

for this model has the following form:

$$\chi\Delta\left(p(x,t) + \lambda_m\frac{\partial p(x,t)}{\partial t}\right) = \frac{\partial p(x,t)}{\partial t}, \tag{102}$$

$$p(x,t) = a(x), \text{while } t = 0, \tag{103}$$

$$p(x,t) = p_1(x,t) \text{ for } x \in \partial\Omega\times[0,T]. \tag{104}$$

After approximation only by time variable t Eq. (102) has the form:

$$\Delta p^{n+1}(x) - a_1 p^{n+1}(x) = f^n(x), \tag{105}$$

where $a_1 = \dfrac{1}{\chi(\Delta\tau + \lambda_m)}$, $f^n(x) = \dfrac{\lambda_m}{\Delta\tau + \lambda_m}\Delta p^n(x) - \dfrac{1}{\chi(\Delta\tau + \lambda_m)}p^n(x)$. Initial and boundary conditions (103)and (104):

$$p^0(x) = a(x), \quad x \in \Omega, \tag{106}$$

$$p^{n+1}(x) = p_1^{n+1}(x), \quad x \in \partial\Omega. \tag{107}$$

Problem (105)–(107) for a fixed time layer $n = 0, 1, \ldots, N-1$ can be considered as a Dirichlet problem for Helmholtz equation.

6.2 *Mathematical Setting of Neumann Problem*

for this model has the following form:

$$\chi\Delta\left(p(x,t) + \lambda_m \frac{\partial p(x,t)}{\partial t}\right) = \frac{\partial p(x,t)}{\partial t}, \tag{108}$$

$$p(x,t) = a(x), \text{ while } t = 0, \tag{109}$$

$$\frac{\partial p(x,t)}{\partial \mathbf{n}} = p_2(x,t) \ \text{ for } x \in \partial\Omega \times [0,T], \tag{110}$$

where $\mathbf{n}$ is an internal normal. After approximation only by time variable t Eq. (108) has the form:

$$\Delta p^{n+1}(x) - a_1 p^{n+1}(x) = f^n(x), \tag{111}$$

where $a_1 = \dfrac{1}{\chi(\Delta\tau + \lambda_m)}$, $f^n(x) = \dfrac{\lambda_m}{\Delta\tau + \lambda_m}\Delta p^n(x) - \dfrac{1}{\chi(\Delta\tau + \lambda_m)}p^n(x)$. Initial and boundary conditions (109) and (110):

$$p^0(x) = a(x), \quad x \in \Omega, \tag{112}$$

$$\frac{\partial p^{n+1}(x)}{\partial \mathbf{n}} = p_2^{n+1}(x), \qquad x \in \partial\Omega. \tag{113}$$

Problem (111)–(113) for a fixed time layer $n = 0, 1, \ldots, N-1$ can be considered as a Neumann problem for Helmholtz equation.

6.3 Mathematical Setting of Mixed Problem

for this model has the following form:

$$\chi\Delta\left(p(x,t) + \lambda_m \frac{\partial p(x,t)}{\partial t}\right) = \frac{\partial p(x,t)}{\partial t}, \tag{114}$$

$$p(x,t) = a(x), \text{while } \ t = 0, \tag{115}$$

$$\alpha_1 p(x,t) + \beta_1 \frac{\partial p(x,t)}{\partial \mathbf{n}} = p_3(x,t) \ \text{ for } \ x \in \partial\Omega \times [0,T], \tag{116}$$

where $\mathbf{n}$ is an internal normal. After approximation only by time variable t Eq. (114) has the form:

$$\Delta p^{n+1}(x) - a_1 p^{n+1}(x) = f^n(x), \tag{117}$$

where $a_1 = \dfrac{1}{\chi(\Delta\tau + \lambda_m)}$, $f^n(x) = \dfrac{\lambda_m}{\Delta\tau + \lambda_m}\Delta p^n(x) - \dfrac{1}{\chi(\Delta\tau + \lambda_m)} p^n(x)$. Initial and boundary conditions (115) and (116):

$$p^0(x) = a(x), \ \ x \in \Omega, \tag{118}$$

$$\alpha_1 p^{n+1}(x) + \beta_2 p^{n+1}(x) = p_3^{n+1}(x), \quad x \in \partial\Omega_\varepsilon, \tag{119}$$

where $\beta_2 = \beta_1\big(d(\partial\Omega)\cdot\nabla\big)$. Problem (117)–(119) for a fixed time layer $n = 0, 1, \ldots, N-1$ can be considered as a mixed problem for Helmholtz equation.

Problems described in 6.1, 6.2 and 6.3, that is Dirichlet, Neumann and mixed problems [(105)–(107), (111)–(113) and (117)–(119)] are solved by Monte Carlo methods algorithms in the same way as problems from 2, as for model **III**—Filtration in relaxationaly-compressed porous environment realized by the linear Darcy law.

Remark 2 In all considered models after evaluating pressure $p(x,t)$ by Monte Carlo methods, the first derivatives of $p(x,t)$ ($\text{grad}_x p(x,t)$) are also evaluated by Monte Carlo methods [4, 6, 8, 17]. Then we can evaluate a rate of filtration $\mathbf{W}(x,t)$ in all models by Monte Carlo methods. For example, for model **II**: vector equation

$$\tau \frac{\partial \mathbf{W}(x,t)}{\partial t} + \mathbf{W}(x,t) = -\frac{\kappa}{\mu}\,\text{grad}_x p(x,t)$$

can be approximated only by t. Then we get:

$$\tau \frac{\mathbf{W}^{n+1}(x) - \mathbf{W}^n(x)}{\Delta\tau} + \mathbf{W}^n(x) = \mathbf{f}_1^n(x)$$

or

$$\mathbf{W}^{n+1}(x) = \left(1 - \frac{\Delta\tau}{\tau}\right)\mathbf{W}^n(x) + \frac{\Delta\tau}{\tau}\mathbf{f}_1^n(x), \quad n = 0, 1, \ldots, N-1,$$

where $\mathbf{W}^0(x)$ is known because of initial condition $\mathbf{W}(x,t) = \tilde{\mathbf{W}}(x)$ for $t = 0$, and function $\mathbf{f}_1^n(x) = -\frac{\kappa}{\mu}\,\mathrm{grad}_x\, p^n(x)$ is also evaluated by Monte Carlo methods function in point x.

Remark 3 If trajectories of the Markov's chains is infinitely long and their amount (of trajectories) is also infinite, then all estimates of solutions in this work convergence to exact solution of the original problem. This obvious fact (at least for those who involved in Monte Carlo methods for partial differential equations) is not written in this article.

References

1. Y.M. Molokovich, P.P. Osipov, *Basics of Relaxation Filtration Theory* (Publisher of Kazan University, Kazan, 1987)
2. M. Müller, Some continuous Monte Carlo methods for the Dirichlet problems. Ann. Math. Stat. **27**(3), 569–589 (1956)
3. A. Haji–Sheikh, E.M. Sparrow, The floating random walk and its application to Monte Carlo solutions of heat equations. J. SIAM Appl. Math. **14**(2), 370–389 (1966)
4. B.S. Elepov, A.A. Kronberg, G.A. Mihailov, K.K. Sabelfeld, *Solution of the Boundary Problems by Monte Carlo Methods* (Nauka, Novosibirsk, 1980)
5. S.M. Ermakov, V.V. Nekrutkin, A.S. Sipin, *Random Process for Solution of Classical Equations of Mathematical Physics* (Nauka, Moscow, 1984)
6. K.K. Shakenov, S. Smagulov, *Monte Carlo Methods in Hydrodynamics and Filtrations* (Publisher of Kazakh University, Almaty, 1999)
7. K.K. Shakenov, Solution of problem for one model of relaxational filtration by probability-difference and Monte Carlo methods. Pol. Acad. Sci. Committee Min. Arch. Min. Sci. **52**(2), 247–255 (2007)
8. S.M. Ermakov, K.K. Shakenov, On the applications of the Monte Carlo method to Navier–Stokes equations. Bulletin of Leningrad State University. Series Mathematics, Mechanics, Astronomy. No.6267-B86. Leningrad, (1986), pp. 1–14
9. S.M. Ermakov, *Monte Carlo Method and Adjacent Questions* (Nauka, Moscow, 1975)
10. K. Shakenov, The solution of the initial mixed boundary value problem for hyperbolic equations by Monte Carlo and probability difference methods, in *Fourier Analysis*. Trend in Mathematics (Springer International Publishing AG, Berlin, 2014), pp. 349–355
11. K.K. Shakenov, N.A. Issabekova, Solution by Monte Carlo methods of the problems for relaxational filtration model described by linear Darcy law. Bull. Kazakh Natl. Univ. Ser. Math. Mech. Inform. **52**(1), 81–95 (2007)
12. H.J. Kushner, *Probability Methods of Approximations in Stochastic Control and for Elliptic Equations* (Academic, New York, 1977)
13. K.K. Shakenov, Solution of mixed problem for elliptic equation by Monte Carlo and probability-difference methods CP1076, in *7th International Summer School and Conference*, vol. 1076, Melville, New York (American Institute of Physics, 2008), pp. 213–218

14. K.K. Shakenov, G.A. Musataeva, Approximation of the solution of finite difference mixed problem for Helmholtz equation with Markov chains. Bull. Kazakh Natl. Univ. Ser. Math. Mech. Inform. **44**(1), 51–59 (2005)
15. K. Shakenov, Solution of the Neumann problem for Helmholtz equation by Monte Carlo methods, in *International Conference on Computational Mathematics. Part 1*, Novosibirsk, 2004, pp. 333–334
16. K. Shakenov, Solution of one problem of linear relaxational filtration by Monte Carlo methods, in *International Conference on Computational Mathematics (ICCM 2002), Part I* (ICM&MG Publisher, Novosibirsk, 2002), pp. 276–280
17. K.K. Shakenov, Solution of the Neumann problems for Helmholtz and Poisson equations and calculation of derivatives of solutions by Monte Carlo methods. Bull. Kazakh Natl. Univ. Ser. Math. Mech. Inform. **26**(3), 25–31 (2001)
18. K. Shakenov, Dispersion of estimation of the solving linearized disturbed system of Navier–Stokes equations. Calculating Technol. **7**(3), 93–97 (2002)
19. K.K. Shakenov, Approximation of the solution of mixed problem for elliptic equation with Markov chains. Bull. Kazakh Natl. Univ. Ser. Math. Mech. Inform. **2**, 78–86 (2005)
20. K. Shakenov, Solution of one mixed problem for equation of a relaxational filtration by Monte Carlo methods, Chap. 71, in *Advances in High Performance Computing and Computational Sciences*. Notes on Numerical Fluid Mechanics and Multidisciplinary Design, vol. 93/2006 (Springer, Berlin, 2006), pp. 205–210